Waste Management Policy for Sustainable (Peri-)Urban Development and the Circular Economy

Waste Management Policy for Sustainable (Peri-)Urban Development and the Circular Economy

Guest Editor

Ioannis Vardopoulos

MDPI

Basel • Beijing • Wuhan • Barcelona • Belgrade • Novi Sad • Cluj • Manchester

Guest Editor
Ioannis Vardopoulos
Open University of Cyprus
Nicosia
Cyprus

Editorial Office
MDPI AG
Grosspeteranlage 5
4052 Basel, Switzerland

This is a reprint of the Special Issue, published open access by the journal *Sustainability* (ISSN 2071-1050), freely accessible at: https://www.mdpi.com/journal/sustainability/special_issues/800KVX4L94.

For citation purposes, cite each article independently as indicated on the article page online and as indicated below:

Lastname, A.A.; Lastname, B.B. Article Title. *Journal Name* **Year**, *Volume Number*, Page Range.

ISBN 978-3-7258-4645-0 (Hbk)
ISBN 978-3-7258-4646-7 (PDF)
https://doi.org/10.3390/books978-3-7258-4646-7

Cover image courtesy of Julia M Cameron

Contents

About the Editor

Ioannis Vardopoulos

Ioannis Vardopoulos studied Civil Engineering and holds two master's degrees, one in Environmental Protection and Management, and the other in Sustainable Development, along with a doctorate in Applied Urban Sustainability. He is currently a Tenured Public Sector Officer, acting as Senior Civil Engineer at the Greek Public Employment Service (DYPA) Directorate of Technical Services for Buildings, where he drives sustainability in public building projects. He is also a Postdoctoral Researcher at the University of Athens and a Visiting Lecturer at the Open University of Cyprus, teaching and conducting high-quality research that addresses key sustainability challenges in the built environment. With over 120 impactful publications and more than 2000 citations, he is featured among the top 2% of the most influential scientists worldwide.

Preface

An interdisciplinary collection of studies exploring the multifaceted dimensions of waste management within the context of (peri-)urban sustainability. This volume delves into the critical role of waste as both a challenge and an opportunity for advancing circular economy principles and fostering resilient urban futures.

By addressing systemic barriers, technological innovations, and the socio-economic forces that shape waste practices, the contributions herewith illuminate pathways for aligning policy, technology, and behavior with sustainable development goals.

The genesis of this reprint was driven by a collective recognition of the urgent need to rethink waste not merely as a by-product of consumption but as a cornerstone of resource efficiency and environmental stewardship. From exploring governance challenges in developing economies to examining advanced technological solutions for material recovery, the included studies offer insights that transcend geographical and disciplinary boundaries. Each work reflects the dedication and expertise of its authors, who represent institutions from Europe, Asia, Africa, and North America, underscoring the global relevance of these themes.

This reprint is intended for researchers, policymakers, urban planners, and sustainability practitioners seeking actionable frameworks and thought-provoking perspectives on waste management and circularity. It also speaks to the broader academic community and stakeholders invested in addressing the complexities of (peri-)urban systems. The contributors' rigorous methodologies and innovative approaches ensure that this collection will serve as both a reference and a catalyst for future research and practice.

The insights presented in this collection envision a future where waste is transformed from an obstacle into an opportunity, and circularity evolves from ambition to reality. This collection explores the challenge of building cities that are not only sustainable but are transformative, dynamic landscapes shaped by innovation and resilience, defining the contours of a sustainable tomorrow.

Ioannis Vardopoulos
Guest Editor

sustainability

MDPI

Editorial

Cities in Flux: Waste, the Shadow of Circularity, and the (Elusive) Promise of Sustainability

Ioannis Vardopoulos [1,2]

[1] Faculty of Pure and Applied Sciences, Open University of Cyprus (OUC), 2231 Nicosia, Cyprus; ivardopoulos@post.com

[2] Department of Agricultural Development, Agri-Food and Natural Resources Management, University of Athens (UoA), 34400 Psachna, Greece

1. Introduction

The transition to a circular economy has emerged as a critical strategy for addressing the escalating challenges posed by resource depletion [1,2], environmental degradation [3,4], and increasing waste generation [5,6]. At the heart of this paradigm shift lies waste management—a sector deeply intertwined with urban sustainability [7,8], resource efficiency [9,10], and public health [11,12]. Moving beyond traditional linear models, where resources are extracted, used, and discarded [13], the circular economy (Figure 1) promotes the continuous use and recovery of materials, aiming to decouple economic growth from environmental harm [14,15].

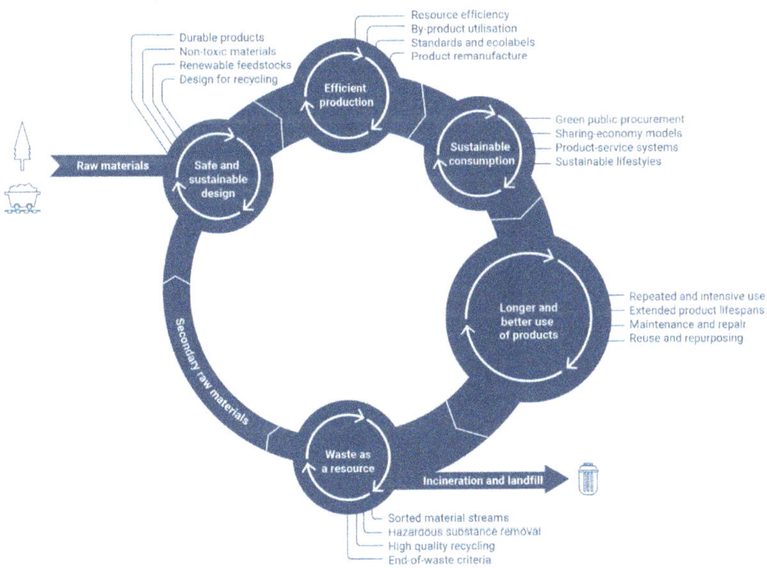

Figure 1. The European Environmental Agency vision for a circular economy in Europe.

Europe has made significant strides in advancing circular economy principles through frameworks such as the European Green Deal [16,17] and the Circular Economy Action Plan [18,19]. These initiatives aim to mitigate climate change [20,21], reduce biodiversity loss [22,23], and foster sustainable economic practices [24,25]. Despite these efforts, key challenges remain. The EU's material circularity rate has stagnated at 11.8% [26], recycling

 https://doi.org/10.3390/su17020705

processes often fail to close the loop effectively [27], and public engagement with waste prevention remains limited overall [28]. Bridging the gap between policy frameworks and tangible outcomes requires evidence-based strategies that integrate technological innovation, public awareness, and regulatory alignment.

Waste management is inherently complex [29], drawing on principles and practices from diverse fields such as environmental science [30], urban planning [31], geography [32], economics [33], social behavior [34], and cultural norms [35]. Its multidimensional nature requires a system-based approach, where technological solutions, policy strategies, spatial considerations, and cultural dynamics are seamlessly integrated to address the challenges of resource efficiency, waste prevention and reduction, and environmental sustainability [36].

The increasing focus on waste management, reflected in evolving policies (such as the "Extended Producer Responsibility" [37], the "Pay-As-You-Throw" [38], the "Ban on Single-Use Plastics" [39], the "Green Public Procurement" [40], etc.), strategic research investments (such as the A2UFood Project (UIA02-115), the LIFE-IP CEI Greece Project (LIFE18 IPE/GR/000013), the GREENHORT Project (PRIN 2022—grant 2022WHTNZT), the BBHORT Project (PRIN 2022 PNRR—grant P2022P52XK), the Horizon Europe RECONMATIC Project (Grant agreement ID: 101058580), the WiedergeBORN Project, etc.) and its prominence in global discussions and high-level forums (such as the "United Nations Environment Assembly", the "Global Waste Management Outlook", the "International Solid Waste Association World Congress", the "International Conference on Sustainable Solid Waste Management", the "RETASTE—Rethink Food Resources, Loses, and Waste Conference", etc.), underscores its critical role in mitigating climate change and advancing environmental justice. For example, at the very recent COP29, the inclusion of waste management in the agenda for the first time emphasized the potential for targeted interventions, such as organic waste diversion and high-quality recycling, to substantially reduce methane emissions while enhancing environmental justice. However, achieving such outcomes requires comprehensive strategies that align policy goals with practical implementation.

The Special Issue entitled "Waste Management Policy for Sustainable (Peri-)Urban Development and the Circular Economy" examines these complexities through 12 diverse contributions, which reveal both persistent gaps and emerging solutions, emphasizing the need for interdisciplinary approaches that balance ambition with feasibility.

Key insights from this collection demonstrate that progress in waste management depends on addressing three interconnected dimensions. First, public awareness must shift beyond recycling to encompass waste prevention and reduction. Second, regulatory frameworks must evolve to support innovation while remaining adaptable to regional and local contexts. Finally, systemic approaches that integrate behavioral, technological, and policy considerations are essential to achieving meaningful and scalable impacts.

This Special Issue contributes to a growing body of knowledge that seeks to transform waste management into a driver of sustainable urban development. By critically examining policy, practice, and innovation, the collected works provide a foundation for advancing circular economy principles in (peri-)urban contexts, offering actionable pathways toward a more resilient and sustainable future.

2. Synopsis of Contributions

This Special Issue brings together 12 contributions authored by 53 researchers representing institutions across Europe (Greece, Cyprus, Germany, Italy, and the United Kingdom), Asia (China and Thailand), Africa (South Africa), and North America (the United States). This geographic diversity reflects the global importance of waste management and the varied approaches to addressing its challenges within different cultural, economic, and environmental contexts. With an average of 4.4 authors per paper, the contributions exemplify the collective effort required to tackle the multifaceted issues associated with waste management.

Spanning 18 months from the initial call for papers to the final publication, this Special Issue highlights the sustained effort and interdisciplinary collaboration that shaped its content. The contributions reflect cutting-edge research and state-of-the-art methodologies, addressing the most pressing challenges and emerging opportunities in waste management and the circular economy. Their innovative approaches and findings provide valuable insights into the current landscape and future directions of sustainable (peri-)urban development.

The current synopsis is organized into four thematic sections. The first addresses policy and implementation challenges, focusing on systemic barriers and regulatory frameworks. The second highlights technological innovations and waste valorization, emphasizing advancements in materials, processes, and decision-support systems. The third explores behavioral and social dynamics, examining public awareness, consumer behavior, and societal engagement. Finally, the Section 2.4 provides systematic reviews and frameworks, offering comprehensive evaluations and methodologies to guide future research and practice.

This thematic organization facilitates a deeper understanding of the Special Issue's diverse contributions, showcasing their collective effort to advance sustainable waste management and the principles of a circular economy.

2.1. Policy and Implementation Challenges

The contributions in this section uncover critical tensions between the ambition of waste management policies and their practical implementation across diverse contexts. They reveal how systemic issues—ranging from governance and funding constraints to regulatory misalignments—undermine efforts to transition toward a circular economy.

One recurring insight is the disconnect between localized successes and their scalability. For instance, the WiedergeBORN project in Germany demonstrated the potential of local circularity initiatives but also highlighted the institutional and regulatory barriers that limit their broader applicability. This suggests that localized experiments are often constrained by fragmented policy environments and insufficient support mechanisms, reflecting broader challenges in harmonizing multi-level governance structures.

Governance emerges as a central obstacle in less resourced contexts, as evidenced by the analysis of the South African municipalities. This study provided insights into how a lack of financial resources, poor institutional coordination, and inadequate infrastructure prevent the effective implementation of integrated waste management plans, stressing the need for robust institutional frameworks that not only allocate resources equitably but also foster collaboration across public and private sectors.

Regulatory misalignments further exacerbate these challenges. For example, an examination of legal barriers in sustainable agriculture highlights how conflicting policies hinder the valorization of agri-food waste and pesticide reduction efforts. Such misalignments indeed reveal deeper structural issues, where short-term economic priorities often overshadow long-term sustainability goals, complicating the operationalization of circular economy principles.

These studies manage to collectively illustrate the complexities of embedding waste management within the broader framework of circular economy transitions. They emphasize the critical need for adaptive policies that are responsive to regional contexts, align with existing regulatory frameworks, and incentivize stakeholder engagement. Without addressing these systemic barriers, the potential of waste management to drive sustainability will remain unrealized, particularly in contexts where institutional and financial capacities are limited.

2.2. Technological Innovations and Waste Valorization

The contributions in this section explore the transformative role of technological advancements and innovative practices in waste management, focusing on their potential to enhance resource recovery, reduce environmental impacts, and support circular economy objectives. These studies critically assess the balance between technological feasibility and practical implementation, providing insights into the opportunities and constraints of integrating innovative solutions into existing systems.

One significant focus is the potential of organic waste valorization. Research on bioplastic production from food waste demonstrates the dual benefits of reducing waste volumes while generating high-value materials. However, the study also highlights critical challenges, including the need for standardized processes and market acceptance of bioplastics, which remain underdeveloped in many contexts. Similarly, an evaluation of the biodegradability of polylactic acid and compostable bags under industrial composting conditions underscores the importance of aligning product design with waste management infrastructure to maximize environmental benefits.

Technological decision-making tools emerge as another key innovation. A study on medical waste management introduces a fuzzy-based decision support system to optimize waste disposal practices, ensuring compliance with environmental standards while addressing safety concerns in high-risk contexts. This research emphasizes the importance of integrating technology into complex waste streams to achieve more effective and sustainable outcomes.

Despite these advancements, the contributions collectively emphasize that technology alone cannot solve systemic waste challenges. The innovative use of Plan-Do-Check-Act processes in (green) university practices, underscores the importance of systematic evaluation and continuous improvement in achieving sustainability benchmarks. The effectiveness of these innovations depends on supportive regulatory frameworks, adequate infrastructure, and public acceptance. For example, while innovations in material recycling and composting hold significant promise, their scalability often hinges on addressing logistical and economic constraints at the local level.

Overall, these studies underline the need for a system-based approach, where technological solutions are embedded within broader waste management strategies. By focusing on the intersection of innovation, regulation, and infrastructure, these contributions provide valuable pathways for advancing the role of technology in achieving circular economy objectives.

2.3. Behavioral and Social Dynamics

The contributions in this section examine the critical role of human behavior, societal norms, and community engagement in shaping waste management practices. They underscore that effective waste management and the transition to a circular economy require not only technological and regulatory solutions but also significant shifts in public awareness, consumer behavior, and collective action.

A central theme is the gap between awareness and behavior. Research from China highlights how policy promotion and internet use influence farmers' and residents' willingness to participate in waste classification, revealing that while awareness is increasing, behavior change remains uneven. This points to the importance of targeted educational campaigns and accessible infrastructure to bridge the gap between knowledge and action.

Public perceptions of waste management policies also emerge as a critical factor. A study on waste classification in Chinese prefecture-level cities shows how age, education, and exposure to policy initiatives influence residents' attitudes and practices. The analysis suggests that younger populations, who are more likely to access information via digital platforms, are more engaged in waste sorting, whereas older groups rely on traditional media and exhibit lower levels of participation. These insights call for tailored communication strategies that consider demographic and cultural differences to maximize impact.

The role of institutional and consumer practices in fostering sustainable behavior is further explored through a study of Germany's mandatory reusable packaging laws. While the policy has created opportunities for more sustainable consumption, the research highlights obstacles such as consumer resistance and operational challenges faced by businesses, emphasizing the need for policies that balance regulatory requirements with practical feasibility.

Collectively, these studies highlight the interconnectedness of behavior, policy, and infrastructure. They demonstrate that fostering meaningful societal engagement requires more than individual behavior change; it demands systemic approaches that integrate public participation with supportive policies and accessible waste management systems. Addressing these behavioral and social dynamics is essential for embedding sustainability into everyday practices and ensuring the success of circular economy initiatives.

2.4. Systematic Reviews and Frameworks

The contributions in this section synthesize existing knowledge, identify persistent gaps, and propose structured approaches to guide future research and policy development.

A notable focus is on the inefficiencies in managing specific waste streams. A systematic review of food waste quantification in the food service industry reveals significant data inconsistencies and underreported waste volumes, which hinder the development of effective reduction strategies. The analysis calls for standardized methodologies and better monitoring systems to address these shortcomings and enhance resource efficiency. Similarly, a comprehensive evaluation of construction and demolition waste management practices highlights the potential for circularity in the sector but emphasizes the need for improved regulatory frameworks and market mechanisms to support the recycling and reuse of materials.

The interaction between legal frameworks and waste management practices is critically explored in a review of agri-food waste valorization. The study identifies regulatory inconsistencies as a key barrier to integrating waste reduction and pesticide use mitigation into agricultural practices. By highlighting the disconnect between policy intentions and practical implementation, the paper underscores the importance of regulatory alignment and policy coherence for advancing sustainable practices.

Frameworks for assessing public engagement in waste management are also examined. A case study of waste classification in Chinese cities evaluates residents' understanding of policies and the effectiveness of public awareness campaigns. The findings provide actionable insights for designing community-focused interventions that foster greater participation and compliance with waste management regulations.

These contributions collectively highlight the necessity of addressing waste management through integrated and multi-dimensional frameworks rather than relying on fragmented or isolated measures. They underscore the importance of aligning policies, standardizing methodologies, and fostering stakeholder collaboration to tackle systemic inefficiencies. By presenting tools for harmonizing data collection, enhancing regulatory coherence, and improving public engagement strategies, these reviews and frameworks provide a concrete foundation for embedding circular economy principles into waste management practices, while ensuring scalability and adaptability across different contexts.

The contributions to this Special Issue underscore the pivotal role of waste management in the broader transition toward a circular economy and sustainable (peri-)urban development. Collectively, they reveal both the challenges and opportunities inherent in aligning policy, technology, behavior, and systemic frameworks to address the complexities of waste generation and resource efficiency.

A key takeaway is the urgent need for a more integrated and adaptive approach to waste management. While advancements in technology and policy frameworks offer promising pathways, their success depends on bridging the gaps between regulatory ambition and practical implementation. The studies highlight the importance of tailoring strategies to specific socio-economic and cultural contexts, fostering public engagement, and ensuring robust infrastructure to support the effective adoption of circular practices.

However, significant challenges remain. The slow progress in material circularity, the limited scalability of localized initiatives, and the persistent gaps in public understanding and behavior indicate that waste management continues to face systemic barriers. Addressing these requires not only improved governance and regulatory coherence but also a commitment to innovation and stakeholder collaboration at all levels.

Future research must prioritize interdisciplinary solutions that integrate technical, behavioral, and policy dimensions. Areas such as digital tools for waste tracking, advanced recycling technologies, and data-driven decision-making frameworks offer substantial potential for transforming waste management systems. Equally, greater attention must be given to the economic and social incentives that drive sustainable practices, ensuring that policies are both equitable and effective.

International collaboration is essential to accelerate progress, as waste management challenges transcend national borders. The lessons learned from diverse contexts in this Special Issue highlight the value of knowledge exchange and cooperative efforts to scale successful models globally. Furthermore, the integration of waste management into global climate agendas, as seen at COP29, provides a critical opportunity to align waste reduction strategies with broader environmental and social goals.

In closing, this Special Issue reaffirms the centrality of waste management in achieving a circular economy and sustainable urban futures. It calls for a concerted effort from policymakers, researchers, and practitioners to move beyond incremental improvements and embrace systemic, scalable solutions. By addressing these challenges with urgency and innovation, waste can be transformed from a liability into a vital resource for a more resilient and equitable world.

Conflicts of Interest: The author declares no conflicts of interest.

List of Contributions

1. Vardopoulos, I.; Abeliotis, K.; Lasaridi, K. A Systematic Informetric Analysis and Literature Review of Food Waste Quantification Studies in the Food Service Industry. *Sustainability* **2025**, *17*, 103. https://doi.org/10.3390/su17010103.
2. Obersteg, A. Successful Project—Limited Transfer: Learnings from the Local Circularity Experiment WiedergeBORN. *Sustainability* **2024**, *16*, 10643. https://doi.org/10.3390/su162310643.
3. Mukwevho, P.; Radzuma, P.; Roos, C. Exploring Barriers to the Effective Implementation of Integrated Waste Management Plans in Developing Economies: Lessons Learned from South African Municipalities. *Sustainability* **2024**, *16*, 9643. https://doi.org/10.3390/su16229643.
4. Nicastro, R.; Papale, M.; Fusco, G.M.; Capone, A.; Morrone, B.; Carillo, P. Legal Barriers in Sustainable Agriculture: Valorization of Agri-Food Waste and Pesticide Use Reduction. *Sustainability* **2024**, *16*, 8677. https://doi.org/10.3390/su16198677.
5. Alhawamdeh, M.; Ferriz-Papi, J.A.; Lee, A. Examining the Drivers to Support Improved Construction and Demolition Waste Management for a Circular Economy: A

Comprehensive Review Using a Systematic Approach. *Sustainability* **2024**, *16*, 6014. https://doi.org/10.3390/su16146014.
6. Synani, K.; Abeliotis, K.; Velonia, K.; Maragkaki, A.; Manios, T.; Lasaridi, K. Environmental Impact and Sustainability of Bioplastic Production from Food Waste. *Sustainability* **2024**, *16*, 5529. https://doi.org/10.3390/su16135529.
7. Theobald, K.; Mich, A.; Hillesheim, S.; Hartard, S.; Rohn, H. One Year of Mandatory Reusable Packaging in Germany: Opportunities and Obstacles from the Perspective of Consumers and Companies. *Sustainability* **2024**, *16*, 5439. https://doi.org/10.3390/su16135439.
8. Phrophayak, J.; Techarungruengsakul, R.; Khotdee, M.; Thuangchon, S.; Ngamsert, R.; Prasanchum, H.; Sivanpheng, O.; Kangrang, A. Enhancing Green University Practices through Effective Waste Management Strategies. *Sustainability* **2024**, *16*, 3346. https://doi.org/10.3390/su16083346.
9. Cui, W.; Chen, Y.; Zeng, H. Can Internet Use Narrow the Gap between Farmers' Willingness and Behavior in Waste Classification? Empirical Evidence from Rural Areas in Jiangsu Province, China. *Sustainability* **2024**, *16*, 2726. https://doi.org/10.3390/su16072726.
10. Kokkinos, K.; Lakioti, E.; Moustakas, K.; Tsanaktsidis, C.; Karayannis, V. Sustainable Medical Waste Management Using an Intuitionistic Fuzzy-Based Decision Support System. *Sustainability* **2023**, *16*, 298. https://doi.org/10.3390/su16010298.
11. Maragkaki, A.; Malliaros, N.G.; Sampathianakis, I.; Lolos, T.; Tsompanidis, C.; Manios, T. Evaluation of Biodegradability of Polylactic Acid and Compostable Bags from Food Waste under Industrial Composting. *Sustainability* **2023**, *15*, 15963. https://doi.org/10.3390/su152215963.
12. Ge-Zhang, S.; Cai, T.; Hu, Z.; Zhu, H.; Mu, P.; Cui, J. Investigation and Suggestions Regarding Residents' Understanding of Waste Classification in Chinese Prefecture-Level Cities—A Case Study of Maanshan City, Anhui Province, China. *Sustainability* **2023**, 15, 11124. https://doi.org/10.3390/su151411124.

References

1. Tan, J.; Tan, F.J.; Ramakrishna, S. Transitioning to a Circular Economy: A Systematic Review of Its Drivers and Barriers. *Sustainability* **2022**, *14*, 1757. [CrossRef]
2. Salvati, L.; Ciaschini, C.; Postigliola, M.; Ermini, B.; Vardopoulos, I. Contextual Differences, Contrasting Trends? Sustainable Development, Economic Performances, and Divergence in Land Resource Depletion. In *Environmental Sustainability and Global Change*; Elsevier: Cambridge, MA, USA, 2025; pp. 107–114.
3. Voukkali, I.; Papamichael, I.; Loizia, P.; Zorpas, A.A. Urbanization and Solid Waste Production: Prospects and Challenges. *Environ. Sci. Pollut. Res.* **2023**, *31*, 17678–17689. [CrossRef]
4. Ramos, A. (Ed.) *Circular Economy Strategies for Sustainable Development: Applications and Impacts*; MDPI: Basel, Switzerland, 2024; ISBN 978-3-7258-1661-3.
5. Zorpas, A.A.; Lekkas, D.F. Nexus Approach and Perspectives in the next Generation Waste Strategies Development. *Sci. Total Environ.* **2024**, *958*, 178183. [CrossRef]
6. Papamichael, I.; Voukkali, I.; Loizia, P.; Stylianou, M.; Economou, F.; Vardopoulos, I.; Klontza, E.E.; Lekkas, D.F.; Zorpas, A.A. Measuring Circularity: Tools for Monitoring a Smooth Transition to Circular Economy. *Sustain. Chem. Pharm.* **2023**, *36*, 101330. [CrossRef]
7. Zorpas, A.A. Promoting Circular Economy: The Transformative Impact of Deposit Refund Systems. *Waste Manag. Res.* **2024**, *42*, 1093–1095. [CrossRef] [PubMed]
8. Vardopoulos, I.; Konstantopoulos, I.; Zorpas, A.A.; Limousy, L.; Bennici, S.; Inglezakis, V.J.; Voukkali, I. Sustainable Metropolitan Areas Perspectives through Assessment of the Existing Waste Management Strategies. *Environ. Sci. Pollut. Res.* **2021**, *28*, 24305–24320. [CrossRef]
9. Wilts, H.; Von Gries, N.; Bahn-Walkowiak, B. From Waste Management to Resource Efficiency—The Need for Policy Mixes. *Sustainability* **2016**, *8*, 622. [CrossRef]

10. Wong, J.; Kaur, G.; Taherzadeh, M.; Pandey, A.; Lasaridi, K. (Eds.) *Current Developments in Biotechnology and Bioengineering. Sustainable Food Waste Management: Resource Recovery and Treatment*; Elsevier: Amsterdam, The Netherlands, 2021; ISBN 9780128191484.
11. Rodríguez-Espinosa, T.; Navarro-Pedreño, J.; Gómez-Lucas, I.; Jordán-Vidal, M.M.; Bech-Borras, J.; Zorpas, A.A. Urban Areas, Human Health and Technosols for the Green Deal. *Environ. Geochem. Health* **2021**, *43*, 5065–5086. [CrossRef] [PubMed]
12. Vinti, G.; Bauza, V.; Clasen, T.; Medlicott, K.; Tudor, T.; Zurbrügg, C.; Vaccari, M. Municipal Solid Waste Management and Adverse Health Outcomes: A Systematic Review. *Int. J. Environ. Res. Public Health* **2021**, *18*, 4331. [CrossRef]
13. Panagiotopoulos, P.; Vardopoulos, I.; Maialetti, M.; Ciaschini, C.; Koundouri, P.; Salvati, L. Reimagining Sustainable Development and Economic Performance Indicators: A Human-Centric Maslow–Bossel Blueprint. *Economies* **2024**, *12*, 338. [CrossRef]
14. Emmanouil, C.; Zorpas, A.A.; Papamichael, I. Editorial: Circular Economy: State of the Art and Future Trends. *Sustain. Chem. Pharm.* **2024**, *39*, 101591. [CrossRef]
15. Scheel, C.; Aguiñaga, E.; Bello, B. Decoupling Economic Development from the Consumption of Finite Resources Using Circular Economy. A Model for Developing Countries. *Sustainability* **2020**, *12*, 1291. [CrossRef]
16. Zorpas, A.A.; Navarro-Pedreño, J.; Jeguirim, M.; Dimitriou, G.; Almendro Candel, M.B.; Argirusis, C.; Vardopoulos, I.; Loizia, P.; Chatziparaskeva, G.; Papamichael, I. Crisis in Leadership vs Waste Management. *Euro-Mediterr. J. Environ. Integr.* **2021**, *6*, 80. [CrossRef]
17. Doukas, Y.E.; Salvati, L.; Vardopoulos, I. Unraveling the European Agricultural Policy Sustainable Development Trajectory. *Land* **2023**, *12*, 1749. [CrossRef]
18. Zorpas, A.A. The Hidden Concept and the Beauty of Multiple "R" in the Framework of Waste Strategies Development Reflecting to Circular Economy Principles. *Sci. Total Environ.* **2024**, *952*, 175508. [CrossRef]
19. Voukkali, I.; Papamichael, I.; Loizia, P.; Lekkas, D.F.; Rodríguez-Espinosa, T.; Navarro-Pedreño, J.; Zorpas, A.A. Waste Metrics in the Framework of Circular Economy. *Waste Manag. Res.* **2023**, *41*, 1741–1753. [CrossRef] [PubMed]
20. Bonoli, A.; Zanni, S.; Serrano-Bernardo, F. Sustainability in Building and Construction within the Framework of Circular Cities and European New Green Deal. The Contribution of Concrete Recycling. *Sustainability* **2021**, *13*, 2139. [CrossRef]
21. Ribeiro Hoffmann, A.; Sandrin, P.; Doukas, Y.E. (Eds.) *Climate Change in Regional Perspective. European Union and Latin American Initiatives, Challenges, and Solutions*; United Nations University Series on Regionalism: Cham, Switzerland; Springer Nature: Berlin/Heidelberg, Germany, 2024; Volume 27, ISBN 978-3-031-49328-7.
22. Smol, M.; Marcinek, P.; Duda, J.; Szołdrowska, D. Importance of Sustainable Mineral Resource Management in Implementing the Circular Economy (CE) Model and the European Green Deal Strategy. *Resources* **2020**, *9*, 55. [CrossRef]
23. Hermoso, V.; Carvalho, S.B.; Giakoumi, S.; Goldsborough, D.; Katsanevakis, S.; Leontiou, S.; Markantonatou, V.; Rumes, B.; Vogiatzakis, I.N.; Yates, K.L. The EU Biodiversity Strategy for 2030: Opportunities and Challenges on the Path towards Biodiversity Recovery. *Environ. Sci. Policy* **2022**, *127*, 263–271. [CrossRef]
24. Peyravi, B.; Peleckis, K.; Limba, T.; Peleckienė, V. The Circular Economy Practices in the European Union: Eco-Innovation and Sustainable Development. *Sustainability* **2024**, *16*, 5473. [CrossRef]
25. Eckert, E.; Kovalevska, O. Sustainability in the European Union: Analyzing the Discourse of the European Green Deal. *J. Risk Financ. Manag.* **2021**, *14*, 80. [CrossRef]
26. Zisopoulos, F.K.; Schraven, D.F.J.; de Jong, M. How Robust Is the Circular Economy in Europe? An Ascendency Analysis with Eurostat Data between 2010 and 2018. *Resour. Conserv. Recycl.* **2022**, *178*, 106032. [CrossRef]
27. Hsu, W.-T.; Domenech, T.; McDowall, W. Closing the Loop on Plastics in Europe: The Role of Data, Information and Knowledge. *Sustain. Prod. Consum.* **2022**, *33*, 942–951. [CrossRef]
28. Zorpas, A.A.; Lasaridi, K.; Pociovalisteanu, D.M.; Loizia, P. Monitoring and Evaluation of Prevention Activities Regarding Household Organics Waste from Insular Communities. *J. Clean. Prod.* **2018**, *172*, 3567–3577. [CrossRef]
29. Zorpas, A.A. Strategy Development in the Framework of Waste Management. *Sci. Total Environ.* **2020**, *716*, 137088. [CrossRef]
30. Zorpas, A.A. Sustainable Waste Management through End-of-Waste Criteria Development. *Environ. Sci. Pollut. Res.* **2016**, *23*, 7376–7389. [CrossRef] [PubMed]
31. Voukkali, I.; Loizia, P.; Navarro Pedreño, J.; Zorpas, A.A. Urban Strategies Evaluation for Waste Management in Coastal Areas in the Framework of Area Metabolism. *Waste Manag. Res.* **2021**, *39*, 448–465. [CrossRef] [PubMed]
32. Chalkias, C.; Lasaridi, K. Benefits from GIS Based Modelling for Municipal Solid Waste Management. In *Integrated Waste Management—Volume I*; Kumar, S., Ed.; InTech: London, UK, 2011; pp. 417–436.
33. Zorpas, A.A.; Voukkali, I.; Loizia, P. Socio Economy Impact in Relation to Waste Prevention. In *Sustainable Economic Development. Green Economy and Green Growth*; Filho, W.L., Pociovalisteanu, D.M., Al-Amin, A.Q., Eds.; Springer: Cham, Switzerland, 2017; pp. 31–48.
34. Zorpas, A.A.; Dimitriou, M.; Voukkali, I. Disposal of Household Pharmaceuticals in Insular Communities: Social Attitude, Behaviour Evaluation and Prevention Activities. *Environ. Sci. Pollut. Res.* **2018**, *25*, 26725–26735. [CrossRef]
35. Zacharatos, T.; Theodoridis, P. From Plate to Bin: Consumer Segments and Food Waste in Greece. *Waste* **2024**, *3*, 2. [CrossRef]

36. Das, S.; Lee, S.-H.; Kumar, P.; Kim, K.-H.; Lee, S.S.; Bhattacharya, S.S. Solid Waste Management: Scope and the Challenge of Sustainability. *J. Clean. Prod.* **2019**, *228*, 658–678. [CrossRef]
37. Leal Filho, W.; Saari, U.; Fedoruk, M.; Iital, A.; Moora, H.; Klöga, M.; Voronova, V. An Overview of the Problems Posed by Plastic Products and the Role of Extended Producer Responsibility in Europe. *J. Clean. Prod.* **2019**, *214*, 550–558. [CrossRef]
38. Emmanouil, C.; Papadopoulou, K.; Papamichael, I.; Zorpas, A.A. Pay-as-You-Throw (PAYT) for Municipal Solid Waste Management in Greece: On Public Opinion and Acceptance. *Sustainability* **2022**, *14*, 15429. [CrossRef]
39. Cowan, E.; Booth, A.M.; Misund, A.; Klun, K.; Rotter, A.; Tiller, R. Single-Use Plastic Bans: Exploring Stakeholder Perspectives on Best Practices for Reducing Plastic Pollution. *Environments* **2021**, *8*, 81. [CrossRef]
40. Orfanidou, V.S.; Dimitriou, D.J.; Rachaniotis, N.P.; Tsoulfas, G.T. Critical Factors for Green Public Procurement: The Case of Greece. *Logistics* **2024**, *8*, 127. [CrossRef]

sustainability

MDPI

Brief Report

Investigation and Suggestions regarding Residents' Understanding of Waste Classification in Chinese Prefecture-Level Cities—A Case Study of Maanshan City, Anhui Province, China

Shangjie Ge-Zhang [1,†], Taoyang Cai [2,†], Zhitao Hu [3], Haotong Zhu [1], Pingxuan Mu [1] and Jingang Cui [1,*]

1 College of Science, Northeast Forestry University, Harbin 150040, China
2 Aulin College, Northeast Forestry University, Harbin 150040, China
3 College of Economics and Management, Northeast Forestry University, Harbin 150040, China
* Correspondence: cjg@nefu.edu.cn
† These authors contributed equally to this work.

Abstract: The implementation of waste-sorting policy is conducive to improving residents' working and living environments, and has positive implications for promoting green development and building in China. As one of the main factors in the implementation of waste-sorting policy, residents' awareness of and behavior regarding waste sorting and disposal affects its promotion, while policies, incentives and infrastructure will affect residents' enthusiasm for implementing waste sorting. Taking Maanshan City in Anhui Province as a case study, this paper discusses the current state of progress towards ecological civilization in China; the influence of policy promotion on residents' perception of it; the correlation between residents' age and their understanding of waste classification; and the correlation between residents' enthusiasm, policies and green infrastructure; and puts forward some suggestions for how to enhance green infrastructure and improve residents' perceptions of waste classification in the future.

Keywords: waste classification; Chinese prefecture-level city; classification consciousness; ecological civilization; residents' enthusiasm

1. Introduction

With the rapid progress of urbanization and modernization in China, the urban population has increased rapidly, and its material quality of life and consumption level have continuously increased, and the output of municipal solid waste has also shown a trend of increasing year by year [1–3]. Landfill and waste incineration consume resources and produce substantial greenhouse gases, which exacerbates environmental issues [4–6]. As an enhancement of the traditional methods of waste collection and disposal, waste classification has strong implications on livelihoods and sustainable development. Through waste classification management and the reuse of resources, the mechanisms of waste disposal can be optimized and the quality of a living environment can be improved [7,8].

The starting point of and the key to effective waste classification is the residents of the area in question. Their willingness to adopt domestic waste classification policies and their approach to it are not simply their personal attitudes, but are also major factors affecting the social environment [9]. Therefore, investigation of residents' understanding of waste classification is an effective means to judge the implementation of waste classification policies [10,11].

Li et al. [12] used the full, extended VBN theory to examine residents' attitudes towards municipal solid waste classification; that is, the higher a waste product's bio value, the more likely residents were to appropriately classify it. Liu et al. [13] investigated the influence of morality and law on urban residents' willingness to sort waste, and found that the perceived efficacy of recycling directly affected attitudes and behavior pertaining to it, and that this was also positively correlated with respondents' level of education and being under 40 years old. Wu et al. [14] used data text mined from Sina Weibo to identify the key factors affecting residents' willingness to participate in waste sorting, and found that residents' positive attitudes were mainly caused by public environmental awareness and government incentives stimulated by publicity and education, while negative emotions came from inadequate infrastructure and unsuitable waste sorting arrangements. Bai et al. [15] studied the data of four randomly selected major cities in China through the ordered Probit model, and came to the conclusion that the proportion of residents who are willing to pay for waste sorting is high, but the amount they are willing to pay is fairly low.

Based on the analysis of the survey data, this paper takes Maanshan City, Anhui Province, China as a case study, explores the progress of residents' understanding of waste classification as part of the waste classification policies of Chinese prefecture-level cities, and puts forward some thoughts and relevant suggestions for the improvement and development of future policies.

2. Survey Objects and Methods

Urban residents of Maanshan City (Yushan District and Huashan District) were selected as the research subjects, and stratified sampling was used to conduct an in-depth investigation by combining online and offline methods. Maanshan, a city in the center of the Yangtze River Delta, is located in Nanjing and Hefei metropolitan areas, between 117°53′–118°52′ east longitude and 31°24′–32°02′ north latitude. Maanshan was chosen because it is a prefecture-level city and one of the 297 cities with widely implemented policies [16]. In addition, Maanshan City has improved its waste sorting policy according to the principle of moderate advance, which has played a role in leading research. The study of Maanshan City can serve as a reference for understanding the future development of other cities [17]. A total of 1504 questionnaires were collected, and after removing invalid questionnaires, such as those with contradictory responses, the number of valid questionnaires reached 1350, with a recovery rate of 89.76%. In order to protect privacy, this study did not ask for the respondents' names, ID cards or other private information. The sample population does not include minors under the age of 16, as they are considered as having no or limited capacity for civil conduct; adults over the age of 18 and minors over the age of 16 whose main source of livelihood is their own labor income were included, as they are considered to have full capacity for civil conduct. After data collection, Origin 2021, Stata 16.0 and SPSS 27 were used to produce images and analyze the data. The influence of policy promotion on residents' initial perception, the correlation between residents' age and their understanding of waste classification, and the correlation between residents' enthusiasm, policy and green development were analyzed and discussed.

The following operations were used to avoid possible survey bias: all the interviewees were interviewed offline, and those among the respondents able to use social media to complete the questionnaire did so online, via a QR code (This is referred to as online collection). People who were not competent with social media used paper questionnaires, and staff then manually input these data. In addition, the online questionnaire stipulated that any one computer/smart phone could only be used once. In order to avoid filling errors, the respondents could enter the questionnaire again to modify the submitted contents and only keep the last submitted data during the time from submitting the questionnaire to the completion of the data export (24:00 on the same day). This online and offline questionnaire collection method served to improve the efficiency of data entry and the accuracy of the data, and to prevent repeat entries.

3. Basic Characteristics of The Respondents

3.1. Geographical Distribution

According to the government's public information, the population ratio between Yushan District and Huashan District is about 44.11:55.89 [18]. The geographical distribution of the respondents should be strictly controlled to make them more suitable for the population distribution in various places when investigating. The ratio of the number of valid questionnaires collected in Yushan District and Huashan District is 44.07:55.93.

3.2. Gender Distribution

A total of 691 valid questionnaires from male respondents and 659 valid questionnaires from female respondents were collected; thus, the gender ratio was 104.86 (for every 100 females). According to public data, the gender ratio of the total population in China is 105.07 (for every 100 females) [19] and that of the permanent population in Maanshan is 104.71 (for every 100 females) [18]. The gender distribution essentially conforms to the gender ratio between the country and the city, and the obtained results are scientific and reasonable.

4. Analysis and Discussion

4.1. Influence of Policy Promotion on Residents' Perception

Residents' understanding of the waste classification policy was the most comprehensive in 2016–2019 (49.78%), followed by 2011–2015 (31.56%), and the number of people who were familiar with waste classification policy before and after 2010 was the lowest. This is positively correlated with the point at which residents first learned about waste classification, which indicates that government policies can positively guide residents' approaches to waste classification. With the increasing publicity, frequency and intensity of government policies related to waste classification, residents' understanding of the main policies of waste classification has gradually improved, and the government and residents have continuously attached importance to the green development.

4.2. Correlation between Residents' Understanding of Waste Classification and Their Age

Residents were assessed on their understanding of waste classification. Among those who passed the assessment, young people (16–40 years old) accounted for the most, comprising 42.33%, followed by mature people (41–60 years old; 33.42%), and people over 60 accounted for only 24.25%. The survey subjects' general knowledge of waste classification showed a downward trend with the increase of age. The means of obtaining information was the key reason for the substantial difference in understanding of waste classification among different age groups. The main channel for people aged 16–40 to obtain information is the internet, and the tools for this are mainly smart phones and computers. Those aged 41–60 years obtain information through the internet and televised news, and occasionally newspapers; the main ways for people over 60 years are watching television, reading newspapers and listening to the radio.

Because online networks can be used to obtain information more widely and rapidly, it is more effective to use the internet to learn about waste classification. Although the internet is gradually becoming one of the main ways for the new generation of residents to obtain information, most residents still use traditional media to obtain information. Therefore, it should be the main method of publicity and promotion in the future, in order to fully integrate traditional media with online media and to reap the combined the benefits of the excellent quality of traditional media and the reach of online media.

4.3. Correlation between Residents' Enthusiasm (RE), Policies (P) and Ecological Civilization (EC)

Table 1 summarizes the respondents' single-choice responses regarding the extent of understanding of waste classification policy (P), the extent of ecological civilization (EC) in China and the level of enthusiasm towards waste classification (RE).

Table 1. Questions, Options and Number of People with Selected Options.

Questions	Options	Number
Recognition degree of waste classification construction in Maanshan city (P)	Have great achievements (Strongly positive)	732
	There are some achievements (Moderately positive)	599
	No results (Not positive)	9
	Unwilling to answer (No evaluation)	10
Possibility of actively participating in waste sorting (RE)	Self-disciplined waste sorting (Strongly positive)	706
	Semi-autonomous waste sorting (Moderately positive)	579
	Refuse to classify waste (Not positive)	60
	Unwilling to answer (No evaluation)	5
Current level of ecological civilization in China (EC)	Make great progress (Strongly positive)	624
	There is some progress (Moderately positive)	678
	No progress (Not positive)	10
	Unwilling to answer (No evaluation)	38

4.3.1. Reliability Analysis and Validity Analysis

The Cronbach α coefficient was used as the measurement index. It can be seen from Table 2 that the reliability coefficient value is 0.953 > 0.9, which shows that the reliability quality of the research data is very high. KMO and Bartlett tests were used to verify the validity, and the KMO value is 0.731, ranging from 0.7 to 0.8, reflecting good validity (Table 3). In addition, the communality values of all research items are higher than 0.4.

Table 2. Reliability Analysis of Scale.

Name	Corrected Item—Total Correlation (CITC)	Cronbach's α if Item Deleted	Cronbach's α Coefficient
RE	0.943	0.898	
P	0.897	0.937	0.953
EC	0.874	0.957	

Table 3. Validity Analysis of Scale.

KMO	Bartlett Test		
	Approx. Chi-Squared	df	*p* Value
0.731	4604.582	3	0.000

4.3.2. Correlation Analysis

The Pearson correlation coefficient is used to study the correlation between RE and P, RE and EC. The correlation coefficient value between RE and P is 0.921, and that between RE and EC is 0.885, and both of them are significant ($p < 0.01$), so there is a significant positive correlation between RE, P and EC.

4.3.3. Regression Analysis and Path Analysis

After the data were centralized, P and EC were taken as independent variables, and RE was taken as a dependent variable for linear regression analysis (Table 4). The model formula is:

$$RE = -0.000 + 0.648 \times P + 0.361 \times EC \quad (1)$$

The square value of model R is 0.897, that is, P and EC can explain the 89.7% change of RE. The model passed the F test (F = 5883.882, $p = 0.000 < 0.05$), which showed that at least one of P and EC would have an influence on RE. The regression coefficient of P is 0.648 (t = 38.728, $p = 0.000 < 0.01$), and that of EC is 0.361 (t = 25.289, $p = 0.000 < 0.01$), that is, both P and EC have significant positive effects on RE.

Table 4. Results of Linear Regression Analysis (n = 1350).

	Unstandardized Coefficients		Standardized Coefficients	t	p	VIF
	B	Std. Error	Beta			
Constant	−0.000	0.005	-	−0.000	1.000	-
P	0.648	0.017	0.599	38.728	0.000 **	3.133
EC	0.361	0.014	0.391	25.289	0.000 **	3.133
R^2			0.897			
Adjusted R Square			0.897			
F			F(2,1347) = 5883.882, p = 0.000			
D-W Value			0.093			

Dependent Variable: RE; ** $p < 0.01$.

As can be seen from Table 4 above, when P affects RE, the normalized path coefficient value is 0.599 > 0, which is significant (z = 38.768, $p = 0.000 < 0.01$), and when EC affects RE, the normalized path coefficient value is 0.391 > 0, which is significant (z = 25.320, p = 0.001). Table 5 shows the covariance relationship between the variables. The standardized path coefficient between EC and P is 0.298 > 0, and this path has a significance of 0.01 (z = 23.384, $p = 0.000 < 0.01$). It can be seen that there is a significant positive covariance correlation between EC and P. According to the above contents, the path diagram is obtained (Figure 1).

Table 5. Covariance Table of Variables.

X	Y	Unstandardized Estimated Coefficient	Std. Error	z	p	Std. Estimate
EC	P	0.298	0.013	23.384	0.000	0.825

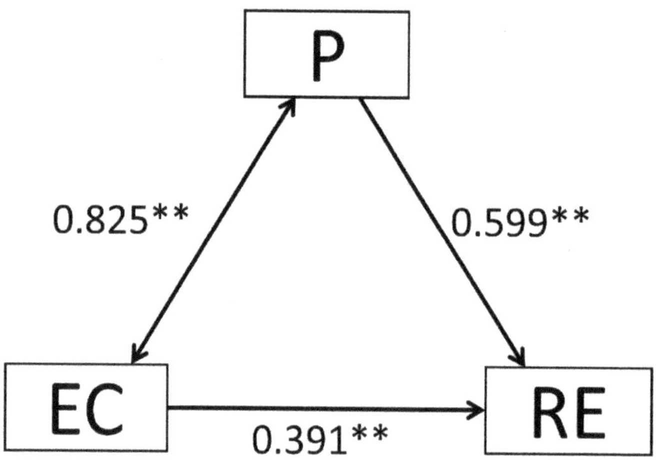

Figure 1. Path analysis diagram of the correlation between residents' enthusiasm (RE), policies (P) and ecological civilization (EC), ** $p < 0.01$.

5. Conclusions and Suggestions

Waste classification is of immense significance for building a modern environmental governance system and promoting environmental governance in China [20]. Taking residents' awareness of waste classification as an example, this research focuses on the factors influencing residents' understanding of waste classification. The results show that: 1. Most residents' understanding of waste classification mainly comes from the frequency of the government's promotion of waste classification policies and the means of promotion.

2. In terms of age division, young and middle-aged residents have a better understanding of waste classification than the elderly because of their ability to obtain and adopt new information via new media platforms. Therefore, we should increase publicity and education regarding waste classification for the elderly in a way that is appropriate for them. 3. In terms of influence, residents' understanding of waste classification policy (P) and the extent of progress towards ecological civilization in China (EC) have a significant positive influence on their approach to waste classification. The P and the EC have a mutual influence on each other. The suggestions are as follows:

1. Expand publicity to improve residents' cognition.

Residents are the main actors in waste classification, and their attitudes and behaviors influence the outcomes of waste classification policy, so it is particularly important to enhance the understanding of waste classification [21]. We should enhance the appreciation of individuals of the importance of proper waste classification, improve the supervision and feedback mechanisms of waste classification, and develop waste classification infrastructure that is comprehensive and integrates recycling. The most feasible way to promote the publicity, popularization and understanding of waste classification policies is to incorporate science and technology, and fully integrate traditional and online media.

At present, the dissemination of relevant policies mainly depends on the government network and TV advertisements. The government can use the network platform to popularize concepts, open official accounts, record relevant videos or write popular science articles, and dispel the limitations of time, space and outreach regarding waste classification education. It should be noted that the interest and the popularity of this content should be optimized in order to encourage people to accept the information. At the same time, in order to better promote the information to children and the elderly, the local government should make in-depth field visits and formulate effective publicity mechanisms according to local conditions.

2. Accelerate the scientific and modern waste sorting facilities.

Popularizing waste classification and fully developing and utilizing waste resources will be the focus of waste classification in the future. As a carrier of waste recycling, waste bins play a vital role in the waste classification and recycling system. Therefore, it is a potential option to post posters of classification instructions next to waste bins to assist residents in classification, or to set up a billboard near waste bins to attract residents' attention by using the slogans in the billboard. With the development of science and technology, intelligent waste bins will emerge as the times require. Users only need to input the waste product's name or display the waste under a camera, and the intelligent waste bin can automatically indicate the waste category and open the corresponding box cover, which reduces the difficulty in remembering categories and improves the classification accuracy.

3. Improve the process of waste classification system

Another reason for citizens' low enthusiasm for participation comes from inadequate government management processes—after waste is classified into waste bins, it is mixed and transported again by waste trucks; after the waste classification is completed, the disposal method is landfill; proper recycling infrastructure is not currently in place. These phenomena have greatly affected the enthusiasm of citizens. Although there are no obvious cases mentioned above in Maanshan, there are still problems in the waste management system, such as low efficiency of packaging and transfer in waste transfer stations, no classification requirements for waste in some areas, and inadequate penalties for littering. Therefore, there is an urgent need to develop relevant recycling policies in strict accordance with the corresponding norms and standards, maximize its efficiency, improve the urban environment, and increase penalties for non-compliance. It is also necessary to optimize the suitability of facilities for waste sorting, continually improve the storage and transportation system for waste sorting, and optimize the waste-sorting process through the supervision of administrative departments.

It should be emphasized that this study only analyzes two significant determinants of residents' enthusiasm, making it a preliminary study. In the follow-up study, we should explore a wider range of determinants, in order to understand the potential causes of each factor and the more complex interaction mechanism, so as to provide a reference for the Chinese government to implement more effective policies. For example, government policies can be differentiated according to target groups, but this requires in-depth research in order to determine the optimal means of differentiation.

Author Contributions: S.G.-Z.: writing—original draft, writing—review and editing, investigation, visualization, methodology; T.C.: writing—original draft, writing—review and editing, investigation; Z.H.: writing—review and editing; H.Z.: writing—review and editing; P.M.: writing—review and editing; J.C.: resources; supervision. All authors have read and agreed to the published version of the manuscript.

Funding: This research was supported by the Fundamental Research Funds for the Central Universities (2572022DJ03).

Institutional Review Board Statement: Not applicable.

Informed Consent Statement: Not applicable.

Data Availability Statement: The data that support the findings of this study are available on request from the first author, Ge-Zhang S., upon reasonable request.

Conflicts of Interest: The authors declare no conflict of interest.

References

1. Ding, Y.; Zhao, J.; Liu, J.W.; Zhou, J.; Cheng, L.; Zhao, J.; Shao, Z.; Iris, Ç.; Pan, B.; Li, X.; et al. A review of China's municipal solid waste (MSW) and comparison with international regions: Management and technologies in treatment and resource utilization. *J. Clean. Prod.* **2021**, *293*, 126144.
2. Kurniawan, T.A.; Liang, X.; O'Callaghan, E.; Goh, H.; Othman, M.H.; Avtar, R.; Kusworo, T.D. Transformation of solid waste management in China: Moving towards sustainability through digitalization-based circular economy. *Sustainability* **2022**, *14*, 2374. [CrossRef]
3. Nanda, S.; Berruti, F. Municipal solid waste management and landfilling technologies: A review. *Environ. Chem. Lett.* **2021**, *19*, 1433–1456. [CrossRef]
4. Lee, R.P.; Meyer, B.; Huang, Q.; Voss, R. Sustainable waste management for zero waste cities in China: Potential, challenges and opportunities. *Clean Energy* **2020**, *4*, 169–201.
5. Kehinde, O.; Ramonu, O.J.; Babaremu, K.O.; Justin, L.D. Plastic wastes: Environmental hazard and instrument for wealth creation in Nigeria. *Heliyon* **2020**, *6*, e05131. [CrossRef]
6. Pan, D.; Su, F.; Liu, C.; Guo, Z. Research progress for plastic waste management and manufacture of value-added products. *Adv. Compos. Hybrid Mater.* **2020**, *3*, 443–461. [CrossRef]
7. Yang, M.; Chen, L.; Wang, J.; Msigwa, G.; Osman, A.I.; Fawzy, S.; Rooney, D.W.; Yap, P.S. Circular economy strategies for combating climate change and other environmental issues. *Environ. Chem. Lett.* **2023**, *21*, 55–80. [CrossRef]
8. Awasthi, S.K.; Sarsaiya, S.; Kumar, V.; Chaturvedi, P.; Sindhu, R.; Binod, P.; Zhang, Z.; Pandey, A.; Awasthi, M.K. Processing of municipal solid waste resources for a circular economy in China: An overview. *Fuel* **2022**, *317*, 123478. [CrossRef]
9. Fu, Y.; Zhu, J. Green design and recycling systems for solving the dilemma of disposable chopsticks waste caused by online food delivery: A review. *BioResources* **2021**, *16*, 8640. [CrossRef]
10. Zhang, S.; Hu, D.; Lin, T.; Li, W.; Zhao, R.; Yang, H.; Pei, Y.; Jiang, L. Determinants affecting residents' waste classification intention and behavior: A study based on TPB and ABC methodology. *J. Environ. Manag.* **2021**, *290*, 112591. [CrossRef] [PubMed]
11. Almulhim, A.I.; Abubakar, I.R. Understanding public environmental awareness and attitudes toward circular economy transition in Saudi Arabia. *Sustainability* **2021**, *13*, 10157. [CrossRef]
12. Li, L.; Yue, G.; Xinquan, G.; Yingmei, Y.; Hua, C.; Jianping, H.; Jian, Z. Exploring the residents' intention to separate MSW in Beijing and understanding the reasons: An explanation by extended VBN theory. *Sustain. Cities Soc.* **2018**, *37*, 637–648. [CrossRef]
13. Liu, C.; Jing, Q.; Cong, J.; Zhang, W. How do integrity level and economic punishment affect residents' willingness and behavior to separate household waste? New evidence from 1293 questionnaires in Jinan. *J. Clean. Prod.* **2022**, *365*, 132713. [CrossRef]
14. Wu, W.; Zhang, M. Exploring the motivations and obstacles of the public's waste classification participation: Evidence from Sina Weibo. *J. Mater. Cycles Waste Manag.* **2023**, *25*, 2049–2062.
15. Bai, R.; Lin, B. Are residents willing to pay for waste recycling: Evidence from a survey in Chinese first-tier cities. *Environ. Impact Assess. Rev.* **2022**, *95*, 106789. [CrossRef]

16. Xinhua News Agency. China Will Basically Achieve Full Coverage of Waste Classification by the End of 2025 [EB/OL]. Chinese Government Network. 24 May 2023. Available online: https://www.gov.cn/govweb/lianbo/bumen/202305/content_6875926.htm (accessed on 26 May 2023).
17. Maanshan Civilization Network. Maanshan: Strive to Basically Achieve Full Coverage of Domestic Waste Classification by the End of Next Year [EB/OL]. China Civilization Network. 15 October 2021. Available online: http://mas.wenming.cn/wmcj/202110/t20211015_7367379.shtml (accessed on 4 May 2023).
18. Dangtu County Bureau of Statistics. Bulletin of the Seventh National Population Census of Maanshan City [EB/OL]. Dangtu County People's Government. 24 May 2021. Available online: https://www.dangtu.gov.cn/xxgk/openness/detail/content/60bf1f3fd1da9beb078b4567.html (accessed on 4 May 2023).
19. National Bureau of Statistics. Bulletin of the Seventh National Census [EB/OL]. Chinese Government Network. 11 May 2021. Available online: http://www.gov.cn/guoqing/2021-05/13/content_5606149.htm (accessed on 4 May 2023).
20. Chen, S.; Liu, N. Research on Citizen Participation in Government Ecological Environment Governance Based on the Research Perspective of "Dual Carbon Target". *J. Environ. Public Health* **2022**, *2022*, 5062620. [PubMed]
21. Wang, S.; Wang, J.; Yang, S.; Li, J.; Zhou, K. From intention to behavior: Comprehending residents' waste sorting intention and behavior formation process. *Waste Manag.* **2020**, *113*, 41–50. [PubMed]

sustainability

MDPI

Article

Evaluation of Biodegradability of Polylactic Acid and Compostable Bags from Food Waste under Industrial Composting

Angeliki Maragkaki [1,*], Nikitas G. Malliaros [2,3], Ioannis Sampathianakis [1], Theofanis Lolos [3], Christos Tsompanidis [3] and Thrassyvoulos Manios [1]

1 Department of Agriculture, School of Agricultural Science, Hellenic Mediterranean University, 71401 Crete, Greece; sabathianakisj@hmu.gr (I.S.); tmanios@hmu.gr (T.M.)
2 Department of Chemistry, Voutes Campus, University of Crete, Crete, 70013 Heraklion, Greece; nmalliaros@uoc.gr
3 ENVIROPLAN SA, 23 Perikleous & Iras St., Gerakas, 15344 Athens, Greece; fl@enviroplan.gr (T.L.); ct@enviroplan.gr (C.T.)
* Correspondence: amaragkaki@hmu.gr

Abstract: To reduce plastic pollution, biodegradable plastics have been introduced to the market to replace petroleum-based plastics. This work investigates the biodegradation/disintegration of pure poly-L-lactic acid ($PLLA_{B2B}$), composed of food waste (FW), and $PLLA_{B2B}$ bags, under industrial composting conditions, in order to determine whether they are compostable and to examine compost quality. In order to study the biodegradation, pure $PLLA_{B2B}$ was degraded in laboratory conditions and bag samples were put into simulation systems using windrow technology. Phytotoxicity tests were carried out for every compost sample and high germination values were found (97–103.8%). The pure PLLA reached an average biodegradation value of 100.2 ± 3.7% and a disintegration value of 100.0 ± 0.0%, resulting in biodegradable mature compost. After a seven week composting trial, the bag samples had fully decomposed. The samples had no influence on compost characteristics and there were no visually perceptible changes to the compost. Therefore, we suggest that food waste bioplastic can be used as a potential eco-material for compostable bags, one which could be used in industrial composting units and which offers degradable active materials with low environmental impact.

Keywords: biodegradation; food waste; polylactic acid; compostable bags; industrial composting

1. Introduction

Plastic waste accounts for global emissions in the range of 19 to 23 million metric tons annually in fresh and sea water [1], with emissions on land also being calculated in the same order. These quantities are expected to double by 2030 [1]. At the same time most of the bioplastics used to remediate plastic pollution are xenobiotic materials and not entirely "green" [2].

Development of bio-based products has been strongly encouraged due to the depletion of fossil fuels, whereas biodegradable and compostable plastic items are developed with the hope of reducing environmental pollution or streamlining the collection of organic waste [3]. The raw materials for bioplastics used in the existing production method are most commonly based on starch (mainly from corn), using resources (water, energy, nutrients, land) and food quality crops for bioplastic production. This environmentally questionable activity will create massive social issues, especially in poorer countries, once the need for bioplastic increases. The main use of bioplastics is in packaging, which accounted for more than 48% (1.15 million tons) of the market for bioplastics in 2021 [4]. The demand for sustainable products from consumers and brands alike is on the rise due to an increased awareness of the effects on the environment and the need to lessen reliance on fossil fuels. This demand is also being fueled by the bioplastics industry's ongoing development of new materials with improved properties and new functionalities. The number of producers, converters, and end users is continually growing along with the quantity of materials, uses, and products. In this context, large relative growth rates are demonstrated by novel and inventive biodegradable plastics, such as polylactic acid (PLA).

On the other hand, municipal organic waste and agro-industrial byproducts have become a primary environmental, social, and economic concern in today's world [5,6]. Through the Waste Framework Directive (Directive 2008/98/EC, 2008), regulatory frameworks have been established throughout the European Union in order to create a sustainable waste management model that balances economic growth, technological advancement, and environmental and human health protection [7,8]. In this context, the circular economy could contribute to recategorizing these waste streams, as either feedstock raw materials or energy, according to the "cascading use principle".

The exploitation of bio-waste and agro-industrial organic waste through integrated processes is a rising area of investigation in the open literature, although studies are limited so far [9,10]. R&D in the area of waste valorization has focused on the production of energy and fuels but not on the production of bio-based products [11,12]. Bio-waste and agro-industrial organic wastes, that are rich in carbohydrates, protein and lipids/fats, are preferred as they can be combined into a waste stream of homogeneous composition, with large volumes, security of supply, possible funds as gate fee and simple logistics (mass transportation is possible) [13].

Due to the coexistence of biotic and abiotic elements in nature, the entire degradation process of a particular substance can be referred to as environmental degradation [14]. The material characteristics of PLA, such as its molecular first-order structure, as well as environmental aspects like humidity, temperature, and catalytic species, have an impact on how PLA degrades in the environment [15]. PLA is brittle, unclear, and opaque at low molecular weights, but it becomes stronger, more transparent, and more resistant to degradation at higher molecular weights [16].

Polymer mineralization happens under aerobic microbial biodegradation conditions [17]. Microorganisms in the compost can break down PLA in a composting setting after 45 to 60 days at 50 to 60 °C [18]. Under composting settings, commercially available PLA bottles and delicatessen containers decomposed noticeably after 30 days, with PLA bottles degrading more slowly due to their higher degree of crystallinity [19]. The term "aerobic biodegradation" describes the bio-degradation process whereby aerobic microbial respiration occurs. Due to the oxygen that is present, aerobic plastic biodegradation typically takes place in natural conditions like those associated with compost and soil [20]. Additionally, the aerobic process uses less energy and releases less methane, a more dangerous greenhouse gas than carbon dioxide, making it more energy-efficient [21].

According to several studies [17,22,23], blending affects how quickly bio-plastic films break down in soil, simulated landfills, marine water, aquatic ecosystems, and industrial composting facilities. The chemical structure of the resin, the activity of the soil and compost, and the depth at which the samples are buried all affect these rates. It has been attempted, in published research on the degradation behavior of blends, to relate the cumulative behavior of biodegradable films and articles, such as total CO_2 evolution and total weight loss, to their degradation and composting rate [24]. As already noted, one method for calculating the rate of total product breakdown under aerobic active microbial conditions is the measurement of evolved CO_2 [25,26].

In this study, the biodegradation/disintegration of $PLLA_{B2B}$ produced from FW was examined under laboratory and real conditions (windrows), in order to assess how these proposed materials will behave after use. Additionally, the impact of the compost produced following the decomposition process was examined in terms of its phytotoxic impact on particular plants. To assess how the two different techniques affected the qualities of the materials, visual observations were carried out for each test at various periods.

2. Materials and Methods

2.1. Raw Materials

The food waste (FW) used in the present study was collected from the students' restaurant at the Hellenic Mediterranean University (HMU), Heraklion. The FW composition was 80% raw-fresh food (vegetables), 10% fruits and 10% salads (on a wet-weight basis). Vegetable waste (VW) was collected from a major supermarket in Crete. The HMU Department of Parks and Gardens provided material from pruning (PR) waste, branches and leaves. The PR was shredded with a Woodchipper Green Mech Abnorist 130. The commercial compostable bag (CB_c) is made of MaterBi designed for delivering organics to household waste collection systems and is compliant with the EN 13432 norm [27]. The compostable bag (CB_{B2B} was made of PBAT with MFI = 4–5 g/10 min, 12% polylactic acid ($PLLA_{B2B}$), produced using FW gathered from the municipality of Heraklion in the context of the B2B Project and various other active ingredients.

The pure $PLLA_{B2B}$ was synthesized through direct azeotropic polycondensation of L-lactic acid. For the production of lactic acid, the approach suggested by Sakai and colleagues [28] was followed and optimized for the creation of a scalable synthetic process. The $PLLA_{B2B}$ was obtained from polymerization as a white powder with a particle size <3 mm and used without pretreatment. For the biodegradability tests the $PLLA_{B2B}$ was cut and milled, obtaining thin, homogeneous pieces < 1 mm. Two fractions of $PLLA_{B2B}$ were used (Figure 1): the polymeric granules 3 mm "$PLLA_{B2B3}$" for the disintegration test and the polymeric powder "$PLLA_{B2BP}$" for the composting process needed to determine the ecotoxic effect.

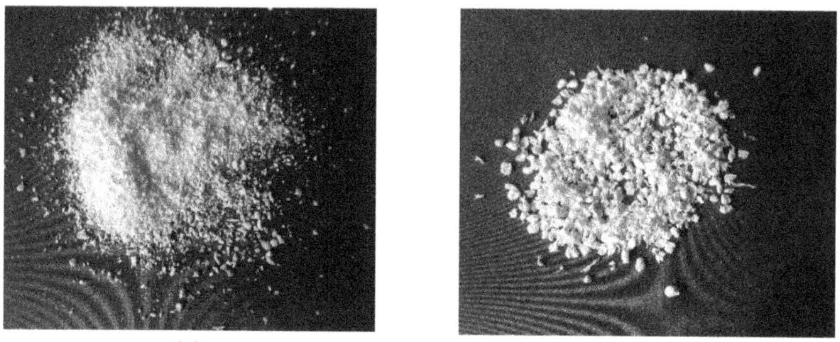

Figure 1. Sample pictures: (**a**) sample $PLLA_{B2BP}$ for ecotoxicity test, (**b**) sample $PLLA_{B2B3}$ for disintegration test.

For the biodegradability tests all the organic material components used to prepare the biowaste (bulking agent (BA) 24%, rabbit feed 46%, sawdust 13%, compost 13%, fruit, vegetable and cereals 3.8% and urea 0.2% dry wt) were reduced to a particle size < 50 mm. The pilot scale test was conducted under defined composting conditions according to ISO 16929:2021 [29].

The inoculum for the biodegradability test was a mature compost obtained from a composting plant treating 70% residues from public green waste management and 30% organic fraction from municipal waste collection. Before using it, the compost was sieved using a 2 mm sieve and, the fraction < 2 mm was used as inoculum. The inoculum characteristics were volatile solids (VS) 37.4%, moisture 41.1%, pH 6.3 and total nitrogen (TN) 1.8%. The mean composition of raw FW, VW, PR, CBc, CB_{B2B}, biowaste and BA is summarized in Table 1. Table 2 provides a summary of the characteristics of the synthetic $PLLA_{B2B}$ used in the current study.

Table 1. Characteristics of raw materials and compostable bags.

Parameter	VW	FW	PR	CB_c	CB_{B2B}	Biowaste	BA
pH	4.2 ± 0.4	4.2 ± 0.1	7.7 ± 0.1	-	-	5.9 ± 0.1	5.3 ± 0.1
Moisture (%)	88.8 ± 0.1	65.9 ± 1.5	59.6 ± 1.0	-	-	64.9 ± 1.2	31.7 ± 1.1
VS (g/kg) (DM)	904 ± 2.1	878.6 ± 11.7	890.3 ± 0.9	968.8 ± 5.5		877.0 ± 2.7	901.0 ± 1.2
N (%)	1.5 ± 0.0	2.5 ± 0.0	1.4 ± 0.0	0.1 ± 0.0	0.1 ± 0.0	1.9 ± 0.0	0.5 ± 0.0
TOC (g/kg)	433.1 ± 4.9	634.1 ± 24.1	566.4 ± 66.1	692 ± 92.0		440.0 ± 2.3	426.2 ± 5.2
P (%)	0.1 ± 0.0	0.2 ± 0.0	0.3 ± 0.2	-	-	0.2 ± 0.0	0.1 ± 0.0

Table 2. Characteristics of produced $PLLA_{B2B}$.

Parameter	$PLLA_{B2B}$
Weight average molecular weight (M_w)	149,452
Number average molecular weight (M_n)	81,146
Polydispersity Đ	1.84

2.2. Determination of the Biodegradability Test

The biodegradability test was performed according to the UNI EN ISO 14855-1:2013 standard [30] by mixing the test sample $PLLA_{B2BP}$ with mature compost to verify the capability of a typical microbial compost environment to biologically convert the organic fraction of the sample into CO_2, water and biomass.

The inoculum consisted of mature compost obtained from an industrial composting plant. The mature compost was sieved to eliminate the coarse fraction. The compost fine fraction obtained represented the test inoculum, while its volatile solids content had to be higher than 30% of the total compost dry solids.

The test sample $PLLA_{B2BP}$ was mixed with the inoculum at a ratio of about 1:6 (dry solids) and introduced into the reactor. The reactors were placed in an incubator and kept at 58 ± 2 °C for the entire duration of the test.

The aerobic condition was maintained by blowing environmental air into the reactors. The aerobic condition permitted the conversion of the sample organic fraction into CO_2 during the test. The gas flow from each reactor was sent to a gas analyzer that determined the CO_2 concentration and the outlet flow rate at regular time intervals. The percentage of biodegradation was determined as the percentage of the initial theoretical carbon in the test compound that was converted into CO_2.

The sample was cut and milled to obtain thin, homogeneous pieces <1 mm. The blank consisted of three glass reactors (3 L capacity) containing testing mixture, while the reference sample consisted of three glass reactors (3 L capacity), each containing about 50 g of microcrystalline cellulose plus testing mixture, and the sample consisted of three glass reactors (3 L capacity), each containing about 50 g of milled sample plus testing mixture. Finally, the testing mixture consisted of mature compost from a composting plant mixed with an inert support (vermiculite, Sigma-Aldrich code 101532822), 300 g compost and 100 g vermiculite (dry weight) for each reactor with humidity at 50 ± 5% throughout the test.

The environmental air was blown through silicone rubber tubes into inlet gas flowmeters regulating the inlet flowrate of the reactors at approximately 10–15 L/h. Aquarium type pumps were used for this purpose. The aerobic condition permits the conversion of the sample organic fraction into CO_2 during the test. The air was previously humidified by bubbling in water at the same temperature as the reactors.

The air at the exit of each reactor was blown through gas-impermeable tubes into a humidity recovery system to eliminate water. Then the dry air was sent to a gas infrared analyzer that determined the CO_2 concentration and the flow rate at regular time intervals. The percentage of biodegradability was calculated as the percentage of CO_2 produced with respect to the total theoretical organic carbon content of the sample. The temperature during the test was 58 ± 2 °C. The CO_2 measurement was carried out twice per day for the first five days of the test, then at least once per day for up to 45 days of the test. Afterwards, the measurement was performed at least five days per week up to the 90th day. The measurements were reduced to 3 times per week when the trial continued for an additional 90 days. Once a week for the whole duration of the test the reactor content was thoroughly mixed, while water was added when necessary. The mixing and water addition operations were recorded.

2.3. Determination of Degree of Disintegration

The purpose of this test was to evaluate the disintegration of the sample in a composting bin in the presence of freshly prepared biowaste. The composting process was monitored regularly and conducted until the compost was fully stabilized (3 months). At the end of the process the compost was sieved, and the disintegration of the sample was carefully measured. The ecotoxic effect of the compost obtained from this process was then evaluated to ensure the absence of any toxic effect on higher plants.

To perform the test, plastic bins of about 60 L capacity were used. The following bins were filled: (a) control bins: 2 bins each containing at least 20 kg of biowaste and (b) sample bins: 2 bins each containing at least 20 kg of biowaste with 10% sample concentration with respect to wet biowaste obtained by adding: 9% ($PLLA_1$: 1.87 kg, $PLLA_2$: 1.87 kg) polymeric granules shredded < 2 mm + 1% ($PLLA_1$: 208.3 g, $PLLA_2$: 206.5 g) polymeric granules with dimensions of 3 mm, to each bin. The two fractions of the sample were well mixed with the biowaste inside the bin. The temperature was measured at regular intervals with probes (every day for the first eight weeks and at least three times a week for the remaining weeks). Periodically the composting material was turned and visually inspected (once a week for the first four weeks and at least every two weeks for the remaining period). Water was added to the bins to restore humidity losses due to evaporation and the water content was maintained around 50–60% during the test, taking care that no free-standing water was present at the bottom of the reactor after turning of the composting material. During turning operations, the smell and appearance of the composting mixture were monitored, as was the appearance of the sample specimens. Sampling of composting material was performed to check pH at 4, 8 and 12 weeks. The reactors were aerated daily for 4 h by fluxing air to ensure that the oxygen content was >10%.

At the end of the 12 weeks of the test all of the bins were cooled. The compost was sieved with 10 mm, 5 mm and 2 mm sieves to isolate fractions of the residual sample. All the compost fractions, except the >10 mm fraction, were then used to determine the quality of the compost.

2.4. Ecotoxic Effect Evaluation

The phytotoxic effect (germination of seeds and growth of the plants) was assessed (a) on compost obtained after exposure to the composting process for 12 weeks, without the presence of the sample, and (b) on compost obtained after exposure to the composting process for 12 weeks, in the presence of the sample "$PLLA_{B2BP}$". Reference compost and sample compost were mixed, before the analysis, with a reference substrate prepared by mixing vermiculite (83%) and peat (17%).

The phytotoxic effect tests were conducted with two different types of seeds: mung bean (Vigna radiata) (100 seeds per pot) and barley (Hordeum vulgare) (50 seeds per pot).

The sowing was conducted in pots filled with an 83% vermiculite/17% peat mixture containing the reference compost or the sample compost, both analyzed at two different concentrations: 25% and 50% (*w*/*w*). At the end of the test, germination (number of plants germinated) and biomass (dry weight of the plants) were evaluated.

Plastic containers of 500 mL volume were filled with 15 g of vermiculite, to which 50 mL of deionized water was added. Then, about 200 g of testing mixture was added and the seeds were distributed on the surface and covered by a thin layer of 83% vermiculite/17% peat mixture. The pots were covered (to preserve humidity) and incubated in the dark till the seeds germinate, when the covering was removed, and the pots were exposed to light/dark cycles with a photoperiod of 16 h for 14 days. The plants were watered as necessary. At the end of the growth period, the plants were cut at the base where they emerge from the soil and the number of plants germinate and plant dry weight were determined.

For each seed type the following pots were set up:

- 3 pots with vermiculite/peat mixture (83% vermiculite and 17% peat), as growth control;
- 3 pots with reference compost 25% *w*/*w*;
- 3 pots with reference compost 50% *w*/*w*;
- 3 pots with sample compost 25% *w*/*w*;
- 3 pots with sample compost 50% *w*/*w*;
- for a total of 15 pots for each seed type.

2.5. Composting in Windrows

The biodegradability study was carried out in windrow technology simulation systems at the Hellenic Mediterranean University in Crete. The composting experiment was conducted in a commercial composter (BioActor, Athens, Greece, Helesi). The composter comprises ten stackable frames. Each frame consists of three integrated layers: (a) a perforated inner ventilation wall, (b) an insulating middle wall, and (c) a perforated outer ventilation wall. The FW, VW and PR were added to the composter in April 2019. PR was added to the material (FW:VW) at a volume ratio of 1:1 (FW & VW:PR). The compostable material was emptied from the composter and turned manually at one week intervals, with 7 turnings in total (7 weeks). In order to maintain moisture levels between 60% and 70% (*w*/*w*), water was occasionally added to the material manually. The temperature in the composter was measured on a daily basis in the core, at a depth of 40–60 cm. During turning, composite samples were taken, one from the top, 20 cm from the top of the system; one from the middle; and one from the bottom, at 10 cm from the bottom of the system. At each turning, the samples were lifted from the compost and all samples were subsequently photographed and assessed. Table 3 provides a summary of the material composition of samples used in the current study.

Table 3. Material composition of samples.

Sample	Type	Description
1	Cellulose (blank)	-
2	Starch	MaterBi EN 13432 norm-industrial compostable
3	CB_{B2B}	PBAT with MFI = 4–5 g/10 min, 12% polylactic acid ($PLLA_{B2B}$) and various other active ingredients

2.6. Analytical Methods

The raw composting waste and outflow were tested for moisture content, pH, electrical conductivity (EC), total organic matter (TOM), volatile solids (VS), total organic carbon (TOC), total nitrogen (TN) and total phosphorus (TP).

The weight loss of the sample after drying at 105 °C allowed us to determine its moisture content [31]. The pH and EC of a 1/1.5 solid/liquid aqueous extract were determined (extraction duration equal to 24 h), and total nitrogen (TN) was quantified using the semi-micro Kjeldahl method. Total phosphorus (TP) was determined by extracting the elements from solid samples and measuring them in the same way as liquid samples. Specifically, the TP was brought to a solution by acidic digestion using HNO_3. For TOC analysis, a solid sample module was utilized (SSM-5000A, Kyoto, Japan, Shimadzu). Each number was calculated as the average of three similar measurements.

The biodegradability test was performed according to the UNI EN ISO 14855-1:2013 standard [30]. The degree of disintegration test was conducted according to ISO 16929:2021 [29].

The ammonium nitrogen and nitric nitrogen were determined according to the method reported in UNI 10780:1998 [32]. An amount of 5 g of the sample is suspended in KCl for 1 h at a 1:10 ratio. After centrifugation, the solution was diluted (ten times) and Devarda's alloy was added. The solution was then distilled in the presence of NaOH boric acid. The captured nitrogen by boric acid was titrated with standard HCl. The evaluation of CO_2 concentration in the outlet gas from the reactors was carried out with a nondispersive infrared (NDIR) detector (Ecocontrol model EC100).

The ecotoxic effect was evaluated according to EN 13432:2000 [27]. The tests were conducted in triplicate. The results obtained from the reference compost and the compost sample were compared at different concentrations. The percentage of germination and the percentage of growth obtained using the sample compost were calculated as a percentage of the values obtained with the corresponding reference compost.

Regarding the germination capacity of the seeds, a layer of cotton and a filter paper disk were piled in a Petri dish and 5 mL of deionized water were added. Twenty seeds were distributed on top of the paper disk, and another previously humidified paper disk was placed on top of the seeds. The Petri dish was covered with the lid and closed by a parafilm strip. The Petri dishes were incubated for 4 days in the dark at room temperature. The germinated seeds were counted, and the percentage of germination (germination capacity) was calculated with respect to the initial seed number. The germination index (GI) was calculated using the formula found in equation [33] (GI): GI (%) = 100 × (average number of seed germination × average length of treatment's roots) (average number of seed germination × average of root length of control).

Arsenic (As), cadmium (Cd), chrome (Cr), mercury (Hg), nickel (Ni), lead (Pb), copper (Cu), selenium (Se), zinc (Zn) and molybdenum (Mo) were determined according to UNI EN 15411:2011 [34]. The levels of heavy metals (Cr, Cu, Ni, Cd, Pb, Zn, Hg) were determined using atomic absorption spectroscopy (Perkin Elmer AAnalyst400, Waltham, MA, USA). An atomic fluorescence spectrometer was used to analyze As (AFS-820; Jitian, Beijing, China).

Fluorine (F) was determined according to EPA 9056A (2007) [35]. The EPA 5050 (1994): Bomb preparation method for solid waste was used for the sample preparation [36]. The sample was oxidized by combustion in a bomb containing oxygen under pressure. The liberated halogen compounds were absorbed in a sodium carbonate/sodium bicarbonate solution. The bomb combustate solution was then analyzed by ion chromatography (IC). IC analyses were conducted using a Dionex ISC-2000 reagent-free ion chromatograph eluent generator cartridge (RFIC EGCII KOH), CD 20 conductivity detector and anion self-regenerating suppressor (ASRS 300, 4 mm, all Dionex, Sunnyvale, CA, USA).

Origin 9 (OriginLab, Northampton, MA, USA) was used to undertake the statistical analysis of the data and the findings of this study (analysis of average values, variance, and standards deviation). Statistical analysis of variance (ANOVA) was used to examine significant differences.

3. Results and Discussion

3.1. Biodegradability Test

Tests on biodegradability are required to assess the environmental effects of plastic materials and to identify strategies with which to prevent the unsettling buildup of polymers after their commercial shelf life. The primary indicator typically used to calculate the rate of composting of biodegradable materials is the measurement of CO_2 evolution during composting of pure and blend samples.

Regarding the validation of the test, the degree of biodegradability of the reference material (microcrystalline cellulose) was 90% (>70% after 45 days). The difference among the percentages of biodegradability of the reference material (microcrystalline cellulose) in the different reactors was 9.2% at the end of the test (<than 20%). The CO_2 production of the compost after 10 days of test was 72.3 (between 50 and 150 mg CO_2/g of volatile solids). This means that the test was validated.

During the test setup, the milled sample was well mixed with compost and vermiculite. After four weeks of testing the sample reactor mixture was very similar to that of the blank and reference reactors.

The evolution of the total cumulative CO_2 production is reported in Figure 2 for the blank, reference and sample reactors. The evolution trend of CO_2 of the sample $PLLA_{B2BP}$ was lower than the microcrystalline cellulose reference sample. After about 20 days it started to increase constantly until, after 50 days of testing, it exceeded the production of CO_2 of the microcrystalline cellulose reference. Table 4 provides the weight of the reactors at 0 and 91 days of test.

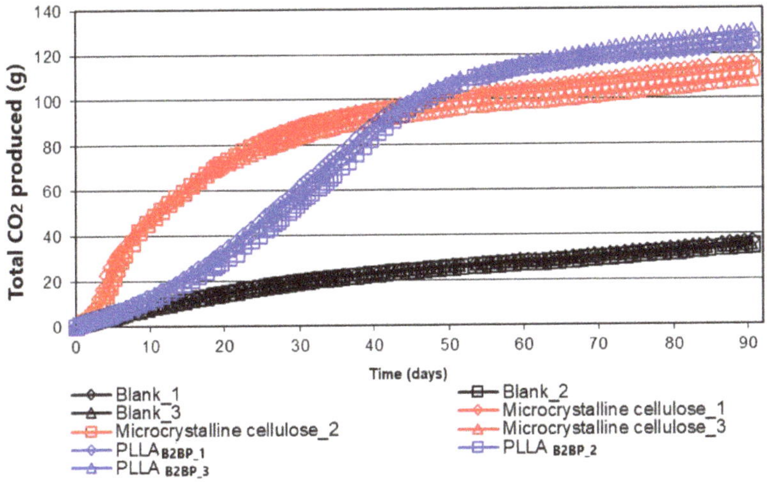

Figure 2. Total CO_2 production of the reference microcrystalline cellulose sample, $PLLA_{B2BP}$ sample and blank.

Table 4. Weight of reactors at 0 and 91 days of test.

Reactor	Initial Weight (0 Day)		Final Weight (91 Days)	
	Gross	Net	Gross	Net
Blank_1	2160	793	2090	723
Blank_2	2152	793	2082	723
Blank_3	2152	796	2056	700
Microcrystalline cellulose_1	2260	896	2117	753
Microcrystalline cellulose_2	2240	877	2182	819
Microcrystalline cellulose_2	2256	886	2162	792
$PLLA_{B2BP\_1}$	2256	888	2186	818
$PLLA_{B2BP\_2}$	2259	888	2215	844
$PLLA_{B2BP\_3}$	2254	885	2184	815

The biodegradability percentage was calculated with respect to the quantity of the total organic carbon initially contained in the samples. Figure 3 shows the evolution of the biodegradation percentages of the different replicates for the sample and the reference. The biodegradation of the sample $PLLA_{B2BP}$ started lower, but after 20 days the biodegradation rate of the sample $PLLA_{B2BP}$ increased quickly and constantly, reaching the biodegradation percentage of the reference.

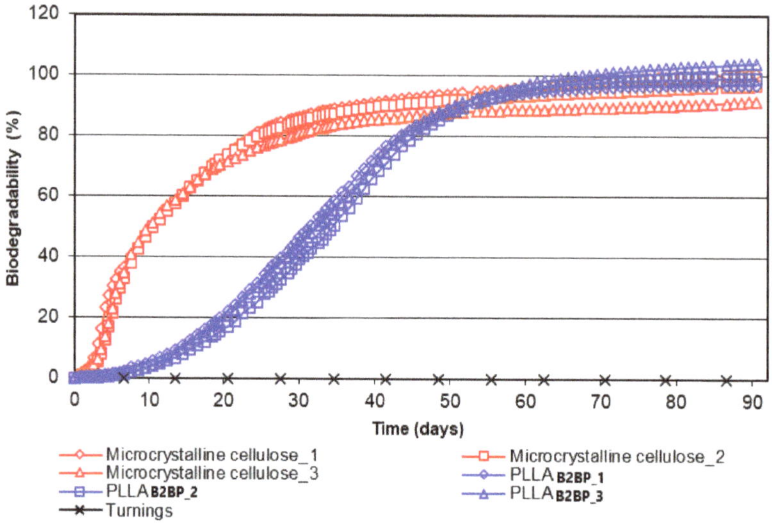

Figure 3. Evolution of aerobic biodegradability under controlled composting conditions for microcrystalline cellulose reference sample and $PLLA_{B2BP}$.

The biodegradability percentage was calculated with respect to the quantity of the total organic carbon initially contained in the samples. At the end of the test an average biodegradation value equal to 96.4 ± 4.6% was reached for microcrystalline cellulose and 100.2 ± 3.7% for $PLLA_{B2BP}$, thus fulfilling the 90% limit required by the EN 13432:2000 norm (limit value required ≥ 90%). Published works on the degradation behavior of blends have tried to relate the cumulative behavior of biodegradable films and articles, e.g., total CO_2 evolution and total weight loss, to their degradation and composting rate.

According to Kalita et al. [37] 80.50% to 94.20% biodegradation was observed in different PLAs at 136 to 139 days. After hydrolytic destruction, PLA oligomers and other intermediates were digested by microorganisms in the compost and turned into carbon dioxide, water, and humus [37]. For $PLLA_{B2B}$, it was found that 85% of the composite fraction decomposed after 50 days as compared to 30 days for blank. Another study reported similar results, but did not include information on the lag phase, which was crucial for comparing the two systems [38,39]. This study confirms our findings on the compostability of PLA in used compost. In the case of Kalita et al. [37] the PLA biodegradation rate was comparatively slow. This could be for a number of reasons, including the material's suitability for composting, molecular structure, and mechanical strength. Additionally, several authors have observed that composite materials, such as PLA blends, disintegrated more slowly when composted [37,40].

3.2. Disintegration Test

In order to monitor the progress of the disintegration during mixing and humidity recovery, the bins were monitored for the odor and the visual appearance of the mixture and the sample under test. The odor and visual inspections evidenced a regular aerobic composting process during the 12 testing weeks overall. No deviations were registered during the test and therefore no corrective operations were adopted.

During the different process phases the following odor evolution for blank bins and sample bins was observed: (a) first 15 days: start of fermentation processes with rancid odor and increasing ammonia smell; (b) from 16 to 40 days: the rancid odor was attenuated until it disappeared completely and was replaced by an ammonia odor that became strong; (c) from 41 to 56 days: ammonia smell decreased and was replaced by fresh soil smell; and (d) from 57 to 84 days: presence of typical odor of mature compost by the end of the test. The visual aspect of the organic waste in the bins evolved, starting from an initial darkening associated with a volume reduction that was much more visible in the sample reactor than in the Blank.

During the second week the growth of white molds was noted, which increased to cover a major part of the biowaste. At the same time, the smell of ammonia significantly increased. The biowaste in the sample reactors showed a significant volume reduction starting from the second week, while the sample mixing with the biowaste increased by disintegration and darkening.

The granules of the sample, both 3 mm and <2 mm, started to become dark and to integrate with the biowaste by the end of the first week. The disintegration process increased in the following days: after only 22 days of the test only small residues of the granules remained in the reactors. At 29 days of the disintegration test the sample completely disappeared. Figure 4 shows the sample during the composting process. Table 5 presents the final product's quality obtained at the end of the test. Table 6 presents the results of the sample $PLLA_{B2B}$ disintegration test.

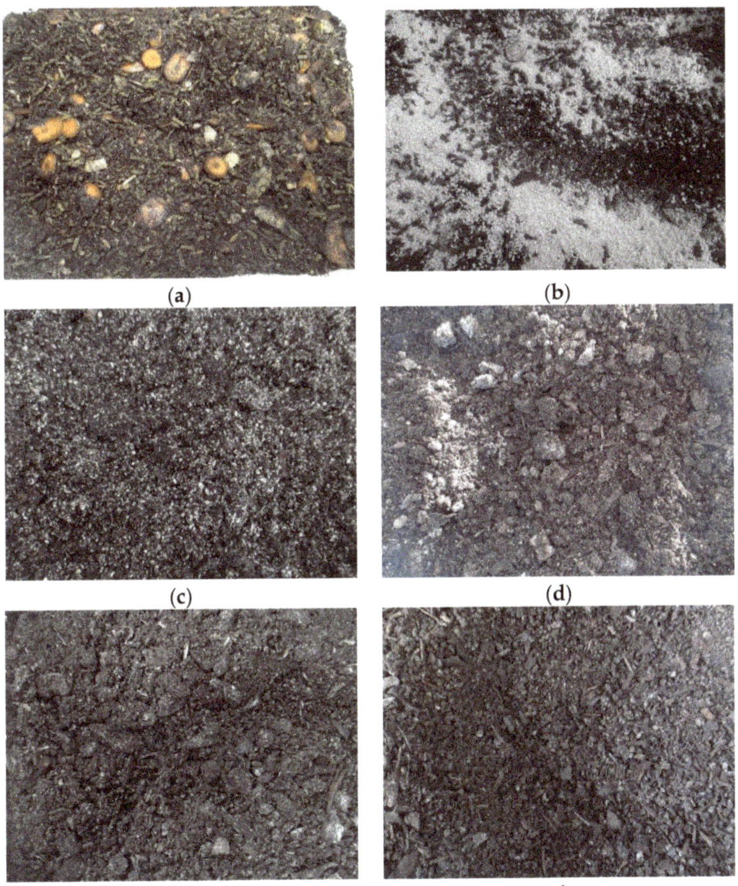

(**a**) (**b**) (**c**) (**d**) (**e**) (**f**)

Figure 4. Photo of the sample during the test, at the end of the test and the final residues of the sample. (**a**) Fresh Biowaste at the beginning of the test, (**b**) sample granules at the beginning of the test, (**c**) disintegration of the granules after 8 days, (**d**) residues of the granules after 22 days, (**e**) final sample compost after 12 weeks, (**f**) final sample compost < 10 mm sieving fraction.

Table 5. Chemical characterization of the compost obtained at the end of the test.

Parameter	$Blank_1$	$Blank_2$	$PLLA_{B2B1}$	$PLLA_{B2B2}$
Moisture (%)	49.2 ± 0.5	39.0 ± 0.3	35.5 ± 0.1	50.1 ± 0.1
Volatile solids (%)	73.8 ± 2.1	72.5 ± 4.5	73.2 ± 2.6	72.5 ± 3.3
pH	6.9 ± 0.1	8.0 ± 0.1	7.9 ± 0.1	7.2 ± 0.1
NH4-N (mg/Kg)	2343 ± 6.1	2145 ± 3.6	3176 ± 2.7	2770 ± 9.3
NOx-N (mg/Kg)	3692 ± 4.9	3951 ± 8.1	3090 ± 21.1	3387 ± 10.2
TN (%)	2.9 ± 0.0	2.8 ± 0.0	2.9 ± 0.0	2.9 ± 0.0
C/N	12.7 ± 3.1	13.2 ± 2.7	12.7 ± 4.2	12.5 ± 0.1

Table 6. Results of the sample $PLLA_{B2B}$ disintegration test.

Sample Bin	Weight of Sample Residues (g) as Dry Weight				Disintegration (%)	Average Disintegration (%)	Disintegration Limit (%)
	Initial	At the End					
		2–10 mm	>10 mm	Total			
$PLLA_{B2B1}$	204.5	0.0	0.0	0.0	100.0	100.0	≥90.0
$PLLA_{B2B2}$	202.7	0.0	0.0	0.0	100.0		

The odor and visual controls evidenced a regular aerobic composting process during the 12 testing weeks overall. At the end of the test the compost obtained from the composting process was mature, as the maximum temperature measured during the self-heating test remained below 30 °C: Rottegrad = V as requested by ISO 16929:2021. No deviations were registered during the test and therefore no corrective operations were adopted. The sample of $PLLA_{B2B}$ polymeric granules with dimensions of 3 mm, disintegrated at 100% value, within the limits specified by the EN 13432:2000/AC:2005. In contrast with earlier studies on thicker samples, the results of $PLLA_{B2B}$ disintegration demonstrate extended stability [41,42]. The sample of polymeric granules with a large surface area and high porosity may be responsible for the rapid breakdown [43,44]. As set out in Table 5, $PLLA_{B2B}$ did not affect the chemical characteristics of the compost obtained at the end of the test.

3.3. *Phytotoxic Effect Test*

Using composts produced from the breakdown of the materials studied in a lab, phytotoxicity tests were conducted to determine the compost quality.

The compost obtained from the disintegration test in the presence of the sample $PLLA_{B2B}$ did not present an inhibiting effect on either the germination or growth of both plants analyzed at the different compost concentrations tested. The results are shown in Table 7. Figure 5 presents photographs of ecotoxic test.

(a)

(b)

Figure 5. Photos of ecotoxic test (**a**) growth of ecotoxic test (4th day), (**b**) end of ecotoxic test (Wastewater and Solid Waste Management Laboratory of the Department of Agriculture).

Table 7. Percentage of germination and growth on the Sample Compost with $PLLA_{B2B}$ respect the values obtained with the Reference Compost.

	Barley		Mung Bean	
Compost (%)	Germination (%)	Grown (%)	Germination (%)	Grown (%)
25	98.5	109.3	103.8	110.6
50	101.5	118.0	97.0	95.3

High germination values were found in the phytotoxicity test: 97.0–118.0%.

3.4. Heavy Metals

The concentrations of heavy metals and other toxic and hazardous substances of sample $PLLA_{B2B}$ were determined within the limits specified by the EN 13432:2000. The results are shown in Table 8.

Table 8. Chemical characterization of the compost obtained at the end of the test.

Substance (mg/kg)	$PLLA_{B2B}$	Limits EN 13432:2000 (mg/kg)
Chrome	0.2 ± 0.1	50
Cobalt	<0.1	38 (a)
Nickel	<0.1	25
Copper	1.9 ± 0.4	50
Zinc	4.7 ± 0.9	150
Arsenic	<0.1	5
Selenium	<0.1	0.75
Molybdenum	<0.1	1
Cadmium	<0.1	0.5
Lead	0.2 ± 0.1	50
Mercury	<0.05	0.5
Fluorine	<50	100

The extended uncertainty values refer to a 95% confidence interval. Coverage factor k = 2. (a) Canadian Certification Program (CAN/BNQ 9011-911-1/2007).

3.5. Composting in Windrows

At the end of the experiment at the HMU, the samples were taken to the Wastewater and Solid Waste Management Laboratory of the Department of Agriculture, where they were subjected to detailed evaluation. In all samples, a visual comparison was made of their initial and final states. In terms of this visual assessment all samples exhibited the highest degree and rate of decomposition. Photographs of the initial condition of the samples, their condition in each week, and the final condition of the samples at the end of the experiment are presented in Figure 6. At the end of the experiment, all samples were decomposed to about 100% of their initial condition. Moisture and temperature were monitored weekly over the course of the experiment (Figure 7).

Due to water absorption and the creation of low molecular weight compounds, the whitening process and formulation opacity are linked to changes in the refractive index [45] while, during degradation, the formation of certain holes on the materials and an induced increase in crystallinity are observed [46].

Figure 7 illustrates the development of composter temperatures and the material moisture during the experimental period. As shown in the graph, the highest composter temperatures were reached in the 2nd turning when the temperature amounted to 70 °C. Composter temperatures in the 3rd and 4th turnings were 67.6 and 47.2 °C, respectively. To determine the compost quality, phytotoxicity tests were carried out using composts resulting from the disintegration of the materials tested under laboratory conditions. The compost GI value was found to be 67% for the 1:10 dilution ratios.

The composition of the blank samples verified the favorable composting conditions. According to Angelica et al. (2020) [47], the ideal moisture content required for composting is between 50 and 60%. The initial moisture content was 61.0, 0.1%, within the 50–60% range recommended in the literature to support the composting process.

(a) (b) (c) (d) (e) (f) (g)

Figure 6. Photos of the sample during the test. (**a**) Fresh FW, VW and PR at the beginning of the test, (**b**) turning 1 after 5 days, (**c**) turning 2 after 7 days and with addition of the samples, (**d**) turning 3 after 7 days, (**e**) turning 4 after 7 days, (**f**) turning 5 after 13 days, and (**g**) turnings 6 and 7 after 7 days.

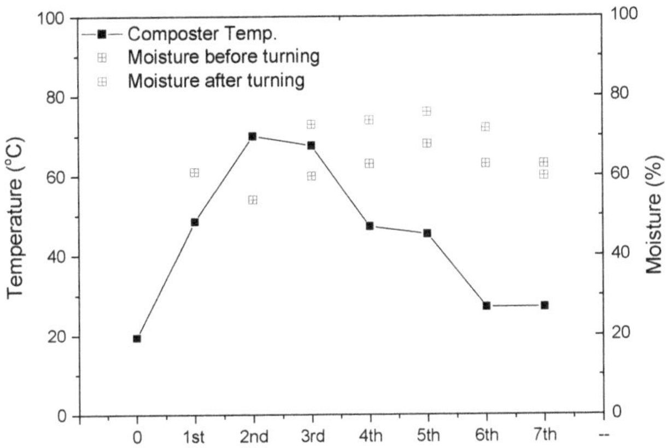

Figure 7. Composter temperature and moisture variation during composting experiment before and after turning of the materials.

Table 9 presents the analysis of materials during the composting process. The pH increased from 5.9 ± 0.0 to 8.6 ± 0.0. The pH tends to rise during the first two to three weeks of composting as ammonia gas is produced from the breakdown of nitrogen, but it tends to fall later as organic acid breaks down into organic matter, according to He et al. (2020) [48]. EC decreased from 4.8 ± 0.0 mS/cm to 3.2 ± 0.0 mS/cm with the composting process. The high salt content in the raw materials presents a significant challenge because the salts would be released into the final compost after composting, and if it were used as fertilizer, it might result in too much salt in the soil, which might prevent the soil from absorbing other salts [49]. The EC was less than 4 mS/cm after the fifth week, which is favorable for plant growth [50]. TN concentration increased (from 1.4% to 2.5%); this was due to an organic mass reduction in the composting process. Composting produced more stable, mature compost and reduced TOC, increased TN, and decreased C/N ratios. The nitrogen content was found to be at an acceptable level when compared with other compost [51].

Table 9. Analysis of materials during the composting process.

Turnings	Days	T (°C)	Moisture B [1] (%)	Moisture A [2] (%)	pH	EC (mS/cm)	TN (%)	TOC (mg/g)	Ash (%)	TP (%)
Beginning	0	19.4		-	-	4.8 ± 0.0	-	561.0	-	-
1st Turning	5	48.5	61.0 ± 0.1	-	5.9 ± 0.0	3.6 ± 0.1	1.4 ± 1.1	545.0	12.3	0.3 ± 0.1
2nd Turning	12	70.0	54.0 ± 1.0	-	7.9 ± 0.0	4.5 ± 0.0	2.2 ± 1.2	508.0	15.5	0.6 ± 0.0
3rd Turning	19	67.6	60.0 ± 0.9	73.0 ± 0.9	8.9 ± 0.0	4.3 ± 0.0	2.1 ± 0.0	495.0	24.6	0.5 ± 0.1
4th Turning	26	47.2	63.0 ± 0.0	74.0 ± 0.0	9.0 ± 0.0	4.1 ± 0.2	2.4 ± 5.2	450.0	29.6	0.9 ± 0.2
5th Turning	39	45.3	68.0 ± 66.1	76.0 ± 66.1	8.6 ± 0.0	3.9 ± 0.1	2.5 ± 0.0	344.0	23.8	2.0 ± 0.0
6th Turning	46	27.0	63.0 ± 0.2	72.0 ± 0.2	8.7 ± 0.0	3.3 ± 0.1	2.4 ± 0.0	440.0	10.1	1.1 ± 0.3
7th Turning	53	27.0	63.0 ± 0.2	60.0 ± 0.2	8.6 ± 0.0	3.2 ± 0.0	2.5 ± 0.0	447.0	6.4	0.9 ± 0.0

[1] before turning; [2] after turning.

The goal of our experiment was to test the decomposition of the above-described samples in real conditions of industrial composting in windrows technology simulation. The temperatures reached over a long term in the compost pile ranged from 50–70 °C. Tibu et al. [52] recorded the same range. Sample 2 (certified as compostable) and 3 were decomposed. Samples 2 and 3 exhibited the highest decomposition rate (before the expiration of the 7 weeks). Sample 1 was a control reference sample with which to confirm that the conditions of decomposition were suitable during the experiment, which was the case. This was to check the potential of biological decomposition in the tested environment. This agreed with other studies [53,54].

Composting seems to be the most promising waste management option for degradable plastics because the composting process is designed to degrade waste. There are, however, obstacles that make many communities reluctant to accept plastic bags for composting. The plant growth phytotoxic test revealed that $PLLA_{B2B}$ had no negative effects on the composting procedure or compost quality. In conclusion, this article describes the biodegradability and compostability of $PLLA_{B2B}$ and indicates that the EN 13432 standards [27] for industrial composting are fulfilled. Additionally, the degradation degree of compostable bags that were tested in a real composting environment was reached at 100%. The knowledge gained in this study will also contribute to the development of policies and assessments for bioplastic waste, as well as provide direction for future bioplastics research and development.

4. Conclusions

This study reports on the biodegradation/disintegration of pure PLLA composed of food waste and the bioplastics produced from this PLLA, ascertaining and their ability to form compostable materials. Only the question of whether the $PLLA_{B2B}$ material is biologically degradable is addressed by the substantiation of degradation in laboratory conditions in accordance with current standards. It does not, however, address whether the bags made from these $PLLA_{B2B}$ are successfully degradable in the setting of an industrial composting operation, so simulation systems using windrow technology were used to study their biodegradation/disintegration. Laboratory tests for $PLLA_{B2B}$ reached a biodegradation level of $100.2 \pm 3.7\%$ with high germination values, and concentrations of heavy metals and hazardous substances which meet the stringent requirements of the European standard EN 13432:2000 for compostability. Furthermore, the results of the tests conducted under actual composting conditions were comparable with laboratory tests. These findings demonstrate the potential for the development of a novel, sustainable bioplastic. Therefore, we suggest food waste bioplastic as a new potential eco-material for compostable bags and their use in industrial composting units, as they can be considered promising degradable active materials with low environmental impact. In conclusion, the results of this study open up new opportunities for material manufacturers, who should prioritize testing in real settings using different raw materials, mainly waste.

Author Contributions: Conceptualization, A.M.; methodology, A.M. and N.G.M.; formal analysis, A.M. and N.G.M.; investigation, A.M. and I.S.; resources, I.S.; data curation, A.M.; writing—original draft preparation, A.M.; writing—review and editing, A.M.; supervision, T.M., T.L. and C.T.; project administration, T.L and C.T. All authors have read and agreed to the published version of the manuscript.

Funding: This research has been co-financed by the European Union and Greek national funds through the Operational Program "Competitiveness, Entrepreneurship and Innovation", under the call RESEARCH–CREATE–INNOVATE (project code: T1EDK-02746).

Institutional Review Board Statement: Not applicable.

Informed Consent Statement: Not applicable.

Data Availability Statement: Data are contained within the article.

Conflicts of Interest: The authors declare no conflict of interest.

Abbreviations

Poly-L-lactic acid (PLLA), biowaste to bioplastic (B2B), total nitrogen (TN), total phosphorus (TP), total organic matter (TOM), volatile solids (VS), total organic carbon (TOC), germination index (GI), polylactic acid (PLA), food waste (FW), pruning (PR), compostable bag (CB), vegetable waste (VW), Hellenic Mediterranean University (HMU).

References

1. Borrelle, S.B.; Ringma, J.; Lavender Law, K.; Monnahan, C.C.; Lebreton, L.; McGivern, A.; Murphy, E.; Jambeck, J.; Leonard, G.H.; Hilleary, M.A.; et al. Predicted Growth in Plastic Waste Exceeds Efforts to Mitigate Plastic Pollution. *Science* **2020**, *369*, 1515–1518. [CrossRef]
2. Sales, A.J. Bioplastics-and-the-Environment. *Electron. J. Biol.* **2017**, *13*, 274–279.
3. Di Bartolo, A.; Infurna, G.; Dintcheva, N.T. A Review of Bioplastics and Their Adoption in the Circular Economy. *Polymers* **2021**, *13*, 1229. [CrossRef]
4. Market—European Bioplastics e.V. Available online: https://www.european-bioplastics.org/market/ (accessed on 20 September 2023).
5. Zorpas, A.A.; Navarro-Pedreño, J.; Jeguirim, M.; Dimitriou, G.; Almendro Candel, M.B.; Argirusis, C.; Vardopoulos, I.; Loizia, P.; Chatziparaskeva, G.; Papamichael, I. Crisis in Leadership vs Waste Management. *Euro-Mediterr. J. Environ. Integr.* **2021**, *6*, 80. [CrossRef]
6. Loizia, P.; Voukkali, I.; Zorpas, A.A.; Navarro Pedreño, J.; Chatziparaskeva, G.; Inglezakis, V.J.; Vardopoulos, I.; Doula, M. Measuring the Level of Environmental Performance in Insular Areas, through Key Performed Indicators, in the Framework of Waste Strategy Development. *Sci. Total Environ.* **2021**, *753*, 141974. [CrossRef] [PubMed]
7. Vardopoulos, I.; Konstantopoulos, I.; Zorpas, A.A.; Limousy, L.; Bennici, S.; Inglezakis, V.J.; Voukkali, I. Sustainable Metropolitan Areas Perspectives through Assessment of the Existing Waste Management Strategies. *Environ. Sci. Pollut. Res.* **2021**, *28*, 24305–24320. [CrossRef] [PubMed]
8. Zorpas, A.A.; Lasaridi, K. Measuring Waste Prevention. *Waste Manag.* **2013**, *33*, 1047–1056. [CrossRef] [PubMed]
9. Maragkaki, A.E.; Vasileiadis, I.; Fountoulakis, M.; Kyriakou, A.; Lasaridi, K.; Manios, T. Improving Biogas Production from Anaerobic Co-Digestion of Sewage Sludge with a Thermal Dried Mixture of Food Waste, Cheese Whey and Olive Mill Wastewater. *Waste Manag.* **2018**, *71*, 644–651. [CrossRef]
10. Orfanoudaki, A.; Makridakis, G.; Maragkaki, A.; Fountoulakis, M.S.; Kallithrakas-Kontos, N.G.; Manios, T. Anaerobic Co-Digestion of Pig Manure and Spent Coffee Grounds for Enhanced Biogas Production. *Waste Biomass Valorization* **2020**, *11*, 4613–4620. [CrossRef]
11. Moussa, R.N.; Moussa, N.; Dionisi, D. Hydrogen Production from Biomass and Organic Waste Using Dark Fermentation: An Analysis of Literature Data on the Effect of Operating Parameters on Process Performance. *Processes* **2022**, *10*, 156. [CrossRef]
12. Thanos, D.; Maragkaki, A.; Venieri, D.; Fountoulakis, M.; Manios, T. Enhanced Biogas Production in Pilot Digesters Treating a Mixture of Olive Mill Wastewater and Agro-Industrial or Agro-Livestock by-Products in Greece. *Waste Biomass Valorization* **2021**, *12*, 135–143. [CrossRef]
13. Bala, S.; Garg, D.; Sridhar, K.; Inbaraj, B.S.; Singh, R.; Kamma, S.; Tripathi, M.; Sharma, M. Transformation of Agro-Waste into Value-Added Bioproducts and Bioactive Compounds: Micro/Nano Formulations and Application in the Agri-Food-Pharma Sector. *Bioengineering* **2023**, *10*, 152. [CrossRef] [PubMed]
14. Madhavan Nampoothiri, K.; Nair, N.R.; John, R.P. An Overview of the Recent Developments in Polylactide (PLA) Research. *Bioresour. Technol.* **2010**, *101*, 8493–8501. [CrossRef]
15. Tsuji, H. Hydrolytic Degradation. In *Poly(Lactic Acid): Synthesis, Structures, Properties, Processing, and Applications*; John Wiley & Sons, Inc.: Hoboken, NJ, USA, 2010; pp. 345–381. [CrossRef]
16. Ho, K.L.G.; Pometto, A.L.; Gadea-Rivas, A.; Briceño, J.A.; Rojas, A. Degradation of Polylactic Acid (PLA) Plastic in Costa Rican Soil and Iowa State University Compost Rows. *J. Environ. Polym. Degrad.* **1999**, *7*, 173–177. [CrossRef]
17. Idris, S.N.; Amelia, T.S.M.; Bhubalan, K.; Lazim, A.M.M.; Zakwan, N.A.M.A.; Jamaluddin, M.I.; Santhanam, R.; Amirul, A.A.A.; Vigneswari, S.; Ramakrishna, S. The Degradation of Single-Use Plastics and Commercially Viable Bioplastics in the Environment: A Review. *Environ. Res.* **2023**, *231*, 115988. [CrossRef] [PubMed]
18. Tokiwa, Y.; Calabia, B.P. Biodegradability and Biodegradation of Poly(Lactide). *Appl. Microbiol. Biotechnol.* **2006**, *72*, 244–251. [CrossRef]
19. Kale, G.; Auras, R.; Singh, S.P. Comparison of the Degradability of Poly(Lactide) Packages in Composting and Ambient Exposure Conditions. *Packag. Technol. Sci.* **2007**, *20*, 49–70. [CrossRef]
20. Alshehrei, F. Biodegradation of Synthetic and Natural Plastic by Microorganisms. *J. Appl. Environ. Microbiol.* **2017**, *5*, 8–19. [CrossRef]
21. Kasar, P.; Sharma, D.K.; Ahmaruzzaman, M. Thermal and Catalytic Decomposition of Waste Plastics and Its Co-Processing with Petroleum Residue through Pyrolysis Process. *J. Clean. Prod.* **2020**, *265*, 121639. [CrossRef]
22. Tabasi, R.Y.; Ajji, A. Selective Degradation of Biodegradable Blends in Simulated Laboratory Composting. *Polym. Degrad. Stab.* **2015**, *120*, 435–442. [CrossRef]
23. Karamanlioglu, M.; Preziosi, R.; Robson, G.D. Abiotic and Biotic Environmental Degradation of the Bioplastic Polymer Poly(Lactic Acid): A Review. *Polym. Degrad. Stab.* **2017**, *137*, 122–130. [CrossRef]
24. Briassoulis, D.; Dejean, C.; Picuno, P. Critical Review of Norms and Standards for Biodegradable Agricultural Plastics Part II: Composting. *J. Polym. Environ.* **2010**, *18*, 364–383. [CrossRef]
25. Cossu, C.; Heuzey, M.C.; Lussier, L.S.; Dubois, C. Early Development of a Biodegradable Energetic Elastomer. *J. Appl. Polym. Sci.* **2011**, *119*, 3645–3657. [CrossRef]

26. Arias, V.; Höglund, A.; Odelius, K.; Albertsson, A.C. Tuning the Degradation Profiles of Poly(L-Lactide)-Based Materials through Miscibility. *Biomacromolecules* **2014**, *15*, 391–402. [CrossRef]
27. *EN 13432:2000*; Packaging. Requirements for Packaging Recoverable through Composting and Biodegradation. Test Scheme and Evaluation Criteria for the Final Acceptance of Packaging. British Standards Institution: London, UK, 2000; pp. 1–26.
28. Sakai, K.; Taniguchi, M.; Miura, S.; Ohara, H.; Matsumoto, T.; Shirai, Y. Making Plastics from Garbage: A Novel Process for Poly-L-Lactate Production from Municipal Food Waste. *J. Ind. Ecol.* **2003**, *7*, 63–74. [CrossRef]
29. *ISO 16929:2019*; Plastics-Determination of the Degree of Disintegration of Plastic Materials under Defined Composting Conditions in a Pilot-Scale Test(E). ISO: Geneva, Switzerland, 2021.
30. *UNE EN ISO 14855-1:2013*; Determination of the Ultimate Aerobic Biodegradability of Plastic Materials under Controlled Composting—Method by Analysis of Evolved Carbon Dioxide—Part 1: General Method (ISO 14855-1:2012). UNE Spanish standardization: Madrid, Spain, 2013. Available online: https://www.en-standard.eu/une-en-iso-14855-1-2013-determination-of-the-ultimate-aerobic-biodegradability-of-plastic-materials-under-controlled-composting-conditions-method-by-analysis-of-evolved-carbon-dioxide-part-1-general-method-iso-14855-1-2012/ (accessed on 25 October 2023).
31. APHA; AWWA; WEF. *Standard Methods for the Examination of Water and Wastewater*; American Public Health Association (APHA): Washington, DC, USA; American Water Works Association (AWWA): Denver, CO, USA; Water Environment Federation (WEF): Alexandria, VA, USA, 2005.
32. *UNI 10780:1998*; Compost—Classification, Requirements and Use Criteria. Ente Nazionale Italiano di Unificazione (UNI): Roma, Italy. Available online: https://infostore.saiglobal.com/en-us/Standards/UNI-10780-1998-1091645_SAIG_UNI_UNI_2541004/ (accessed on 4 October 2023).
33. Evaluating Toxicity of Immature Compost—ScienceOpen. Available online: https://www.scienceopen.com/document?vid=d003a586-348f-4d40-b89d-24d98889ac5e (accessed on 7 September 2023).
34. *EN 15411:2011*; Solid Recovered Fuels—Methods for the Determination of the Content of Trace. British Standards Institution: London, UK, 2011. Available online: https://standards.iteh.ai/catalog/standards/cen/108a8614-5b62-4295-b150-ea56279b8828/en-15411-2011 (accessed on 4 October 2023).
35. NEMI Method Summary—9056A. Available online: https://www.nemi.gov/methods/method_summary/4734/ (accessed on 4 October 2023).
36. CD-ROM 5050-2 Revision 0. 1994. Available online: https://19january2017snapshot.epa.gov/sites/production/files/2015-12/documents/5050.pdf (accessed on 4 October 2023).
37. Kumar Kalita, N.; Sarmah, A.; Bhasney, S.M.; Kalamdhad, A.; Katiyar, V. Demonstrating an Ideal Compostable Plastic Using Biodegradability Kinetics of Poly(Lactic Acid) (PLA) Based Green Biocomposite Films under Aerobic Composting Conditions. *Environ. Chall.* **2021**, *3*, 100030. [CrossRef]
38. Kale, G.; Kijchavengkul, T.; Auras, R.; Rubino, M.; Selke, S.E.; Singh, S.P. Compostability of Bioplastic Packaging Materials: An Overview. *Macromol. Biosci.* **2007**, *7*, 255–277. [CrossRef]
39. Rudeekit: Determining Biodegradability of Polylactic...—Designer. Google. Available online: https://scholar.google.com/scholar_lookup?title=Determining%20biodegradability%20of%20polylactic%20acid%20under%20different%20environments&publication_year=2008&author=Y.%20Rudeekit&author=J.%20Numnoi&author=M.%20Tajan&author=P.%20Chaiwutthinan&author=T.%20Leejarkpai (accessed on 9 October 2023).
40. Bitinis, N.; Verdejo, R.; Bras, J.; Fortunati, E.; Kenny, J.M.; Torre, L.; López-Manchado, M.A. Undefined Poly (Lactic Acid)/Natural Rubber/Cellulose Nanocrystal Bionanocomposites Part I. In *Processing and Morphology*; Elsevier: Amsterdam, The Netherlands, 2013.
41. Wang, Y.; Pan, J.; Han, X.; Sinka, C.; Ding, L. A Phenomenological Model for the Degradation of Biodegradable Polymers. *Biomaterials* **2008**, *29*, 3393–3401. [CrossRef]
42. Kale, G.; Auras, R.; Singh, S.P.; Narayan, R. Biodegradability of Polylactide Bottles in Real and Simulated Composting Conditions. *Polym. Test.* **2007**, *26*, 1049–1061. [CrossRef]
43. Ruggero, F.; Onderwater, R.C.A.; Carretti, E.; Roosa, S.; Benali, S.; Raquez, J.M.; Gori, R.; Lubello, C.; Wattiez, R. Degradation of Film and Rigid Bioplastics During the Thermophilic Phase and the Maturation Phase of Simulated Composting. *J. Polym. Environ.* **2021**, *29*, 3015–3028. [CrossRef]
44. Czigány, T. Editorial Corner–a Personal View Disposable or Single-Use Plastics? Neither! Recyclable or Reusable Plastics! *Express Polym. Lett.* **2020**, *14*, 1. [CrossRef]
45. Bitinis, N.; Fortunati, E.; Verdejo, R.; Bras, J.; Kenny, J.M.; Torre, L.; López-Manchado, M.A. Poly(Lactic Acid)/Natural Rubber/Cellulose Nanocrystal Bionanocomposites. Part II: Properties Evaluation. *Carbohydr. Polym.* **2013**, *96*, 621–627. [CrossRef]
46. Paul, M.A.; Delcourt, C.; Alexandre, M.; Degée, P.; Monteverde, F.; Dubois, P. Polylactide/Montmorillonite Nanocomposites: Study of the Hydrolytic Degradation. *Polym. Degrad. Stab.* **2005**, *87*, 535–542. [CrossRef]
47. Hernández-Gómez, A.; Calderón, A.; Medina, C.; Sanchez-Torres, V.; Oviedo-Ocaña, E.R. Implementation of Strategies to Optimize the Co-Composting of Green Waste and Food Waste in Developing Countries. A Case Study: Colombia. *Environ. Sci. Pollut. Res. Int.* **2021**, *28*, 24321–24327. [CrossRef]
48. He, Z.; Li, Q.; Zeng, X.; Tian, K.; Kong, X.; Tian, X. Impacts of Peat on Nitrogen Conservation and Fungal Community Composition Dynamics during Food Waste Composting. *Appl. Biol. Chem.* **2020**, *63*, 72. [CrossRef]

49. Lee, M.H.; Han, S.J.; Lee, Y.K.; Ike, I.A.; Ok, Y.S.; Hur, J. Enhancing Copper Binding Property of Compost-Derived Humic Substances by Biochar Amendment: Further Insight from Two-Dimensional Correlation Spectroscopy. *J. Hazard. Mater.* **2020**, *390*, 121128. [CrossRef] [PubMed]
50. Maragkaki, A.; Gamvroudis, C.; Lountou, C.; Stamatiadis, P.; Sampathianakis, I.; Papadaki, A.; Manios, T. Autonomous Home Composting Units for Urban Areas in Greece: The Case Study of the Municipality of Rhodes. *Sustainability* **2022**, *14*, 12362. [CrossRef]
51. Zorpas, A.A.; Lasaridi, K.; Pociovalisteanu, D.M.; Loizia, P. Monitoring and Evaluation of Prevention Activities Regarding Household Organics Waste from Insular Communities. *J. Clean. Prod.* **2018**, *172*, 3567–3577. [CrossRef]
52. Tibu, C.; Annang, T.Y.; Solomon, N.; Yirenya-Tawiah, D. Effect of the Composting Process on Physicochemical Properties and Concentration of Heavy Metals in Market Waste with Additive Materials in the Ga West Municipality, Ghana. *Int. J. Recycl. Org. Waste Agric.* **2019**, *8*, 393–403. [CrossRef]
53. Vaverková, M.; Adamcová, D.; Kotovicová, J.; Toman, F. Evaluation of Biodegradability of Plastics Bags in Composting Conditions. *Ecol. Chem. Eng. S* **2014**, *21*, 45–57. [CrossRef]
54. Adamcová, D.; Zloch, J.; Brtnický, M.; Vaverková, M.D. Biodegradation/Disintegration of Selected Range of Polymers: Impact on the Compost Quality. *J. Polym. Environ.* **2019**, *27*, 892–899. [CrossRef]

sustainability

MDPI

Article

Sustainable Medical Waste Management Using an Intuitionistic Fuzzy-Based Decision Support System

Konstantinos Kokkinos [1], Evangelia Lakioti [2,*], Konstantinos Moustakas [3], Constantinos Tsanaktsidis [4] and Vayos Karayannis [4]

[1] Department of Digital Systems, University of Thessaly, 41500 Larissa, Greece; kokkinos@uth.gr
[2] School of Health Sciences, University of Thessaly, 42100 Larissa, Greece
[3] School of Chemical Engineering, National Technical University of Athens, 15780 Athens, Greece; konmoust@central.ntua.gr
[4] Department of Chemical Engineering, University of Western Macedonia, 50100 Kozani, Greece; ktsanaktsidis@uowm.gr (C.T.); vkarayannis@uowm.gr (V.K.)
* Correspondence: elakioti@uth.gr

Abstract: The growing urban population and increased use of healthcare services have brought significant attention to the safe and sustainable management of medical waste. Selecting the proper technology in medical waste management (MWM) represents one of the most critical challenges for decision-makers to ensure public health. In order to evaluate and choose the best MWM methodology, the current research provides a novel multi-criteria decision-making (MCDM) strategy for a variety of social stakeholders, to compute criteria weights, decision-making weights, and alternative ranking algorithms. The suggested structure addresses uncertain assessments of alternatives by extending weighting and ranking methods to acquire the decision-making weight and rank the MWM alternatives based on uncertain conditions. It also uses 'intuitionistic fuzzy' linguistic variables to indicate criteria weights. To assess all the factors pertaining to the sustainability of MWM actions, this study suggests the creation of a decision support system (DSS). Our DSS is built upon a novel strategy that utilizes a collection of MCDM models that are grounded on contemporary intuitionistic fuzzy logic methodologies. Alternative scenarios have been assessed for the instance of Greece, after specialists in the healthcare management field imposed 17 criteria and sub-criteria. The IF-MCDM methodologies used were the Intuitionistic Fuzzy DEMATEL, TOPSIS, and CORPAS. The alternative scenarios ranged from the prioritizing of safety laws and regulations to public acceptance and awareness, with the handling of hazardous risks and transportation playing a crucial part in the process. All ensemble methods produced the same ranking of the alternatives, demonstrating that safety and risk avoidance is the most significant scenario for sustainable urban development and public health.

Keywords: medical (healthcare) waste; sustainable management; decision support system; intuitionistic fuzzy logic; urban development; public health

1. Introduction

As the global population, the number of healthcare facilities, and the demand for medical services have all grown over the last decade, there has been a corresponding increase in the generation of healthcare wastes (HCWs) on a global scale. Simultaneously, trends such as the production of medical devices with multiple uses and a shift towards safer, single-use alternatives have contributed to the rise in HCW production. Consequently, managing HCW has become a challenging task for municipalities and regions, particularly in European countries that have enforced stringent regulations governing the safe handling, transport, and management of municipal and HCW waste [1]. Given that the improper disposal of HCWs poses a serious risk of infection or injury to healthcare workers and the general public, it is now a top priority in global public health and environmental concerns.

These factors have fueled a growing interest in HCW management within academic and professional circles [2].

The intricate and complex process of HCW management involves waste collection, the careful selection of transfer routes to minimize hazards, choosing appropriate disposal plant sites, selecting suitable HCW treatment technologies to minimize pollution, and, ultimately, recovering some of the lost energy in the entire chain. Numerous socio-cultural factors associated with these processes concurrently lead local societies to erect barriers and impose bureaucratic procedures ('red tape') to mitigate the adverse impacts resulting from improper HCW management. The selection of the best and most efficient treatment technique for HCWs has garnered significant scientific interest due to its far-reaching ecological and economic consequences. In general, the use of the multi-criteria decision-making (MCDM) tool can be of considerable assistance in evaluating HCW treatment options, addressing this complex and significant challenge.

2. Related Work

MCDM assists various decision-makers and businesses in reaching reasonable conclusions, both in quantitative and qualitative terms. It offers suggestions for improving options in the direction of a particular objective or utility, making use of a variety of analytical techniques [3] and instruments to assess how all pertinent data affect the decision-making process. MCDM is a goal-oriented procedure that offers the assistance required to accomplish this aim. While the traditional approach aims to find a single optimal solution, real-life problems often involve multiple options and evaluation criteria, making this goal unattainable [4]. In practice, many problems require the consideration of numerous criteria before a final decision can be reached. Unfortunately, as the number of assessment criteria increases, the decision-making process becomes more challenging. To address this issue, decision-makers often employ ensembles of various techniques in order to achieve near optimal performance in the system. The ensemble approaches have proven advantageous in improving the effectiveness and reliability of the produced DSS. Another challenge may also refer to deciding which scenario is beneficial and/or which approach to follow amongst alternative choices [5]. Simultaneously, when criteria evaluation is expressed through linguistic terms, fuzzy and intuitionistic fuzzy MCDM (IFMCDM) have demonstrated their effectiveness. Among various IFMCDM methodologies [6], intuitionistic fuzzy TOPSIS (IFTOPSIS) stands out as the most popular. According to [7], IFTOPSIS has been shown to yield results close to optimal when ordering criteria in multi-objective constrained waste management optimization problems [8].

Studies on Medical/Hazardous Waste Management via Intuitionistic Fuzzy MCDM Methodologies

Healthcare waste (HCW) can be classified as either hazardous or non-hazardous based on specific characteristics it exhibits. Medical waste is inherently categorized as hazardous due to its potential inclusion of unwanted flammable, oxidizing, poisonous, radioactive, toxic, or explosive materials. The improper handling of such waste poses significant threats to both humans and the environment. Extensive research has been conducted on multiple-criteria decision-making (MCDM) methodologies for handling hazardous waste, including medical waste. However, there has been limited research utilizing intuitionistic fuzzy methodologies in this field until the occurrence of the COVID-19 pandemic. Due to the exponential growth in medical waste generated post-pandemic, there has been a rapid increase in research employing intuitionistic fuzzy methodologies to establish a practical hierarchy of criteria for effective waste management. The most recent and up-to-date insights on technologies for the treatment of solid medical waste have been presented by [9]. The classification of medical waste, as outlined by [7], is depicted in Figure 1 and is representative of the current state in the Greek region in terms of hospital-produced waste. The authors list incineration, autoclaving, chemical disinfection, microwaving, irradiation, biological treatment, and encapsulation as the most popular methodologies for treating medical waste from a chemical-engineering perspective. The nature and categorization of medical waste, along with local laws and regulations, determine the most effective treatment procedure. To ensure the efficient and secure disposal of medical waste, some facilities may additionally employ a variety of treatment techniques.

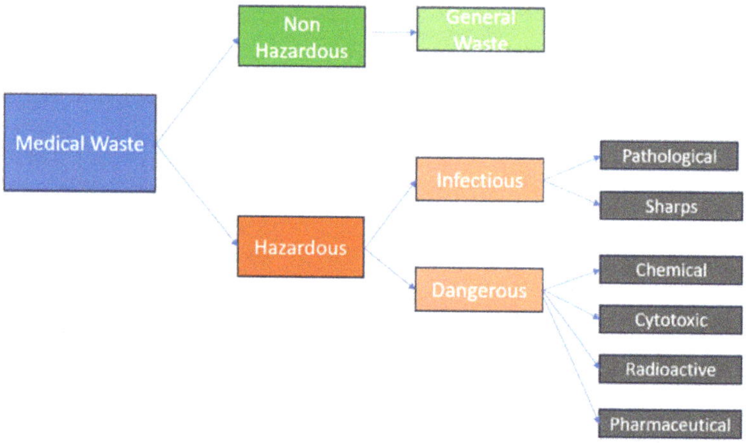

Figure 1. Classification of medical waste (inspired by [7]).

In our original work, we explore the issue of medical waste treatment using an innovative holistic approach that encompasses social, transportation, management, and policy aspects. These include collection, storage, segregation, transportation, accident avoidance during transportation, safety practices of health workers, waste treatment, sanitation processes, disposal, capacity building of appointed companies, public awareness, monitoring processes, and overall safety. Our research on healthcare waste treatment and management is unique due to its unique approach to holistically treating the significant importance of related criteria using a repertoire of methodologies. This research is unique as it combines fuzzy methodologies with a holistic methodology, making it a unique study. Additionally, our research is the first to relate HCW holistic treatment methodologies with the region of Thessaly, Greece, making it a unique attempt in this area. Furthermore, our work successfully connects results from similar methodologies like IF-CORPAS and IF-TOPSIS with alternative objectives and scope methodologies like IF-DEMATEL, making it a unique comparative approach in the literature.

Furthermore, our focus is on advanced fuzzy logic methodologies to address the multi-criteria decision making involved in the aforementioned aspects. In this novel approach, we propose using an ensemble of such methodologies as the primary decision-making engine in a unique decision support system (DSS). Table 1 summarizes the research on IF approaches used for this multi-objective issue.

Table 1. HCW management/treatment research based on IF methodologies.

Public. Year	References	Main Topic
2022	[10]	Novel approach based on intuitionistic fuzzy sets (IFSs) and the VIKOR method to establish recycling channel capacity evaluation index system that improves efficiency. The process incorporates the best–worst method and entropy to comprehensively determine decision-makers' weighting. New aggregation operator is used to define capacity rankings of hospitals.
2023	[11]	Intuitionistic fuzzy multi-criteria decision-making (IFMCDM) methods to determine the most effective hospital for medical waste management. The methodology includes intuitionistic fuzzy weighted averaging (IFWA), IF analytical hierarchy process (IFAHP), IF technique for order preference by similarity to ideal solution (IFTOPSIS), and one-dimensional sensitivity analysis. The most important criteria are qualified personnel, health institution infrastructure, and waste control.

Table 1. *Cont.*

Public. Year	References	Main Topic
2021	[12]	It investigates landfill sites' evaluation using fuzzy multi-criteria decision making. A novel circular intuitionistic fuzzy MCDM approach is developed, extending the use of circular intuitionistic fuzzy sets (C-IFS) and proposing defuzzification functions. The proposed method aims to guide future C-IFS MCDM methodologies.
2023	[13]	It presents the IF-EM-SWARA-TOPSIS technique, which assesses HCW based on social, environmental, economic, and technical criteria in an intuitionistic fuzzy environment. The technique uses an entropy measure to evaluate objective weights of criteria, while the distance measure constructs a technique for order performance by similarity to ideal solution (TOPSIS). The stepwise weight assessment ratio analysis (SWARA) technique obtains subjective weights for criteria. The sustainability and stability of the proposed framework are demonstrated through comparative studies and sensitivity assessments.
2016	[14]	A two-level hierarchical multicriteria decision model for HCW disposal method selection. It assesses methods against intuitionistic fuzzy preference relations and real values, using cross-ratio-based values. The model uses a weighted geometric operator, a cross-ratio-based bipolar 0.1–0.9 scale, and a two-level hierarchical structure to handle multicriteria decision problems with intuitionistic preference relations. The framework is applied to HCW management DM to demonstrate feasibility and effectiveness.
2020	[15]	Evaluation based on distance from average solution framework using parametric divergence measures and IF sets. The framework evaluates four HCW disposal alternatives, including incineration, steam sterilization, microwave, and landfill disposal. The framework's validity is demonstrated through comparisons with existing approaches.
2017	[16]	A systematic, integrated MCDM approach based on the TOPSIS method in an intuitionistic fuzzy (IF) environment. The group decision-making approach minimizes bias and manages uncertainties. A real case application from Turkey illustrates the use of this methodology.
2020	[17]	The framework investigates different disposal methods based on social acceptance, technology, environmental protection, cost, noise, and health risk. The hesitant MCDM method is used to select the best treatment method. The HF-SOWIA and HF-MOOSRA novel methods are proposed. The evaluation results show autoclaving is the best alternative for HCW disposal treatment. Sensitivity analysis is conducted to observe the difference in alternative ranking when subjective and objective weights change.
2022	[18]	Proposition of a novel strategy using IF-MOORA in spherical fuzzy sets (SFSs) to counterbalance FMEA's disadvantages. The method identifies barriers using FMEA, assigns risk factors, and prioritizes them in the SF area. The proposed methodology is compared to IF-MOORA and TOPSIS for reliability and correlation.
2021	[19]	Pythagorean fuzzy-based decision-making methodology is developed to rank HCW alternatives and develop weight-determining procedures. A case study demonstrates the effectiveness of the proposed framework, with autoclaving technology being the most suitable alternative. A sensitivity analysis and comparative study highlight the methodology's strength and validity.
2021	[20]	Proposition of a novel approach to selecting the optimal landfill for medical waste disposal using MCDM methods and SFS. The criteria include environmental, economic, and social factors. The weights of the criteria were computed using SFSWARA, and alternatives were analyzed and ranked using SFWASPAS. The results were compared with the IF-SWARA-WASPAS method, showing more reliability and accuracy.
2023	[21]	IF-MCDM model using the CRITIC method for objective criterion weights and the WASPAS method for ranking alternatives. The methodology is developed in single and interval-valued spherical fuzzy environments, allowing users to model membership, non-membership, and hesitancy parameters independently.
2022	[22]	Proposition of a new MCDM approach to compute criteria weights, DMs' weights, and alternative ranking methods for assessing and selecting the best HCW treatment technology. The model uses IF variables and extends weighting and ranking methods for uncertain conditions. Results show the first alternative and DM have a high preference, and the final ranking results are reliable with changing criteria weights.

Table 1. *Cont.*

Public. Year	References	Main Topic
2023	[23]	A robust integrated methodology using SWARA and COPRAS techniques is developed. The methodology is applied to a large-scale international hospital chain, revealing that storage conditions are the most important criterion for selecting the best disposal and logistics firm. The implications provide managerial insights and decision aids to improve the sustainability of the healthcare industry.
2023	[24]	Optimizing treatment procedures involves complex multi-criteria decision making. This study extends the DNMA approach with interval-valued intuitionistic fuzzy sets (IVIFSs) to address this problem. The proposed approach efficiently tackles HCW treatment selection, with steam sterilization being the optimal choice. The prioritization options obtained are reliable and suitable.
2023	[25]	It presents an integrated MAGDM methodology for selecting the most capable organization to manage HCW using IVFFS data. The study combines PROMETHEE II and SWARA methods, estimating attribute weights and determining preference order of options. Results demonstrate the bio-chemistry lab as the best management option in terms of stability and reliability.
2023	[26]	The framework uses MCDM to identify the best and worst medical waste management techniques. Results show that incineration, microwave, pyrolysis, autoclave chemical, vaporized hydrogen peroxide, dry heat, ozone, and ultraviolet light were the most effective methods for disposing of medical waste during the pandemic. Ozone was identified as the most suitable technique for the circular economy of HCW. Integration of MCDM with other techniques including AHP, LDFN-FDOSM, and ANN to generate more robust results.
2021	[27]	A new MCDM approach is introduced under IF conditions to evaluate the importance of criteria and decision-makers. The approach uses an integrated weighting criteria method (OWA), aggregating subjective and objective criterion weights, and a ranking approach with ideal and anti-ideal distance-based methods. The IF condition ensures high performance in making appropriate decisions.
2021	[28]	This IF TOPSIS-Sort approach achieves classification into low-, moderate-, and high-exposure classes, providing insights for planning and policy formulation. Comparative analysis with other distance-based MCS methods revealed that the IF TOPSIS-Sort is pessimistic and may underestimate assignments, and is potentially counterintuitive in public health applications. The novel IF TOPSIS-Sort is suitable for many domain-sorting applications.
2021	[29]	This study presents a novel framework for assessing socioeconomic phenomena using survey data, focusing on complex phenomena like quality of life and services. The intuitionistic fuzzy TOPSIS (IF-TOPSIS) method is proposed, considering the uncertainty in object assessments and the use of continuous data from public statistics. This approach addresses the challenges of traditional multicriteria methods in evaluating these complex phenomena.

3. Preliminaries

Since real-world HCW management issues are, by their very nature, multi-objective, we use intuitionistic fuzzy sets (IFS). The use of IFS guarantees the coverage of all participating criteria in the process and the inclusion of categorical data (linguistic expert responses) when needed. In support of that, many associated methodologies have been developed lately such as interval-valued [26], IF-DEMATEL [27], Pythagorean IFS [30], etc. to address the HCW-related problems. A membership function and a non-membership function are used to represent an IFS of this type. This feature makes IFS superior compared to conventional fuzzy systems since it enables them to integrate reluctance and uncertainty at the same time. The fundamental definitions of IFS are provided in this section, while subsequent subsections provide thorough descriptions of the particular IFS approaches used in this study.

3.1. IF Sets and Relevant Terminology

Definition 1 ([30]). *Assuming a nonempty set X, and the existence of mapping as shown in the following function definition:*

$$
\begin{aligned}
f_A &: X \to [0,1] \times [0,1] \\
x &\mapsto (\mu_A(x),\ v_A(x))
\end{aligned}
\tag{1}
$$

where the IFS of X, designated as A, is activated by f_A *with* $A = \{\langle x, \mu_A(x), v_A(x)\rangle \mid x \in X\}$. *Furthermore, let* $\mu_A(x)$ *be the membership function, displaying the quantity of membership of x to the IFS A, and* $v_A(x)$ *be the non-membership quantity which both indicate the level of membership of x to the IFS A, presuming the restriction that* $0 \leq \mu_A(x) + v_A(x) \leq 1$.

Definition 2 ([30]). *Let* $\pi(x)$ *be the uncertainty degree with* $\pi(x) = 1 - \mu_A(x) - v_A(x)$ *that introduces the hesitation principle so that IFSs are to be [resented as tuples of three elements* $(\mu_A(x), v_A(x),\ \pi(x))$.

Definition 3 ([30,31]). *For an IF number* $\alpha = (\mu_A(x), v_A(x))$, *we now define the Score and the Accuracy functions given by the following equations. Let us note that this initial definition of Score and Accuracy has been updated (update is indicated by *) and shown in Equation (2b):*

$$
\begin{aligned}
S(a) &= \mu - v \quad \text{where } S(\alpha) \in [-1,1] \\
H(a) &= \mu + v \quad \text{where } H(\alpha) \in [-0,1]
\end{aligned}
\tag{2a}
$$

$$
\begin{aligned}
S^*(a) &= \tfrac{S(\alpha)+1}{2} \quad \text{where } S^*(\alpha) \in [0,1] \\
H^*(a) &= \tfrac{\mu+v}{2} \quad \text{where } H^*(\alpha) \in [-0,1]
\end{aligned}
\tag{2b}
$$

Definition 4. *Let* $Q = \{\alpha_i = (\mu_i,\ v_i) \mid i = 1 \ldots n\}$ *be a set of IF numbers. Moreover, let* $w = (w_1,\ w_2,\ \ldots\ldots, w_n)^T$ *be the ordered set of their weights assuming an imposed multi-objective and multi-criteria decision-making methodology. For the set Q, we call* $IFWAA_w(Q)$ *the IF weighted average operator which is illustrated below:*

$$
IFWAA_w(Q) = \bigoplus_{i=1}^{n} w_i \alpha_i = \left(1 - \prod_{i=1}^{n} (1-\mu_i)^{w_i}, \prod_{i=1}^{n} v_i^{\,w_i}\right) \tag{3}
$$

3.2. Intuitionistic Fuzzy MCDM Methodology

Many components in an MCDM process cannot be precisely evaluated due to the heterogeneity of information and the inherent fuzziness in defining criteria weights by stakeholders. As an alternative, a significant number of decision-makers and stakeholders express their evaluations and assessments using linguistic phrases. Table 2 below presents the most commonly used rating scale, when evaluating events in fuzzy theory. This rating scale is also commonly used in evaluating and/or rating criteria in real-life questionnaire and focus group responses. Similarly, in Table 3, we display the most frequent ratings for the expertise level of each evaluator in the criteria evaluation process.

Table 2. IF rating terminology, equivalent notation, and equivalent IF numbers [30].

Linguistic Terms	Rate Notation	Equivalent IFN
Extremely Good Efficiency (EGE)	s_9	(1.00, 0.00,0.00)
Very Very Good Efficiency (VVGE)	s_8	(0.95, 0.05, 0.05)
Very Good Efficiency (VGE)	s_7	(0.85, 0.10, 0.05)
Good Efficiency (GE)	s_6	(0.75, 0.20, 0.05)
Medium Efficiency (ME)	s_5	(0.65, 0.30, 0.05)
Fair Efficiency (FE)	s_4	(0.55, 0.40, 0.05)
Medium Bad Efficiency (MBE)	s_3	(0.40, 0.50, 0.10)
Bad Efficiency (BE)	s_2	(0.30, 0.60, 0.10)
Very Bad Efficiency (VBE)	s_1	(0.20, 0.70, 0.10)
Extremely Bad Efficiency (EBE)	s_0	(0.10, 0.80, 0.10)

Table 3. Fuzzy linguistic terms for the levels of significance of experts and equivalent IFNs.

Linguistic Terms	Equivalent IFN
Extremely Significant (ES)	(1.00, 0.00, 0.00)
Very Significant (VS)	(0.90, 0.05, 0.05)
Significant (S)	(0.70, 0.20, 0.10)
Moderate (M)	(0.60, 0.30, 0.10)
Insignificant (I)	(0.40, 0.50, 0.10)
Very Insignificant (VI)	(0.30, 0.60, 0.10)
Extremely Insignificant (EI)	(0.10, 0.80, 0.10)

This work addresses the challenge of selecting the optimal alternative solution for healthcare waste (HCW) treatment by introducing a decision-making technique along with an illustrative use-case solution. HCW and other healthcare professionals, as well as municipal staff members focusing on public safety and regional planning, are considered experts in the field. We are particularly interested in the treatment of HCW, as it has the potential to significantly impact municipality operators, local and regional couriers, small- and medium-sized enterprises (SMEs), and the broader society.

Any produced DSS must facilitate the IFWAA operator of Equation (3) to implement an ensemble of IF methodologies. In this research work, we utilize this operator in conjunction with the IF-TOPSIS and IF-CORPAS methods [32]. The linguistic terms are converted to IF- numbers. This is predicated on the existence of plausible employment scenarios for HCW. The overall methodology begins with an extensive literature review to identify all the criteria involved in examining and evaluating the best possible solution for HCW treatment. This list of criteria, along with the metrics and performance indicators needed for evaluating the DSS under construction, is finalized through verification by field experts. Simultaneously, we establish an intuitionistic fuzzy linguistic term set for all criteria to obtain expert responses, thus organizing the process of expert evaluation.

Given a set of criteria for evaluation and a set of alternative solutions, the relative importance decision matrix for the criteria ratings is formed. It is important to note that, for each alternative solution and each overall scenario of development, different matrices must be created and computed. Moreover, the model is immune to scalability issues since the number of criteria used for the ranking process does not affect the outcome, thus model resilience is preserved. The selected methodology builds an ensemble DSS holistically using three models for criteria evaluation and ranking. Each model, when run, indicates which alternative solution scenario, out of all the candidates, has the best potential. We first give a detailed explanation of each IF methodology, and then we illustrate the logical architecture of this methodology in Figure 2.

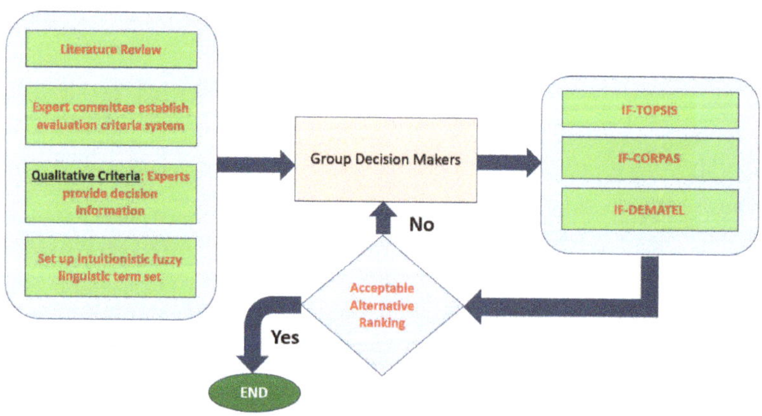

Figure 2. The logical architecture of the DSS based on an ensemble of IF models.

3.3. The IFTOPSIS Model [33]

Input:

Given a set of experts in the HCW management and treatment technology research, denoted as k, let $\vec{W} = \{\vec{w}_1, \vec{w}_2, \ldots\ldots \vec{w}_k\}$ with $\sum_{d=1}^{k} w_d = 1$ be the weights vector of k indicating the importance of their opinion. Moreover, let $C = \{1, 2, \ldots n\}$ be the set of criteria under the process of evaluation and ranking. For this set, let us also denote this with $W_C = \{w_{1c}, w_{2c}, \ldots w_{nc}\}$ and $\sum_{i=1}^{n} w_{ic} = 1$.

Step 1: Initially the discrete experts linguistically rate the criteria C using the rating scale of Table 1. Immediately afterwards, we calculate the IF decision matrix $A[k \times n] = \{\alpha_{i,j} | i = 1 \ldots k, j = 1 \ldots n\}$ with $\alpha_{i,j} = (\mu_{\alpha_{ij}}, v_{\alpha_{ij}}, \pi_{\alpha_{ij}})$ to be the corresponding IFN number and $\pi_{i,j} = (1 - \mu_{\alpha_{ij}} - v_{\alpha_{ij}})$.

Step 2: Based on the results of Step 1, we compute the aggregated IF decision matrix $\hat{A}$. This is achieved by calculating the corresponding IFWAA operators. In the computation process of $\hat{A}$, we basically compute each matrix element using the two equations below:

$$\begin{aligned} \hat{A}_{[k\times n]} &= IFWAA_w(p_{1,1}, p_{1,2}, \ldots p_{2,1} \ldots p_{i,j} \ldots p_{k,n}) \\ &= \left(1 - \prod_{i=1}^{k} (1 - \mu_{p_i})^{\varphi_i}, \prod_{i=1}^{k} (1 - v_{p_i})^{\varphi_i}, \prod_{i=1}^{k} (1 - \mu_{p_i})^{w_j} - \prod_{i=1}^{k} (1 - \mu_{p_i})^{\varphi_i}\right) \end{aligned} \quad (4)$$

where

$$\varphi_i = \frac{\left(\mu_i + \pi_i\left(\frac{\mu_i}{\mu_i + v_i}\right)\right)}{\sum_{i=1}^{k}\left(\mu_i + \pi_i\left(\frac{\mu_i}{\mu_i + v_i}\right)\right)} \quad (5)$$

Step 3: In this step, we compute the ideal solutions denoted as τ_1 and τ_2, respectively. Using τ_1 (beneficial) and τ_2 (non-beneficial). We compute the IF positive ideal solution (IFPIS) and the IF negative ideal solution (IFNIS). respectively (which are denoted as θ_1 and θ_2). according to the following equations:

$$\tilde{r}_{ij} = w_j . p_{i,j} \,\Big|\, i = 1 \ldots k, \; j = 1 \ldots n \quad (6)$$

$$\theta_1 = \left\{\left(\left(\max_i \tilde{r}_{ij}\right) | j \in \tau_1\right), \left(\left(\min_i \tilde{r}_{ij}\right) | j \in \tau_2\right) | i = 1 \ldots k\right\} = \{r_1^+, r_2^+, \ldots r_n^+\} \quad (7a)$$

$$\theta_2 = \left\{\left(\left(\min_i \tilde{r}_{ij}\right) | j \in \tau_1\right), \left(\left(\max_i \tilde{r}_{ij}\right) | j \in \tau_2\right) | i = 1 \ldots k\right\} = \{r_1^-, r_2^-, \ldots r_n^-\} \quad (7b)$$

Step 4: Using the values θ_1 and θ_2 above, we now calculate the distances from the positive and negative ideal solution IFFPIS and IFNIS. respectively. These are calculated in the following equations, respectively, using the notion of the IF normalized Euclidean distance:

$$\delta_i^+ = \sqrt{\sum_{j=1}^{n}\left(\tilde{r}_{ij} - r_j^+\right)^2} \quad (i = 1, 2, \ldots k) \quad \delta_i^- = \sqrt{\sum_{j=1}^{n}\left(\tilde{r}_{ij} - r_j^-\right)^2} \quad (i = 1, 2, \ldots k) \quad (8)$$

Step 5: Using the two equations above, we calculate the closeness coefficient values (CCVs). Note that, for each criterion C_i, the corresponding CCV, which, from now on, is denoted as $\hat{C}_i$, is found by the following equation:

$$\hat{C}_i = \frac{\delta_i^-}{\delta_i^- + \delta_i^+} \text{ where } 0 \leq \hat{C}_i \leq 1 \text{ and } i = 1, 2, \ldots n \quad (9)$$

Step 6: Finally, we compute the criteria weights using the CCV where we presume that the total of the weights is equal to 1. In the opposite case, we repeat the normalization process to achieve the aforementioned summation.

3.4. *The IF-COPRAS Model [34]*

Input:

The aggregated IF decision matrix $\hat{A}$ calculated in Equation (4) and in step 2 of the IF-TOPSIS model process is as follows

Step 1: We first calculate the weighted normalized values for each matrix cell of $\hat{A}$. In order to do that, we initially calculate the value $\overline{x}_{im}$. This is computed either for the normalized case by Equation (10) or for the weighted normalized case by Equation (11) as shown below:

$$\overline{x}_{im} = \frac{x_{im}}{\sum_{i=1}^{n} x_{im}}, \tag{10}$$

$$\overline{x}_{im} = \frac{w_m * (x_{im})}{\sum_{i=1}^{n} x_{im}} \tag{11}$$

Step 2: Calculate the total weighted normalized value for each expertly recommended alternative in relation to the beneficial (BC) and non-beneficial (NBC) criteria. respectively. The following formulae illustrate this:

$\delta_p^+ = \sum_{m \in BC} \overline{x}_{im}$ for the beneficial criteria and $\delta_p^- = \sum_{m \in NBC} \overline{x}_{im}$ for the non-beneficial criteria.

Step 3: From the aforementioned δ_p^+ and δ_p^- values, we calculate the relative importance for each expertly recommended alternative solution w_p. In addition, using the value of w_p, we then compute the utility degree α_p as follows:

$$w_p = \delta_p^+ + \frac{\sum_{p=1}^{n} \delta_p^-}{\delta_p^- \sum_{p=1}^{n} \frac{1}{\delta_p^-}} \tag{12}$$

$$\alpha_p = \left[\frac{w_p}{w_{max}}\right] \times 100 \tag{13}$$

Step 4: Either the value of w_p or the value of α_p can equally help determine the ranking of the expertly recommended alternative solutions.

3.5. *Algorithm Intuitionistic Fuzzy DEMATEL [35]*

Step 1: Construct a direct-relation matrix where the entries are based on the obtained linguistic scale data.

Ratings given by k-th experts on the interrelationship between criteria are collected and used to construct a direct-relation matrix where entries are rating scales.

Step 2: Construct a direct-relation matrix where entries are IFNs.

The IFNs defined in Table 3 are substituted according to their linguistic terms into the direct-relation matrix.

$$A^k = \left[a_{ij}^k\right] = \begin{bmatrix} a_{11}^k & a_{12}^k & \cdots & a_{1n}^k \\ a_{21}^k & a_{22}^k & \cdots & a_{2n}^k \\ \vdots & \vdots & \ddots & \vdots \\ a_{m1}^k & a_{m2}^k & \cdots & a_{mn}^k \end{bmatrix} \tag{14}$$

where $a_{ij}^k = \left[\mu_{ij}^k, v_{ij}^k, \pi_{ij}^k\right], i = 1, 2, \cdots, m, j = 1, 2, \ldots, n$ is the rating for each one of the experts that participate on the criteria interrelationship evaluation process.

Step 3: Expert weight determination.

According to their level of expertise, the evaluators are assigned specific weights (levels of expertise). This is in accordance with their job experience and knowledge in the field of HCW and medical waste management. The determination of the weight for each expert is given in Equation (15) and it is based on the study of [33]. More specifically, if $a_{ij}^{k} = \left[\mu_{ij}^{k}, v_{ij}^{k}, \pi_{ij}^{k}\right], i = 1, 2, \cdots, m, j = 1, 2, \ldots, n$, the rating of the kth expert weight φ_k is:

$$\varphi_k = \frac{\left(\mu_k + \pi_k\left(\frac{\mu_k}{\mu_k + v_k}\right)\right)}{\sum_{k=1}^{l}\left(\mu_k + \pi_k\left(\frac{\mu_k}{\mu_k + v_k}\right)\right)} \quad \text{where} \quad \sum_{k=1}^{l} \varphi_k = 1 \tag{15}$$

Step 4: Find the aggregated intuitionistic fuzzy decision matrix.

To construct the aggregated intuitionistic fuzzy decision matrix, the intuitionistic fuzzy weighted average (IFWA) operator is used to fuse the weights of all decision-makers into one evaluation matrix of decision-makers:

$$\begin{aligned} \hat{A}_{[k \times n]} &= IFWAA_w(\alpha_{1,1}, \alpha_{1,2}, \ldots \alpha_{2,1} \ldots \alpha_{i,j} \ldots \alpha_{k,n}) \\ &= \left(1 - \prod_{i=1}^{k}(1 - \mu_{\alpha_i})^{\varphi_i}, \prod_{i=1}^{k}(1 - v_{\alpha_i})^{\varphi_i}, \prod_{i=1}^{k}(1 - \mu_{\alpha_i})^{w_j} - \prod_{i=1}^{k}(1 - \mu_{\alpha_i})^{\varphi_i}\right) \end{aligned} \tag{16}$$

Step 5: Transform the aggregated matrix into its numeric equivalent according to [34]. Then, the crisp values are computed for the total relationship matrix using the following equation:

$$p(x) = \frac{\mu - v + 1}{2} \tag{17}$$

using the same terminology for the membership and non-membership as in Equation (2a) and in Equation (2b).

Step 6: Construct the normalized direct-relation matrix X using the crisp matrix A and the constant k, $X = k \times A$ where the equation is as follows:

$$k = min\left(\frac{1}{max\sum_{j=1}^{n} a_{ij}}, \frac{1}{max\sum_{i=1}^{n} a_{ij}}\right) \tag{18}$$

Step 7: Construct the total-relation matrix T from the following equation:

$$T = X(I - X)^{-1} \tag{19}$$

with I to be the identity matrix.

Step 8: Determine the sum of rows and the sum of columns from the matrix T. The sum of the rows and the sum of the columns (denoted by D and R, respectively) can be calculated by Equations (20) and (21):

$$R = \left[\sum_{i=1}^{n} t_{ij}\right]_{1 \times n} = [t_j]_{1 \times n} \tag{20}$$

$$D = \left[\sum_{j=1}^{n} t_{ij}\right]_{n \times 1} = [t_i]_{n \times 1} \tag{21}$$

Step 9: Find the quantities $(D + R)$ and $(D - R)$, respectively. With the calculation of $(D + R)$, we can obtain the Cartesian co-ordinate values with respect to both axes.

Step 10: Draw the causal diagram.

Once the values of $(D + R)$ and $(D - R)$ are determined, then the two co-ordinates of each one of the criteria are determined and they can be mapped on the plane. The values of $(D + R)$ represent the *Prominence Vector*. Using this vector, we can visualize the importance of the criteria. On the other hand, the values of $(D - R)$ represent the *Relation Vector*. Depending on the values of $(D - R)$, we can then classify the criteria either in the *Cause Group* or in the *Effect Group* if the values of the Relation Vector are positive or negative, respectively.

Step 11: The construction of the network relationship map (NRM) using the threshold is as follows: the threshold value is computed, finding the mean of all cells of the total-relation matrix.

4. Applying the IF-Models of IFTOPSIS, IFCORPAS, and IFDEMATEL for the Thessaly Case Study: Detailed Analysis and Results

Optimizing the use of effective healthcare waste (HCW) treatment methods is crucial for promoting a sustainable and environmentally friendly urban environment. Concurrently, effective HCW management has the potential to generate energy through processes like anaerobic digestion and incineration. This promotes the production of sustainable energy while also helping to reduce waste. Therefore, in order to maximize the benefits of HCW management, competent authorities and regional decision-makers are very interested in choosing the most effective treatment alternative.

Our research focuses on identifying the best alternative methodologies for HCW treatment based on various related criteria and aspects in the Thessaly region of Greece. Thessaly, one of the thirteen regions of Greece, is located in the central part of the country and is known for its diverse landscapes, including mountains, plains, and coastal areas. Within Thessaly, several cities and towns, including Larissa (the regional capital), Volos, and Trikala, house hospitals and healthcare facilities to serve the local population. Medical or clinical waste produced from these facilities, including needles, syringes, bandages, expired medicines, and other biomedical materials, is predominantly non-hazardous, although the existence of hazardous medical waste remains significant. Proper HCW management is crucial in order to prevent the spread of diseases and protect both healthcare workers and the general public.

The volume of HCW produced is closely related to population density, the number of healthcare facilities, the types of medical services provided, and healthcare practices in the region. As depicted in the thematic map in Figure 3, Larissa has the greatest concentration of facilities, with three other smaller cities contributing with medium-sized hospital facilities. Healthcare waste management in Thessaly adheres to national regulations and guidelines, outlining proper segregation, collection, treatment, and disposal methods for HCW. Common methods include incineration, autoclaving (sterilization), and, sometimes, landfilling for non-hazardous biomedical waste. Given the importance of environmental sustainability, efforts may also be in place to promote eco-friendly HCW treatment practices, such as waste-to-energy technologies or the recycling of certain medical materials.

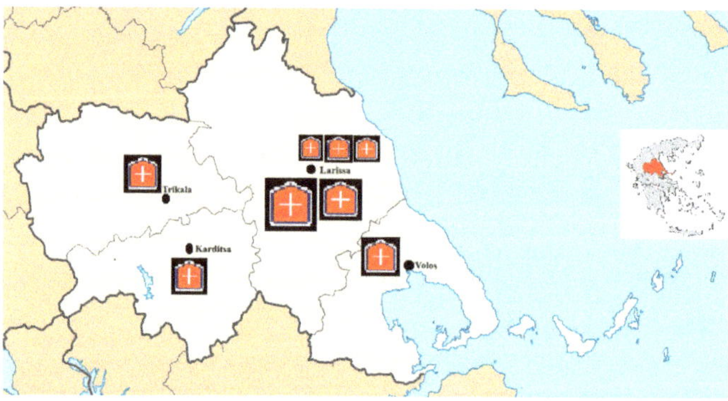

Figure 3. Thematic map of the HCW locations for the Thessaly region, Greece.

The following actions are necessary to address the problem of determining the most practical and effective waste treatment for HCW in a large area with dispersed healthcare facilities, like Thessaly, Greece. The steps involved in this process include: (a) selecting the most relevant impact factors (criteria); (b) involving experts to evaluate and rank the importance of these factors/criteria; (c) selecting the most pertinent treatment alternatives based again on expert opinions and the regional priorities; and (d) compiling the responses from all decision-makers to establish a hierarchy of the criteria and compare the significance of the criteria for each proposed solution.

Various indicators, including environmental and economic costs, sensitivity to the environment, and social aspects, demonstrate that the introduction of organized infrastructures and facilities, along with integrated HCW treatment methodologies, may offer solutions to numerous challenges in hazardous material transportation, management, and treatment. To address the complex problem of choosing the best methodology or creating a comprehensive policy for HCW treatment, multiple IF methodologies can be integrated. As seen in Table 4, we present 17 criteria grouped into six main categories: environmental, economic, social, organizational, waste-related, and legislation. The selection of the criteria is based on relevant literature, the professional judgment of hospital staff, capable authorities, and experts in the field of hazardous material treatment facilities:

Table 4. Categories of aspects, criteria, and justification of usage.

Criteria ID	Aspect	Criteria Name	Justification
C_1	Environmental	Hazard/risk potential	Proper waste management strategy is needed to ensure avoidance of any risk factors that affect public health [35,36].
C_2		Distance of storing facility in relation to society	Regional society avoids having hazardous-material-storing facilities nearby residential areas (Not In My Back Yard (NIMBY) phenomenon) [37,38].
C_3		Segregation, categorization, and transportation of HCW	The type of treatment, as well as the transportation of hazardous medical waste, is of great consideration [39–41].
C_4		Occupational health and safety	Investigation of health risks to staff from various hazards and suggestion of control measures to prevent recurrence of accidents [22,42]
C_5		Water and residue disposal and emission control	Control incineration of HCW residues via thermal treatment and improve volatilization to minimize the effects from the residue disposal in HCW treatment [43–46].
C_6	Economic	Cost of the waste treatment methodology	Waste management planning to ensure socio-economic and environmental sustainability [47].
C_7		Energy and other resource utilization	Beneficial and effective biomedical waste treatment technologies generate byproducts that can be further utilized as useful inputs in other industries, reinforcing the circular economy concept [48–50].
C_8	Social	Qualified personnel availability	The management of HCW involves a substantial amount of effort by healthcare professionals who are involved in daily care and other healthcare tasks [51,52]
C_9		Sensitivity to the environment, existence of the NIMBY phenomenon	Because of NIMBYism, waste facilities have to be located farther away from inhabited areas, which results in significant cost savings as trash management efficiency rises [53,54].
C_{10}	Organizational	HCW plan management substance and execution	Emphasize the role that medical waste management plays in creating a sustainable, environmentally friendly HCW setting [55].
C_{11}		Supply chain management and finance	A sustainable model for medical waste supply chain network is needed [56].
C_{12}		Health institution resources and infrastructure	Exploit the learnings from COVID-19 to identify any potential opportunities in post-pandemic waste infrastructure [57].
C_{13}	Waste Related	HCW type and hazard	Extra vigilance is necessary in order to reduce the danger of injury, cross-contamination, and infection; as a result, waste management professionals and healthcare personnel must adhere to the legal safety precautions [58].
C_{14}		HCW material	Material may affect generation of energy but, at the same time, may be dangerous for public health [59].
C_{15}		HCW treatment monitoring and control	Managing medical waste has become crucial, given its potential environmental and public health risks [60].

Table 4. *Cont.*

Criteria ID	Aspect	Criteria Name	Justification
C_{16}	**Authorities and Legislation**	Municipality services, competence, and treatment facility functions	Establishing a sustainable network for the handling of HCW under the supervision of appropriate authorities and municipal oversight [61]
C_{17}		Personnel competences in the treating facilities	Personnel inability to properly handle the hazardous material leads to disastrous accidents and many risks for public health [62,63]

In order to maximize the environmental, economic, and social sustainability of the regional societies, public safety and health, and regional development, we employ the IFTOPSIS, IFCORPAS, and IFDEMATEL IF methodologies in this research to choose the best alternative HCW treatment methodology. Four HCW treatment options are available as stated:

- **Incineration** is a method to treat medical waste, including hazardous materials and sharps. The process involves segregation and collection, receiving and inspection, loading the incinerator, and burning the waste at high temperatures. At the same time, air pollution control devices are used to capture and neutralize harmful emissions. After incineration, the remaining ash is inspected and disposed of in designated landfills or undergoes further treatment if necessary. We denote this alternative solution as ALT1.
- **HCW Storage** is a methodology that includes a series of steps: Initially, HCW is segregated into categories like sharps, infectious waste, pharmaceutical waste, and chemical waste, and placed in designated containers. Containers must be sturdy and leak-proof to prevent leaks. Proper labeling and documentation are essential for proper tracking and disposal. Medical waste is stored on-site in secure areas, with storage times regulated by local, state, and national regulations. Refrigeration is necessary for certain wastes, such as biological samples and certain pharmaceuticals. Healthcare staff are trained in proper protocols, and medical waste is prepared for transportation to treatment facilities. Facilities have emergency response and contingency plans in place to mitigate risks. Regulatory compliance is essential, with regular inspections and adherence to guidelines ensuring healthcare facilities maintain a safe environment for patients and staff. We denote this alternative solution as ALT2.
- **HCW Recycling** is a complex process that converts HCW material into reusable products, reducing the environmental impact and strain on natural resources. It involves the identification and segregation of waste, decontamination to remove harmful substances, and shredded or crushed materials for transportation and processing. Specialized facilities handle processed HCW using advanced technologies. Methods include mechanical recycling, chemical recycling, and energy recovery. Quality control checks ensure materials meet required standards, especially for medical devices or sensitive applications. Regulatory compliance is essential throughout the recycling process to ensure safety and environmental standards are met. However, there are limited methodologies for treating medical waste, particularly for biohazardous materials. We denote this alternative solution as ALT3.
- **HCW Segregation and Landfilling** is a methodology that involves identifying waste types, using color-coded containers, and training healthcare staff on proper segregation. Segregated waste is stored securely in containers, labeled with information about the type, date of generation, and facility name. On-site storage is also used to prevent environmental contamination. Packaging is carried out using double-bagged or leak-proof containers, while manifests document the type and quantity of waste being transported. Landfills are lined with protective liners and monitoring systems to prevent the leaching of hazardous materials. The waste is compacted to reduce volume and minimize environmental impact. Covering is carried out daily to minimize odor

and disease transmission. Regulatory compliance ensures the safe management of medical waste and public health. We call this alternative scenario ALT4.

Under the following scenarios, experts assess the alternatives: (a) public safety; (b) environmental safety, efficiency, and pollution prevention; and (c) risk assessment for the use of HCW treatment. It is also important to note that safety cannot be quantified, meaning it cannot be assigned a weight value. Consequently, even with interactive modeling approaches, it is impossible to compare the two scenarios. The categorical phrases that the experts used to compare the 17 criteria in Table 4 with the rating scale are listed in Table 5:

Table 5. Criteria evaluation terms and their corresponding IFN.

Linguistic Term	Equivalent IFN Triplet
Very Important (VI)	(0.88, 0.08, 0.04)
Important (I)	(0.75, 0.20, 0.05)
Medium (M)	(0.50, 0.45, 0.05)
Unimportant (UI)	(0.35, 0.60, 0.05)
Very Unimportant (VUI)	(0.08, 0.88, 0.04)

4.1. IFTOPSIS Analysis

Expert evaluators E_1, E_2, E_3, and E_4 responded, evaluating the pertinent criteria. Their findings in terms of criteria importance rates are presented in Table 3. It is important to emphasize that E_1, E_2 are competent authorities' officials, whilst the other two are hospital-focused personnel in different hospitals of the region. Additionally, we assume that each expert response is equally important, with an importance weight of 0.250.

Table 6 displays the linguistic responses from the expert evaluators E_1, E_2, E_3, and E_4 according to the criteria importance rates. We must also emphasize that E_1, E_2 are competent authorities' officials and the other two are hospital-focused personnel. Furthermore, we assume a 0.250 importance weight for each expert response, meaning that they are all equally important.

Table 6. Coded linguistic expert responses.

Crit. Code	Crit. Category	E_1	E_2	E_3	E_4
C_1	Environmental	M	M	M	I
C_2		I	M	M	M
C_3		I	M	M	M
C_4		VI	I	VI	VI
C_5		M	M	VI	VI
C_6	Economic	VI	I	M	I
C_7		I	I	I	I
C_8	Social	I	M	M	M
C_9		I	I	VI	UI
C_{10}	Organizational	VI	VI	UI	UI
C_{11}		VI	VI	I	M
C_{12}		VI	I	M	M
C_{13}	Waste Related	VI	I	M	M
C_{14}		VI	VI	I	I
C_{15}		M	I	M	I
C_{16}	Authorities and Legislation	VI	VI	I	I
C_{17}		M	I	M	I

The stated expert criteria evaluations and objectives and the aims of the under-consideration alternatives serve as the primary guides in the selection and formulation of assessment criteria. A hierarchical criteria tree could be created, but, since the weights of the experts' opinions have equal weight and importance for all respondents, the hierarchical tree may not provide any new information. Furthermore, unlike a multi-criteria analysis, which often uses the results or effects of the activities themselves as assessment criteria, MCDM procedures use the goals and objectives of the experts when these serve as evaluators of criteria. If all pertinent experts/stakeholders are included, these consequences will automatically show in the aims of the experts. Thus, the experts are responsible for providing non-redundant and legibility data.

Equations (4) and (5) of the IF-TOPSIS method drive the calculations. Table 5 shows the outcomes of this process including the membership, non-membership, and the uncertainty elements. It is imperative to acknowledge that these results align with the initial scenario, which pertains to environmental safety, efficiency, and pollution prevention. Two other sets of tables similar to Tables 6 and 7, respectively, have been produced and they are related to the public safety and the risk evaluation for the use of HCW treatment scenarios. However, due to space limitation reasons, they are not included but can be submitted in the supplementary material. In Table 8, for each criterion. We indicate whether it is a beneficial or not beneficial (B/NB), the calculated values columns μ, v. and π from the aggregated intuitionistic fuzzy decision matrix, and the values δ_i^+ and δ_i^-, as well as $\hat{C}_i$.

Table 7. Coded linguistic responses/valuations of experts for each alternative solution.

	ALT-1				ALT-2				ALT-3				ALT-4			
	E_1	E_2	E_3	E_4	E_1	E_2	E_3	E_4	E_1	E_2	E_3	E_4	E_1	E_2	E_3	E_4
C_1	VGE	GE	FE	GE	ME	FE	ME	GE	EGE	VVGE	VVGE	VVGE	GE	ME	VVGE	VVGE
C_2	EGE	VGE	FE	GE	ME	ME	VGE	VGE	VGE	VGE	VVGE	VVGE	EGE	EGE	EGE	EGE
C_3	VVGE	VGE	GE	GE	ME	ME	GE	GE	GE	VGE	VGE	EGE	VVGE	VVGE	GE	GE
C_4	EGE	EGE	EGE	VGE	EGE	EGE	EGE	EGE	ME	FE	ME	ME	VGE	VVGE	VVGE	VGE
C_5	EGE	EGE	EGE	EGE	EGE	EGE	EGE	EGE	GE	ME	VGE	GE	EGE	EGE	EGE	EGE
C_6	GE	ME	VVGE	VVGE	ME	FE	VGE	GE	VGE	VGE	VVGE	VVGE	EGE	EGE	VGE	GE
C_7	EGE	EGE	EGE	EGE	ME	FE	GE	GE	GE	VGE	VVGE	GE	EGE	EGE	VVGE	VVGE
C_8	VVGE	VVGE	GE	GE	ME	ME	EGE	EGE	GE	ME	ME	VGE	EGE	VVGE	VVGE	VGE
C_9	VGE	VVGE	VVGE	VGE	ME	ME	GE	VGE	VGE	VVGE	GE	GE	ME	GE	VGE	VGE
C_{10}	EGE	EGE	EGE	EGE	ME	ME	VGE	VGE	VGE	ME	ME	GE	VGE	GE	EGE	EGE
C_{11}	EGE	EGE	VGE	GE	GE	VGE	VGE	GE	VGE	GE	GE	VGE	ME	FE	ME	GE
C_{12}	EGE	EGE	VVGE	VVGE	GE	GE	VGE	VVGE	EGE	EGE	ME	ME	ME	ME	VGE	VGE
C_{13}	EGE	VVGE	VVGE	VGE	GE	GE	VGE	VVGE	VGE	GE	FE	GE	ME	ME	GE	GE
C_{14}	EGE	EGE	ME	ME	ME	FE	ME	GE	EGE	VGE	FE	GE	EGE	EGE	EGE	EGE
C_{15}	EGE	VVGE	VVGE	VGE	GE	GE	VGE	VGE	VVGE	VGE	GE	GE	EGE	EGE	EGE	EGE
C_{16}	ME	GE	VGE	VGE	ME	FE	ME	GE	EGE	EGE	EGE	VGE	ME	FE	VGE	GE
C_{17}	VGE	GE	EGE	EGE	VGE	GE	GE	GE	EGE	EGE	EGE	EGE	ME	FE	GE	GE

Table 8. Intuitionistic fuzzy decision matrix δ_i^+, δ_i^- and $\hat{C}_i$ (IFTOPSIS)—Scenario (a).

Criteria ID	B/NB	μ	v	π	δ_i^+	δ_i^-	$\hat{C}_i$
C_1	NB	0.811	0.132	0.057	0.140	0.934	0.916
C_2	B	0.742	0.194	0.064	0.113	0.947	0.927
C_3	B	0.764	0.176	0.06	0.042	1.004	0.955
C_4	B	0.857	0.102	0.041	0.196	0.912	0.886
C_5	NB	0.699	0.236	0.065	0.217	0.907	0.872
C_6	B	0.684	0.252	0.064	0.165	0.924	0.902
C_7	B	0.723	0.216	0.061	0.109	0.949	0.942
C_8	B	0.768	0.175	0.057	0.086	0.961	0.950
C_9	B	0.788	0.153	0.059	0.354	0.917	0.766
C_{10}	B	0.587	0.358	0.055	0.150	0.931	0.914
C_{11}	B	0.735	0.207	0.058	0.200	0.912	0.882
C_{12}	B	0.697	0.242	0.061	0.140	0.934	0.921
C_{13}	NB	0.811	0.132	0.057	0.113	0.947	0.936
C_{14}	NB	0.857	0.102	0.041	0.200	0.912	0.877
C_{15}	B	0.699	0.236	0.065	0.042	1.004	0.957
C_{16}	B	0.684	0.252	0.064	0.109	0.949	0.939
C_{17}	B	0.723	0.216	0.061	0.150	0.931	0.906

Using the findings from Table 8, we can compute the IFPIS and IFNIS solutions. Applying the corresponding equations, we find that $\theta_1 = \{0.857, 0.132, 0.011\}$ and

$\theta_2 = \{0.587, 0.242, 0.171\}$, respectively. We then calculate the relative assessment matrix. For each matrix cell, there exists a threshold function σ in order to recognize and break alternatives of equal distance from the ideal solution. As it has been proposed in [64], this threshold is allowed to obtain values between 0.01 and 0.05. The ranking of the four alternatives is shown in Table 9:

Table 9. Relative assessment matrix ranking.

	ALT_1	ALT_2	ALT_3	ALT_4	$\sum q_{ij}$	Ranking
ALT_1	0.000	0.317	0.247	0.189	0.753	1
ALT_2	−0.317	0.000	−0.295	−0.224	−0.836	4
ALT_3	−0.247	0.295	0.000	−0.175	−0.127	3
ALT_4	−0.189	0.224	0.175	0.000	0.210	2

The above ranking indicates that the incineration alternative (ALT1) solution is the most preferable, with the ALT4 (segregation and landfill) to follow. The third preferable is ALT3, which refers to HCW recycling, with the ALT2 (HCW storage) being the least preferable.

4.2. IFCORPAS Analysis

The intuitionistic fuzzy decision matrix, which comprises all expert responses for all alternatives and is displayed in Table 5, is the input used by the approach. We then compute the weighted decision matrix in a way that each cell is the aggregated outcome of all 17 criteria and all four alternative solutions. This is achieved by the use of Equations (10) and (11) in order to obtain the normalized value $\overline{x}_{im}$. The IFCORPAS algorithm's first step includes both procedures. Table 10 displays the acquired results.

Table 10. Normalized and weighted normalized decision matrix for IFCORPAS—Scenario (a).

		Normalized				Weighted Normalized			
Criteria ID	B/NB	ALT_1	ALT_2	ALT_3	ALT_4	ALT_1	ALT_2	ALT_3	ALT_4
C_1	NB	0.274	0.329	0.297	0.253	0.379	0.346	0.359	0.357
C_2	B	0.315	0.362	0.404	0.295	0.345	0.407	0.305	0.331
C_3	B	3.891	4.592	5.004	3.718	0.329	0.297	0.253	0.303
C_4	B	0.658	0.617	0.566	0.583	0.362	0.404	0.295	0.339
C_5	NB	0.387	0.391	0.402	0.338	4.592	5.004	3.718	4.387
C_6	B	0.331	0.329	0.357	0.300	0.617	0.566	0.583	0.547
C_7	B	0.267	0.273	0.255	0.251	0.391	0.402	0.338	0.341
C_8	B	0.372	0.479	0.518	0.354	0.329	0.357	0.300	0.298
C_9	B	0.324	0.317	0.366	0.248	0.273	0.255	0.251	0.256
C_{10}	B	0.311	0.308	0.350	0.284	0.479	0.518	0.354	0.455
C_{11}	B	0.334	0.329	0.289	0.295	0.317	0.366	0.248	0.243
C_{12}	B	0.388	0.374	0.398	0.362	0.308	0.350	0.284	0.282
C_{13}	NB	0.436	0.361	0.378	0.420	0.329	0.289	0.295	0.290
C_{14}	NB	0.367	0.381	0.379	0.346	0.374	0.398	0.362	0.349
C_{15}	B	0.424	0.318	0.345	0.407	0.638	0.658	0.617	0.566
C_{16}	B	0.322	0.274	0.329	0.297	0.461	0.387	0.391	0.402
C_{17}	B	0.431	0.315	0.362	0.404	0.394	0.331	0.329	0.357

We calculate the overall weighted normalized value for Step 2 of the IFCORPAS model in relation to the non-beneficial criteria (NBC) and the beneficial criteria (BC). The last four columns of Table 10 can be immediately utilized to obtain the results for δ_p^+ and δ_p^-, respectively. Lastly, using Step 3, we determine each alternative's relative importance and utility degree, providing us with a ranking of the alternatives. Table 11 displays the acquired results. There is agreement between the IFTOPSIS rankings as seen in Section 4.1.

Table 11. Distances for the ideal solution and ranking of the alternatives in the IFCORPAS model for scenario (a).

	δ_p^+	δ_p^-	a_p	**Ranking**
ALT_1	8.416	1.962	1.000	1
ALT_2	6.492	1.613	0.681	4
ALT_3	7.127	1.672	0.743	3
ALT_4	7.437	1.707	0.814	2

4.3. IF-DEMATEL Analysis

We employed the intuitionistic fuzzy DEMATEL method for capturing the complex relationships among the 17 criteria and the four alternative options for HCW treatment. Incorporating the experts' responses in terms of the comparative linguistic rating of the 17 criteria, as well as their rating of the criteria within each one of the alternatives $ALT_1 - ALT_4$, we first calculate the direct-relation matrix where the entries are IFNs. At the next step, we compute the intuitionistic fuzzy decision matrix. The transformation of this matrix into crisp values used the Equation (17) of the IF-DEMATEL methodology. Table 12 shows the crisp values of the intuitionistic fuzzy transform. In this table, we included Step 8 of the methodology, determining the sum of the rows and the sum of the columns to compute R and D from Equation (20) and Equation (21), respectively. Finally, in Table 13, we construct the total-relation matrix T.

Table 12. Relative importance of alternatives and ranking of IF DEMATEL for Scenario (a).

	C_1	C_2	C_3	C_4	C_5	C_6	C_7	C_8	C_9	C_{10}	C_{11}	C_{12}	C_{13}	C_{14}	C_{15}	C_{16}	C_{17}	D
C_1	0.5904	0.5737	0.5582	0.6272	0.5709	0.6707	0.5911	0.5737	0.6177	0.6606	0.5709	0.6190	0.5911	0.6278	0.6177	0.6349	0.6606	**10.3562**
C_2	0.5916	0.5550	0.5569	0.6247	0.5500	0.6594	0.6196	0.5550	0.6245	0.6698	0.5500	0.6203	0.6196	0.6296	0.6245	0.6428	0.6698	**10.3631**
C_3	0.6046	0.5735	0.5636	0.6054	0.5734	0.6622	0.5828	0.5735	0.6061	0.6643	0.5734	0.6308	0.5828	0.6306	0.6061	0.6239	0.6643	**10.3213**
C_4	0.5803	0.5363	0.5217	0.5611	0.5168	0.5913	0.5093	0.5363	0.6046	0.5735	0.5636	0.6054	0.5734	0.6622	0.5828	0.5482	0.5935	**9.6603**
C_5	0.6280	0.6098	0.6077	0.6727	0.6180	0.7192	0.6374	0.6098	0.5311	0.4934	0.5064	0.5644	0.5521	0.5890	0.5310	0.6659	0.7034	**10.2393**
C_6	0.5963	0.5730	0.5479	0.6056	0.5289	0.6783	0.5780	0.5730	0.6030	0.5552	0.5616	0.6174	0.5487	0.6494	0.5978	0.6150	0.6261	**10.0552**
C_7	0.5036	0.5570	0.5115	0.5516	0.4925	0.5912	0.5418	0.5570	0.6262	0.5978	0.5804	0.6700	0.5802	0.6675	0.6574	0.5525	0.5558	**9.794**
C_8	0.5718	0.5517	0.5920	0.6037	0.5552	0.6374	0.5542	0.5517	0.5913	0.5843	0.5732	0.6393	0.5879	0.6633	0.5969	0.5943	0.6195	**10.0677**
C_9	0.5622	0.5620	0.5468	0.6260	0.5044	0.6065	0.5676	0.5620	0.5730	0.5479	0.6056	0.5289	0.6783	0.5780	0.5730	0.5504	0.5980	**9.7706**
C_{10}	0.6046	0.5735	0.5636	0.6054	0.5734	0.6622	0.5828	0.5735	0.5570	0.5115	0.5516	0.4925	0.5912	0.5418	0.5570	0.6239	0.6643	**9.8298**
C_{11}	0.5311	0.4934	0.5064	0.5644	0.5521	0.5890	0.5310	0.4934	0.6245	0.6698	0.5500	0.6203	0.6196	0.6296	0.6245	0.6428	0.6698	**9.9117**
C_{12}	0.6030	0.5552	0.5616	0.6174	0.5487	0.6494	0.5978	0.5552	0.6061	0.6643	0.5734	0.6308	0.5828	0.6306	0.6061	0.6239	0.6643	**10.2706**
C_{13}	0.6262	0.5978	0.5804	0.6700	0.5802	0.6675	0.6574	0.5978	0.6046	0.5735	0.5636	0.6054	0.5734	0.6622	0.5828	0.5482	0.5935	**10.2845**
C_{14}	0.5913	0.5843	0.5732	0.6393	0.5879	0.6633	0.5969	0.5843	0.5311	0.4934	0.5064	0.5644	0.5521	0.5890	0.5310	0.6659	0.7034	**9.9572**
C_{15}	0.6088	0.5881	0.5854	0.6419	0.5801	0.6862	0.6044	0.5881	0.6030	0.5552	0.5616	0.6174	0.5487	0.6494	0.5978	0.6150	0.6261	**10.2572**
C_{16}	0.5904	0.5737	0.5582	0.6272	0.5709	0.6707	0.5911	0.5737	0.6262	0.5978	0.5804	0.6700	0.5802	0.6675	0.6574	0.5525	0.5558	**10.2437**
C_{17}	0.5567	0.5180	0.5213	0.5923	0.5304	0.6238	0.5708	0.5180	0.5913	0.5843	0.5732	0.6393	0.5879	0.6633	0.5969	0.5943	0.6195	**9.8813**
R	**9.9409**	**9.5760**	**9.4564**	**10.4359**	**9.4338**	**11.0283**	**9.9140**	**9.5760**	**10.1213**	**9.9966**	**9.5453**	**10.3356**	**9.9500**	**10.7308**	**10.1407**	**10.2944**	**10.7877**	**0.5926**

Table 13. Total-relation matrix T for IF DEMATEL for Scenario (a).

	C_1	C_2	C_3	C_4	C_5	C_6	C_7	C_8	C_9	C_{10}	C_{11}	C_{12}	C_{13}	C_{14}	C_{15}	C_{16}	C_{17}
C_1	0.0000	0.0000	0.0000	0.6272	0.0000	0.6707	0.0000	0.0000	0.6177	0.6606	0.0000	0.6190	0.0000	0.6278	0.6177	0.6349	0.6606
C_2	0.0000	0.0000	0.0000	0.6247	0.0000	0.6594	0.6196	0.0000	0.6245	0.6698	0.0000	0.6203	0.6196	0.6296	0.6245	0.6428	0.6698
C_3	0.6046	0.0000	0.0000	0.6054	0.0000	0.6622	0.0000	0.0000	0.6061	0.6643	0.0000	0.6308	0.0000	0.6306	0.6061	0.6239	0.6643
C_4	0.0000	0.0000	0.0000	0.0000	0.0000	0.0000	0.0000	0.0000	0.6046	0.0000	0.0000	0.6054	0.0000	0.6622	0.0000	0.0000	0.5935
C_5	0.6280	0.6098	0.6077	0.6727	0.6180	0.7192	0.6374	0.6098	0.0000	0.0000	0.0000	0.0000	0.0000	0.0000	0.0000	0.6659	0.7034
C_6	0.5963	0.0000	0.0000	0.6056	0.0000	0.6783	0.0000	0.0000	0.6030	0.0000	0.0000	0.6174	0.0000	0.6494	0.5978	0.6150	0.6261
C_7	0.0000	0.0000	0.0000	0.0000	0.0000	0.0000	0.0000	0.0000	0.6262	0.5978	0.0000	0.6700	0.0000	0.6675	0.6574	0.0000	0.0000
C_8	0.0000	0.0000	0.0000	0.6037	0.0000	0.6374	0.0000	0.0000	0.0000	0.0000	0.0000	0.6393	0.0000	0.6633	0.5969	0.5943	0.6195
C_9	0.0000	0.0000	0.0000	0.6260	0.0000	0.6065	0.0000	0.0000	0.0000	0.0000	0.6056	0.0000	0.6783	0.0000	0.0000	0.0000	0.5980
C_{10}	0.6046	0.0000	0.0000	0.6054	0.0000	0.6622	0.0000	0.0000	0.0000	0.0000	0.0000	0.0000	0.0000	0.0000	0.0000	0.6239	0.6643
C_{11}	0.0000	0.0000	0.0000	0.0000	0.0000	0.0000	0.0000	0.0000	0.6245	0.6698	0.0000	0.6203	0.6196	0.6296	0.6245	0.6428	0.6698
C_{12}	0.6030	0.0000	0.0000	0.6174	0.0000	0.6494	0.5978	0.0000	0.6061	0.6643	0.0000	0.6308	0.0000	0.6306	0.6061	0.6239	0.6643
C_{13}	0.6262	0.5978	0.0000	0.6700	0.0000	0.6675	0.6574	0.5978	0.6046	0.0000	0.0000	0.6054	0.0000	0.6622	0.0000	0.0000	0.5935
C_{14}	0.0000	0.0000	0.0000	0.6393	0.0000	0.6633	0.5969	0.0000	0.0000	0.0000	0.0000	0.0000	0.0000	0.0000	0.0000	0.6659	0.7034
C_{15}	0.6088	0.0000	0.0000	0.6419	0.0000	0.6862	0.6044	0.0000	0.6030	0.0000	0.0000	0.6174	0.0000	0.6494	0.5978	0.6150	0.6261
C_{16}	0.0000	0.0000	0.0000	0.6272	0.0000	0.6707	0.0000	0.0000	0.6262	0.5978	0.0000	0.6700	0.0000	0.6675	0.6574	0.0000	0.0000
C_{17}	0.0000	0.0000	0.0000	0.0000	0.0000	0.6238	0.0000	0.0000	0.0000	0.0000	0.0000	0.6393	0.0000	0.6633	0.5969	0.5943	0.6195

We obtained the cause-and-effect diagram (Figure 4) by mapping a dataset of $(D + R,\ D - R)$ where the horizontal axis $(D + R)$ represents the importance of the 17 criteria and $(D - R)$ classifies the identified challenges into each of the cause groups as shown in Table 14.

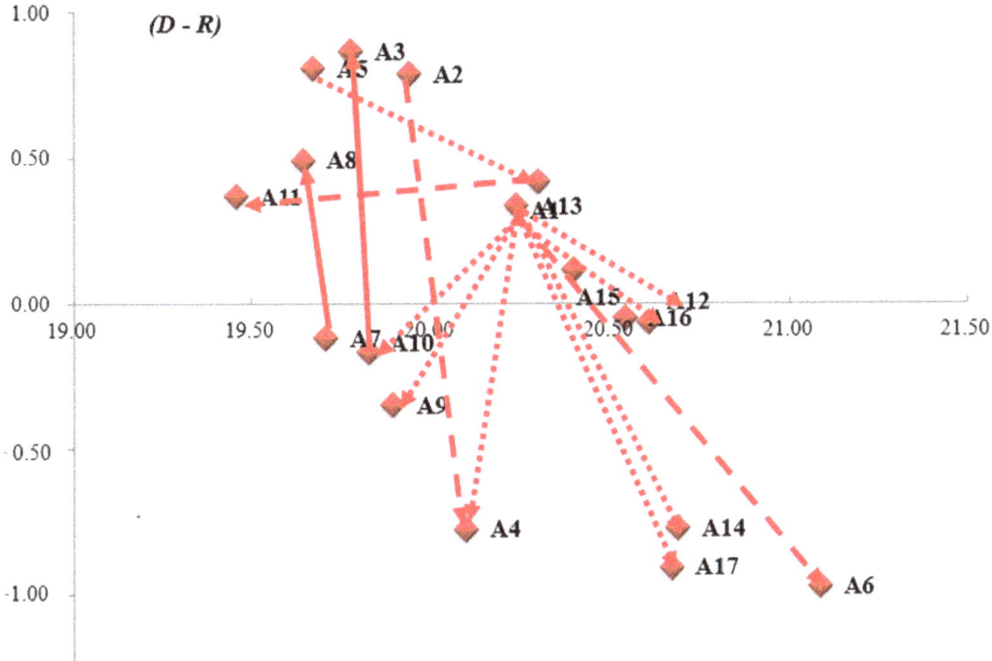

Figure 4. Causal relationships of the criteria for Scenario (a).

Table 14. Criteria (network) relationship map using the (D–R) and the mean threshold of the causal relationships of the 17 criteria in Scenario (a).

Crit. Code	Crit. Category	*R*	*D*	*D+R*	*D−R*
C_1	Environmental	9.9409	10.3562	20.2971	0.4153
C_2		9.576	10.3631	19.9391	0.7871
C_3		9.4564	10.3213	19.7777	0.8649
C_4		10.4359	9.6603	20.0962	**−0.7756**
C_5		9.4338	10.2393	19.6731	0.8055
C_6	Economic	11.0283	10.0552	21.0835	**−0.9731**
C_7		9.914	9.794	19.708	**−0.12**
C_8	Social	9.576	10.0677	19.6437	0.4917
C_9		10.1213	9.7706	19.8919	**−0.3507**
C_{10}	Organizational	9.9966	9.8298	19.8264	**−0.1668**
C_{11}		9.5453	9.9117	19.457	0.3664
C_{12}		10.3356	10.2706	20.6062	**−0.065**
C_{13}	Waste Related	9.95	10.2845	20.2345	0.3345
C_{14}		10.7308	9.9572	20.688	**−0.7736**
C_{15}		10.1407	10.2572	20.3979	0.1165
C_{16}	Authorities and Legislation	10.2944	10.2437	20.5381	**−0.0507**
C_{17}		10.7877	9.8813	20.669	**−0.9064**

The ranking of the four alternatives is calculated indirectly from their relative importance in terms of the scenarios. The IF-DEMATEL method is supportive of the ranking process since the ranking of alternatives can be calculated the same way as IF-TOPSIS from the intuitionistic fuzzy decision matrix of the 17 criteria. The ranking of the criteria is shown in Table 15.

Table 15. Relative criteria ranking from IF-DEMATEL for Scenario (a).

Crit. ID	Aspect	Criteria Name	B/NB	Closeness Coefficient	Ranking
C_1	Environmental	Hazard/risk potential	NB	0.916	9
C_2		Distance of storing facility in relation to society	B	0.927	7
C_3		Segregation, categorization, and transportation of HCW	B	0.955	2
C_4		Occupational health and safety	B	0.886	13
C_5		Water and residue disposal and emission control	NB	0.872	16
C_6	Economic	Cost of the waste treatment methodology	B	0.902	12
C_7		Energy and other resource utilization	B	0.942	4
C_8	Social	Qualified personnel availability	B	0.950	3
C_9		Sensitivity to the environment, existence of the NIMBY phenomenon	B	0.766	17
C_{10}	Organizational	HCW plan management substance and execution	B	0.914	10
C_{11}		Supply chain management and finance	B	0.882	14
C_{12}		Health institution resources and infrastructure	B	0.921	8
C_{13}	Waste-Related	HCW type and hazard	NB	0.936	6
C_{14}		HCW material	NB	0.877	15
C_{15}		HCW treatment monitoring and control	B	0.957	1
C_{16}	Authorities and Legislation	Municipality services, competence and treatment facility functions	B	0.939	5
C_{17}		Personnel competences in the treating facilities	B	0.906	11

For that reason, the results of IF-DEMATEL are identical to IF-TOPSIS as shown in Table 16:

Table 16. Relative importance of alternatives and ranking of IFCORPAS for Scenario (a).

	ALT_1	ALT_2	ALT_3	ALT_4	$\sum q_{ij}$	Ranking
ALT_1	0.000	0.317	0.247	0.189	0.753	1
ALT_2	−0.317	0.000	−0.295	−0.224	−0.836	4
ALT_3	−0.247	0.295	0.000	−0.175	−0.127	3
ALT_4	−0.189	0.224	0.175	0.000	0.210	2

5. Discussion of Results

Discussing the qualitative data results, as opposed to the quantitative analysis of Section 4, we now focus on the chosen models used in the evaluation/ranking process, prioritizing the ranking of the alternatives. For frameworks utilized in the formulation of policies for sustainable urban development, this is always important [65]. As a result, it is critical to evaluate the framework in its entirety as a tool to assist the policy-maker in supporting their decision. The results show that the three alternatives were ranked identically by all models. When the above methodologies are used on the same dataset, this demonstrates their consistency. However, if all the participating models yield the same outcome, then this fact invites the question of whether the comparative analysis is necessary for the study. To make such an argument, one must, however, further demonstrate that every issue is incorporated in this framework, which is not obviously the case. On the contrary, there are many other aspects, criteria, and parameters which have not been included in the study because they have not indicated critical significance relative to the 17 already-taken criteria. But the factor of the direct relevance of these criteria with the region of study is the one that justifies the existence of alternative methodologies working on the same aspect to produce justifiable results. We collected all data pertaining to the ranking outcomes of IF-TOPSIS, IF-CORPAS, and IF-DEMATEL, respectively, from the relative tables. More specifically, the exact relevance weights that each approach allocated to each of the different alternatives are of greater interest to us. Only IF-CORPAS gives normalized significance weights among the methodologies used. The other approaches have been normalized by us so that their performance is consistent across the board. The outcomes are displayed in Table 17:

Table 17. Relative performance matrix for all methodologies for Scenario (a).

	Initial			Normalized		
Alternatives	**IF-Topsis**	**IF-Corpas**	**IF-Dematel**	**IF-Topsis**	**IF-Corpas**	**IF-Dematel**
ALT_1	7.257	1.000	0.349	0.300149	0.281136	0.262209
ALT_2	5.732	0.814	0.327	0.237075	0.228845	0.24568
ALT_3	4.958	0.781	0.321	0.205062	0.219567	0.241172
ALT_4	6.231	0.962	0.334	0.257714	0.270453	0.250939

Incineration is ranked as the most preferable solution due to its ability to safely dispose of a wide range of medical waste types. When operated correctly, modern incinerators can efficiently burn medical waste at high temperatures, minimizing the risk of contamination and emissions. The method is considered effective for medical waste disposal because it reduces the volume of waste significantly and can destroy pathogens and harmful chemicals.

On the other hand, the storing of HCW is ranked as the least preferable alternative. This is justified by the fact that storing HCW for prolonged periods can lead to the risk of contamination, environmental pollution, and potential health hazards for workers and nearby communities. At the same time, the method does not address the disposal issue, and the waste may pose increased risks over time.

Segregation and landfill is the second most preferable alternative for the case of the Thessaly region. Experts indicated that segregating medical waste before landfilling can help reduce the volume of hazardous waste going into landfills, making it easier to manage. Properly engineered landfills can contain medical waste and prevent environmental contamination if managed effectively. On the other hand, there is still a great hazard imposed due to possible soil contamination because of the landfills.

Finally, recycling was ranked as the third important alternative HCW treatment for Thessaly. Experts indicated that recycling certain types of medical waste, such as plastics and glass, can reduce the demand for new raw materials and lower the environmental impact. However, recycling HCW is also challenging due to the risk of contamination. Not all medical waste materials are suitable for recycling, especially those contaminated with biohazardous materials. Therefore, while it is environmentally friendly, it is limited in scope due to the specific nature of medical waste. Recycling can be a part of the waste management process but cannot be the sole solution due to its limitations in handling all types of medical waste.

6. Conclusions

Rapid global urbanization and population growth have presented municipalities with a significant challenge in managing healthcare waste (HCW), attracting increased attention in recent years. Selecting the appropriate HCW treatment technology is a complex task, involving conflicting criteria and multiple social stakeholders. HCW management systems often grapple with imprecise parameters expressed through uncertain linguistic expressions. The use of fuzzy membership functions in decision making can sometimes result in the loss of valuable information.

In this study, we introduced a comprehensive decision-making framework to facilitate optimal HCW treatment technology selection, considering various stakeholders' perspectives and employing a range of fuzzy intuitionistic methodologies, including IF-TOPSIS, IF-CORPAS, and IF-DEMATEL. We utilized an ensemble of these models to handle the linguistic assessments of experts in the field, focusing on the use case of the Thessaly region, Greece. Our approach objectively calculates decision-makers' weights, addressing their uncertainty and divergence levels. We demonstrated our methodology through a practical case study and conducted a comparative analysis to validate its effectiveness. The study showed that our decision-making approach effectively resolves HCW disposal technology selection problems in uncertain and complex environments, thereby strengthening public health. We used the aforementioned three methods to identify the best alternative HCW treatment methods for environmental, economic, and social sustainability. Four options are identified: incineration, storage, recycling, and segregation and landfilling. We also assumed that regulatory compliance ensures the safe management of medical waste and public health. By employing distances from the ideal solution, two techniques—IF-TOPSIS and IF-CORPAS—basically distinguish criteria and alternative solutions, respectively. Distance is the basis of the IF-DEMATEL model. Every model managed to generate an identical rating amongst the alternatives, regardless of the model's core. For the incineration alternative which was the most preferred by all IF-methodologies, IF-TOPSIS presents the highest critical significance of ~0.30, with IF-CORPAS in second place with ~0.28, and IF-DEMATEL in third place, computing ~0.26 out of the whole critical importance vector. The second-best ranking alternative is HCW segregation and landfilling. Even though, in other studies, this alternative is not popular, the Greek study showed an augmented preference. However, the most popular choice was the incineration of HCW because it can securely dispose of various kinds of medical waste. Modern incinerators can effectively burn medical waste at high temperatures while limiting the danger of contamination and pollution when used properly. The safety criterion was intended to be non-negotiable across all participants, avoiding competition with other options. We must mention that our study compares the most dominant technologies for HCW treatment in a holistic way (multiple categories of criteria have been used). Even though the results could support the medical waste hierarchy, we strongly believe that life-cycle assessment is a crucial tool in waste management practice, but little research has been performed on the evaluation of medical waste treatment from a life-cycle perspective.

Future research will focus on expanding this integrated approach to address HCW management issues involving diverse interrelationships among evaluation criteria. Moreover, methods for objectively determining criteria weights should be developed, especially in real-world situations where subjective estimation is challenging. Another avenue for future research involves creating a computer-based application system to streamline the implementation of our proposed HCW treatment technology selection algorithm and enhance human–machine interaction for presenting and analyzing selection results. This framework can be made more flexible by adding more models for more comprehensive evaluation procedures. Furthermore, further research is needed to examine the characteristics of ensemble-based systems in scenarios with inconsistent ranking outcomes. Inter-val-valued, Pythagorean, or interval type-2 fuzzy sets can also be added to a more complete framework to improve the current approach and guarantee public health.

Author Contributions: Conceptualization, K.K. and V.K.; methodology, K.K. and E.L.; software, K.K.; validation, K.K. and E.L.; investigation, K.K., E.L. and V.K.; resources, K.M. and C.T.; data curation, K.K., E.L. and C.T.; writing—original draft preparation, K.K.; writing—review and editing, E.L., C.T. and V.K.; visualization, K.K., K.M. and C.T.; supervision, K.M. and V.K.; project administration, K.M. All authors have read and agreed to the published version of the manuscript.

Funding: This research received no external funding.

Institutional Review Board Statement: Not applicable.

Informed Consent Statement: Informed consent was obtained from all subjects involved in the study.

Data Availability Statement: The data presented in this study are available on request from the corresponding author.

Conflicts of Interest: The authors declare no conflict of interest.

References

1. Mondal, R.; Mishra, S.; Pillai, J.S.K.; Sahoo, M.C. COVID-19 Pandemic and biomedical waste management practices in healthcare system. *J. Fam. Med. Prim. Care* **2022**, *11*, 439–446. [CrossRef] [PubMed]
2. Ranjbari, M.; Esfandabadi, Z.S.; Shevchenko, T.; Chassagnon-Haned, N.; Peng, W.; Tabatabaei, M.; Aghbashlo, M. Mapping healthcare waste management research: Past evolution, current challenges, and future perspectives towards a circular economy transition. *J. Hazard. Mater.* **2022**, *422*, 126724. [CrossRef] [PubMed]
3. Ervural, B.Ç.; Öner, S.C.; Çoban, V.; Kahraman, C. A novel Multiple Attribute Group Decision Making methodology based on Intuitionistic Fuzzy TOPSIS. In Proceedings of the 2015 IEEE International Conference on Fuzzy Systems (FUZZ-IEEE), Istanbul, Turkey, 2–5 August 2015; pp. 1–6. [CrossRef]
4. Alvarez, P.A.; Ishizaka, A.; Martínez, L. Multiple-criteria decision-making sorting methods: A survey. *Expert Syst. Appl.* **2021**, *183*, 115368. [CrossRef]
5. Kaya, G.K.; Ozturk, F. A Comparison of the Multi-criteria Decision-Making Methods for the Selection of Researchers. In *Industrial Engineering in the Internet-of-Things World*; Calisir, F., Ed.; Lecture Notes in Management and Industrial Engineering; Springer International Publishing: Cham, Switzerland, 2022; pp. 147–159. [CrossRef]
6. Schitea, D.; Deveci, M.; Iordache, M.; Bilgili, K.; Akyurt, İ.Z.; Iordache, I. Hydrogen mobility roll-up site selection using intuitionistic fuzzy sets based WASPAS, COPRAS and EDAS. *Int. J. Hydrogen Energy* **2019**, *44*, 8585–8600. [CrossRef]
7. Chhibber, D.; Srivastava, P.K.; Bisht, D.C.S. Intuitionistic fuzzy TOPSIS for non-linear multi-objective transportation and manufacturing problem. *Expert Syst. Appl.* **2022**, *210*, 118357. [CrossRef]
8. Sharma, K.; Singh, V.P.; Ebrahimnejad, A.; Chakraborty, D. Solving a multi-objective chance constrained hierarchical optimization problem under intuitionistic fuzzy environment with its application. *Expert Syst. Appl.* **2023**, *217*, 119595. [CrossRef]
9. Mazzei, H.G.; Specchia, S. Latest insights on technologies for the treatment of solid medical waste: A review. *J. Environ. Chem. Eng.* **2023**, *11*, 109309. [CrossRef]
10. Liu, S.; Zhang, J.; Niu, B.; Liu, L.; He, X. A novel hybrid multi-criteria group decision-making approach with intuitionistic fuzzy sets to design reverse supply chains for COVID-19 medical waste recycling channels. *Comput. Ind. Eng.* **2022**, *169*, 108228. [CrossRef]
11. Çelik, S.; Peker, İ.; Gök-Kısa, A.C.; Büyüközkan, G. Multi-criteria evaluation of medical waste management process under intuitionistic fuzzy environment: A case study on hospitals in Turkey. *Socio-Econ. Plan. Sci.* **2023**, *86*, 101499. [CrossRef]
12. Çakir, E.; Taş, M.A.; Ulukan, Z. A New Circular Intuitionistic Fuzzy MCDM: A Case of COVID-19 Medical Waste Landfill Site Evaluation. In Proceedings of the 2021 IEEE 21st International Symposium on Computational Intelligence and Informatics (CINTI), Budapest, Hungary, 18–20 November 2021; pp. 143–148. [CrossRef]
13. Patel, A.; Jana, S.; Mahanta, J. Intuitionistic fuzzy EM-SWARA-TOPSIS approach based on new distance measure to assess the medical waste treatment techniques. *Appl. Soft Comput.* **2023**, *144*, 110521. [CrossRef]
14. Qian, W.; Wang, Z.-J.; Li, K.W. Medical Waste Disposal Method Selection Based on a Hierarchical Decision Model with Intuitionistic Fuzzy Relations. *Int. J. Environ. Res. Public Health* **2016**, *13*, 896. [CrossRef] [PubMed]
15. Mishra, A.R.; Mardani, A.; Rani, P.; Zavadskas, E.K. A novel EDAS approach on intuitionistic fuzzy set for assessment of health-care waste disposal technology using new parametric divergence measures. *J. Clean. Prod.* **2020**, *272*, 122807. [CrossRef]
16. Büyüközkan, G.; Gocer, F. An Intuitionistic Fuzzy MCDM Approach for Effective Hazardous Waste Management. In *Intelligence Systems in Environmental Management: Theory and Applications*; Kahraman, C., Sari, İ.U., Eds.; Intelligent Systems Reference Library; Springer International Publishing: Cham, Switzerland, 2017; pp. 21–40. [CrossRef]
17. Narayanamoorthy, S.; Annapoorani, V.; Kang, D.; Baleanu, D.; Jeon, J.; Kureethara, J.V.; Ramya, L. A novel assessment of bio-medical waste disposal methods using integrating weighting approach and hesitant fuzzy MOOSRA. *J. Clean. Prod.* **2020**, *275*, 122587. [CrossRef]
18. Ghoushchi, S.J.; Ghiaci, A.M.; Bonab, S.R.; Ranjbarzadeh, R. Barriers to circular economy implementation in designing of sustainable medical waste management systems using a new extended decision-making and FMEA models. *Environ. Sci. Pollut. Res.* **2022**, *29*, 79735–79753. [CrossRef] [PubMed]
19. Liu, P.; Rani, P.; Mishra, A.R. A novel Pythagorean fuzzy combined compromise solution framework for the assessment of medical waste treatment technology. *J. Clean. Prod.* **2021**, *292*, 126047. [CrossRef]
20. Ghoushchi, S.J.; Bonab, S.R.; Ghiaci, A.M.; Haseli, G.; Tomaskova, H.; Hajiaghaei-Keshteli, M. Landfill Site Selection for Medical Waste Using an Integrated SWARA-WASPAS Framework Based on Spherical Fuzzy Set. *Sustainability* **2021**, *13*, 13950. [CrossRef]
21. Menekşe, A.; Akdağ, H.C. Medical waste disposal planning for healthcare units using spherical fuzzy CRITIC-WASPAS. *Appl. Soft Comput.* **2023**, *144*, 110480. [CrossRef]
22. Salimian, S.; Mousavi, S.M. The selection of healthcare waste treatment technologies by a multi-criteria group decision-making method with intuitionistic fuzzy sets. *J. Ind. Syst. Eng.* **2022**, *14*, 205–220.
23. Görçün, Ö.F.; Aytekin, A.; Korucuk, S.; Tirkolaee, E.B. Evaluating and selecting sustainable logistics service providers for medical waste disposal treatment in the healthcare industry. *J. Clean. Prod.* **2023**, *408*, 137194. [CrossRef]
24. Al-Barakati, A.; Rani, P. Assessment of healthcare waste treatment methods using an interval-valued intuitionistic fuzzy double normalization-based multiple aggregation approach. *Environ. Dev. Sustain.* **2023**, *1*, 1–28. [CrossRef]
25. Seikh, M.R.; Mandal, U. Interval-valued Fermatean fuzzy Dombi aggregation operators and SWARA based PROMETHEE II method to bio-medical waste management. *Expert Syst. Appl.* **2023**, *226*, 120082. [CrossRef]

26. Chew, X.; Khaw, K.W.; Alnoor, A.; Ferasso, M.; Al Halbusi, H.; Muhsen, Y.R. Circular economy of medical waste: Novel intelligent medical waste management framework based on extension linear Diophantine fuzzy FDOSM and neural network approach. *Environ. Sci. Pollut. Res.* **2023**, *30*, 60473–60499. [CrossRef] [PubMed]
27. Salimian, S.; Mousavi, S.M. Healthcare waste disposal location selection by a multi-criteria decision-making method with intuitionistic fuzzy sets. *J. Qual. Eng. Prod. Optim.* **2021**, *6*, 143–156.
28. Ocampo, L.; Tanaid, R.A.; Tiu, A.M.; Selerio, E., Jr.; Yamagishi, K. Classifying the degree of exposure of customers to COVID-19 in the restaurant industry: A novel intuitionistic fuzzy set extension of the TOPSIS-Sort. *Appl. Soft Comput.* **2021**, *113*, 107906. [CrossRef]
29. Roszkowska, E.; Kusterka-Jefmańska, M.; Jefmański, B. Intuitionistic fuzzy TOPSIS as a method for assessing socioeconomic phenomena on the basis of survey data. *Entropy* **2021**, *23*, 563. [CrossRef]
30. Atanassov, K.T. More on intuitionistic fuzzy sets. *Fuzzy Sets Syst.* **1989**, *33*, 37–45. [CrossRef]
31. Xu, Z.-S. Methods for aggregating interval-valued intuitionistic fuzzy information and their application to decision making. *Control Decis.* **2007**, *2*, 215–219.
32. Beheshtinia, M.A.; Kahoo, S.J.; Fathi, M. Prioritizing healthcare waste disposal methods considering environmental health using an enhanced multi-criteria decision-making method. *Environ. Pollut. Bioavailab.* **2023**, *35*, 2218568. [CrossRef]
33. Boran, F.E.; Genç, S.; Kurt, M.; Akay, D. A multi-criteria intuitionistic fuzzy group decision making for supplier selection with TOPSIS method. *Expert Syst. Appl.* **2009**, *36*, 11363–11368. [CrossRef]
34. Xu, Z. Intuitionistic Fuzzy Aggregation Operators. *IEEE Trans. Fuzzy Syst.* **2007**, *15*, 1179–1187. [CrossRef]
35. Akter, N. Medical Waste Management: A Review. January 2000. Available online: http://dspace.bracu.ac.bd/xmlui/handle/10361/12964 (accessed on 9 October 2023).
36. Marinković, N.; Vitale, K.; Holcer, N.J.; Džakula, A.; Pavić, T. Management of hazardous medical waste in Croatia. *Waste Manag.* **2008**, *28*, 1049–1056. [CrossRef]
37. Kargar, S.; Paydar, M.M.; Safaei, A.S. A reverse supply chain for medical waste: A case study in Babol healthcare sector. *Waste Manag.* **2020**, *113*, 197–209. [CrossRef] [PubMed]
38. Luo, X.; Liao, W. Collaborative Reverse Logistics Network for Infectious Medical Waste Management during the COVID-19 Outbreak. *Int. J. Environ. Res. Public Health* **2022**, *19*, 9735. [CrossRef] [PubMed]
39. Chand, S.; Shastry, C.S.; Hiremath, S.; Joel, J.J.; Krishnabhat, C.H.; Mateti, U.V. Updates on biomedical waste management during COVID-19: The Indian scenario. *Clin. Epidemiol. Glob. Health* **2021**, *11*, 100715. [CrossRef] [PubMed]
40. Lemma, H.; Asefa, L.; Gemeda, T.; Dhengesu, D. Infectious medical waste management during the COVID-19 pandemic in public hospitals of West Guji zone, southern Ethiopia. *Clin. Epidemiol. Glob. Health* **2022**, *15*, 101037. [CrossRef] [PubMed]
41. Kokkinos, K.; Papadopoulos, E.; Samaras, N.; Chaikalis, K. An Integrated Modeling Framework for Routing of Hazardous Materials. In Proceedings of the 2012 IEEE 21st International Workshop on Enabling Technologies: Infrastructure for Collaborative Enterprises, Toulouse, France, 25–27 June 2012; Available online: https://ieeexplore.ieee.org/abstract/document/6269733 (accessed on 9 October 2023).
42. Eren, E.; Tuzkaya, U.R. Occupational health and safety-oriented medical waste management: A case study of Istanbul. *Waste Manag. Res.* **2019**, *37*, 876–884. [CrossRef] [PubMed]
43. Yang, S.; Long, R.; Wu, M.; Chen, H.; Li, Q. Research progress and frontier of global solid waste management based on bibliometrics. *Environ. Dev.* **2023**, *48*, 100922. [CrossRef]
44. Kokkinos, K.; Karayannis, V.; Moustakas, K. Optimizing Microalgal Biomass Feedstock Selection for Nanocatalytic Conversion Into Biofuel Clean Energy, Using Fuzzy Multi-Criteria Decision Making Processes. *Front. Energy Res.* **2021**, *8*, 622210. [CrossRef]
45. Hendalianpour, A.; Razmi, J.; Fakhrabadi, M.; Kokkinos, K.; Papageorgiou, E.I. A linguistic multi-objective mixed integer programming model for multi-echelon supply chain network at bio-refinery. *EuroMed J. Manag.* **2018**, *2*, 329–355. [CrossRef]
46. Shen, W.; Zhu, N.; Xi, Y.; Huang, J.; Li, F.; Wu, P.; Dang, Z. Effects of medical waste incineration fly ash on the promotion of heavy metal chlorination volatilization from incineration residues. *J. Hazard. Mater.* **2022**, *4251*, 128037. [CrossRef]
47. Govindan, K.; Nosrati-Abarghooee, S.; Nasiri, M.M.; Jolai, F. Green reverse logistics network design for medical waste management: A circular economy transition through case approach. *J. Environ. Manag.* **2022**, *322*, 115888. [CrossRef]
48. Wang, P.; Gu, X.; Xue, M.; Li, Y.; Dong, S.; Chen, G.; Zhang, J. Resource utilization of medical waste under COVID-19: Waste mask used as crude oil fluidity improver. *J. Clean. Prod.* **2022**, *358*, 131903. [CrossRef] [PubMed]
49. Kokkinos, K.; Karayannis, V.; Samaras, N.; Moustakas, K. Multi-scenario analysis on hydrogen production development using PESTEL and FCM models. *J. Clean. Prod.* **2023**, *419*, 138251. [CrossRef]
50. Kokkinos, K.; Karayannis, V.; Nathanail, E.; Moustakas, K. A comparative analysis of Statistical and Computational Intelligence methodologies for the prediction of traffic-induced fine particulate matter and NO_2. *J. Clean. Prod.* **2021**, *328*, 129500. [CrossRef]
51. Thirunavukkarasu, A.; Al-Hazmi, A.H.; Dar, U.F.; Alruwaili, A.M.; Alsharari, S.D.; Alazmi, F.A.; Alruwaili, S.F.; Alarjan, A.M. Knowledge, attitude and practice towards bio-medical waste management among healthcare workers: A northern Saudi study. *PeerJ* **2022**, *10*, e13773. [CrossRef] [PubMed]
52. Al-Omran, K.; Abahussain, A.; Khan, E. Integrated Environmental Assessment of Medical Waste Management in the Kingdom of Bahrain. *Sustainability* **2023**, *15*, 2397. [CrossRef]
53. Karimi, N.; Ng, K.T.W.; Richter, A. Development of a regional solid waste management framework and its application to a prairie province in central Canada. *Sustain. Cities Soc.* **2022**, *82*, 103904. [CrossRef]

54. Ho, P. The Costs and Environmental Justice Concerns of NIMBY in Solid Waste Disposal. *J. Assoc. Environ. Resour. Econ.* **2023**, *10*, 607–654. [CrossRef]
55. Lee, S.M.; Lee, D. Effective Medical Waste Management for Sustainable Green Healthcare. *Int. J. Environ. Res. Public Health* **2022**, *19*, 14820. [CrossRef]
56. Goodarzian, F.; Ghasemi, P.; Gunasekaran, A.; Labib, A. A fuzzy sustainable model for COVID-19 medical waste supply chain network. *Fuzzy Optim. Decis. Mak.* **2023**, *1*, 1–35. [CrossRef]
57. Jayasinghe, P.A.; Jalilzadeh, H.; Hettiaratchi, P. The Impact of COVID-19 on Waste Infrastructure: Lessons Learned and Opportunities for a Sustainable Future. *Int. J. Environ. Res. Public Health* **2023**, *20*, 4310. [CrossRef] [PubMed]
58. Attrah, M.; Elmanadely, A.; Akter, D.; Rene, E.R. A Review on Medical Waste Management: Treatment, Recycling, and Disposal Options. *Environments* **2022**, *9*, 146. [CrossRef]
59. Huda, M.N.; Hailemariam, T.G.; Hossain, S.Z.; Malo, J.S.; Khan, S.; Hadisuyatmana, S.; Ferdous, A.; Akombi-Inyang, B.; Islam, R.M.; Renzaho, A.M. Medical waste management-related factors affecting health and experiences of health risks among medical waste handlers in low and middle-income countries: A systematic review protocol of qualitative studies. *BMJ Open* **2022**, *12*, e056037. Available online: https://bmjopen.bmj.com/content/12/3/e056037.abstract (accessed on 9 October 2023). [CrossRef] [PubMed]
60. Yoon, C.-W.; Kim, M.-J.; Park, Y.-S.; Jeon, T.-W.; Lee, M.-Y. A Review of Medical Waste Management Systems in the Republic of Korea for Hospital and Medical Waste Generated from the COVID-19 Pandemic. *Sustainability* **2022**, *14*, 3678. [CrossRef]
61. Joneghani, N.M.; Zarrinpoor, N.; Eghtesadifard, M. A mathematical model for designing a network of sustainable medical waste management under uncertainty. *Comput. Ind. Eng.* **2022**, *171*, 108372. [CrossRef]
62. Chtioui, A.; Bouhaddou, I.; Benabdella, A.C.; Benghabrit, A. A Multi-Agent System for the Optimization of Medical Waste Management. In *Artificial Intelligence and Industrial Applications*; Masrour, T., Ramchoun, H., Hajji, T., Hosni, M., Eds.; Lecture Notes in Networks and Systems; Springer Nature: Cham, Switzerland, 2023; pp. 134–149. [CrossRef]
63. Fadaei, A. Comparison of medical waste management methods in different countries: A systematic review. *Rev. Environ. Health* **2023**, *38*, 339–348. [CrossRef] [PubMed]
64. Ghorabaee, M.K.; Zavadskas, E.K.; Turskis, Z.; Antucheviciene, J. A new combinative distance-based assessment (CODAS) method for multi-criteria decision-making. *Econ. Comput. Econ. Cybern. Stud. Res.* **2016**, *50*, 25–44.
65. Vardopoulos, I. Critical sustainable development factors in the adaptive reuse of urban industrial buildings. *A fuzzy DEMATEL approach. Sustain. Cities Soc.* **2019**, *50*, 101684. [CrossRef]

sustainability

MDPI

Article

Can Internet Use Narrow the Gap between Farmers' Willingness and Behavior in Waste Classification? Empirical Evidence from Rural Areas in Jiangsu Province, China

Wenchao Cui [1], Yanjun Chen [2,*] and Hengyuan Zeng [1]

1 School of Economics and Management, Nanjing Forestry University, Nanjing 210037, China; cuiwenchao@njfu.edu.cn (W.C.); hengyuan@njfu.edu.cn (H.Z.)
2 School of Economics and Management, Shanxi Normal University, Taiyuan 030000, China
* Correspondence: 222101013@sxnu.edu.cn; Tel.: +86-138-8327-3881

Abstract: Household waste classification and treatment are important for environmental protection and sustainable development. The Logit model is used to analyze differences in farmers' willingness and behavior regarding waste classification based on data from the China Land Economic Survey. Key findings include the following: (1) There is an evident discrepancy between waste classification willingness and action among rural residents. Despite nearly 90% of the sampled farmers expressing a willingness, nearly 40% do not practice waste classification. (2) Internet usage significantly reduces the discrepancy between farmers' willingness and actual waste classification practices. This observation is valid even after robustness checks and endogeneity discussions. (3) There is a partial mediating effect between ecological awareness and knowledge perception on farmers' willingness and behavior differences, which accounts for 12.9% and 52.6%, respectively, of the total impact. Notably, institutional limitations amplify the negative influence of Internet use on this discrepancy. (4) According to heterogeneity analysis, Internet use has a greater negative impact on farmers' willingness and behavior in waste classification in suburban villages and villages with environmental governance projects. This study proposes policy suggestions such as strengthening the construction of digital infrastructure in rural areas, enhancing the promotion and training of waste classification among farmers, and improving incentives and restraint mechanisms for rural household waste classification.

Keywords: internet; waste classification; deviation; willingness; behavior; logit model; environmental management

1. Introduction

Classifying household garbage plays a crucial role in transforming lifestyles to be more eco-friendly [1], and is important for environmental protection and achieving sustainable development. Since 2004, China has been the world's largest producer of household garbage [2]. In recent years, the Chinese government is increasingly focused on issues around the classification of household garbage [3]. For rural China, the effective management of garbage is related to the livelihood and well-being of nearly 500 million farmers and influences the environmental improvement of over 90% of the land area. However, with increases in farmers' economic incomes, consumption levels rise, resulting in a sharp increase in rural China's garbage production [4]. Research has shown that rural China produces approximately 0.8 kg of household garbage per person per day [5], with an annual growth rate of 8% to 10% [6]. Based on the estimated rural population of around 498 million at the end of 2022, rural China produces 145 million tons of household garbage annually. Without such a vast amount of garbage being scientifically classified and resourcefully utilized, the ecological environment could face severe threats.

Theory and practice have shown that classified collection and treatment are effective ways to treat household garbage. Farmers are the main producers of rural household garbage and the main executors of waste classification. Their willingness and behavior influence the effectiveness of household garbage management [7]. In the current research, scholars have addressed issues around rural household waste classification and governance from macro and microdimensions. (1) The effectiveness of waste management policies in different countries or regions have been studied, and scholars have compared the characteristics of different waste management models [8–10]. Other scholars have explored the impact of different macropolicies such as the institutional environment, reward and punishment measures, and promotional policies on waste classification [6,11,12]. For example, Zhang and Zhao [13] found that monitoring combined with reward and punishment systems can effectively promote residents' participation in waste classification. They also found that combining reward and punishment systems proves more effective. (2) Researchers have primarily studied residents' willingness, behaviors, and the influencing factors for household waste classification [14,15], along with potential strategies for improving household waste classification and governance [16,17]. In addition, some researchers have found that psychological factors such as environmental awareness, environmental skills, and group identity significantly influence residents' willingness and behavior toward waste classification [5,18,19].

Although the classification of household garbage is crucial, the actual implementation of relevant policies faces many challenges. In recent years, Chinese residents have shown improved environmental awareness. However, many lack the necessary knowledge and skills to classify garbage effectively, leading to a poor implementation of classification policies [20]. A particularly noteworthy phenomenon is that most Chinese residents have a willingness to classify garbage, but a considerable portion of them do not carry out waste classification in their daily lives. A survey of urban Chinese residents revealed that while 82.5% of participants expressed a willingness to classify garbage, only 13% did so [21]. Surveys targeting rural residents reported similar findings [22].

Studies indicate that Internet use influences the willingness and behavior of rural residents when it comes to waste classification [23]. China's digital expansion in rural areas has been rapidly growing, enhancing the online accessibility for farmers. This widespread Internet use has revolutionized farmers' information access, production methods, and lifestyle habits [24,25], reshaping their environmental awareness. Farmers can acquire garbage classification-related knowledge and learn classification-related skills through words, pictures, videos, and other means on the Internet, which can facilitate the implementation of garbage classification. Moreover, the Internet-embedded governance field can effectively provide technology empowerment and improve governance effectiveness, and has played a positive role in the implementation, supervision, publicity, and other processes of waste classification. The Internet brings efficient policy publicity, which is conducive to the formation of public awareness of environmental protection [26]. Current research primarily explores the impact of Internet use on farmers in areas like labor employment, agricultural production, and welfare levels [27–29]. However, as ecological and environmental concerns in rural areas grow, researchers are increasingly exploring how Internet use affects farmers' attitudes and actions towards waste classification [8,30]. For example, Liu et al. [31] found that using the Internet can positively influence farmers' willingness to classify garbage, though different Internet types on farmers' willingness to classify garbage is heterogeneous. Zhou et al. [32] and Xu et al. [22] also found that the use of the Internet has significantly improved farmers' willingness and action on waste classification.

Existing research has provided an important theoretical basis for this article, but there remain gaps in understanding. First, in the existing research on the willingness and behavior of household waste classification among farmers, most of the literature treats classification willingness and classification behavior as two distinct dependent variables, and emphasizes that willingness has a direct determining effect on behavior. However, only a few researchers have examined the deviation between farmers' willingness and behavior in waste classification as a separate dependent variable. Second, while some researchers have confirmed the discrepancy between farmers' willingness and behavior in waste classification [33], they have not delved deeply into the mechanisms of this inconsistency. Furthermore, more evidence is needed to determine if, and how, Internet use influences farmers' attitudes and behaviors around waste classification [21]. Third, while some studies have considered how institutional constraints enhance farmers' willingness and behavior, the potential moderating effect of these constraints on the relationship between Internet use and the gap in farmers' willingness and behavior remains untested. Given these gaps, we draw from the China Land Economy Survey data and target farmers willing to classify household waste. We aim to determine whether Internet use can reduce the discrepancy between farmers' intent and actions concerning waste classification. Additionally, we evaluate the mediating roles of ecological awareness and knowledge perception and the moderating effects of institutional constraints. This study's findings help address certain limitations in the existing theoretical and empirical approaches.

2. Materials and Methods

Recently, China has rapidly developed its rural information infrastructure. The influx of information from the Internet inevitably influences farmers' production and lifestyles through mediums like education and publicity [29,34,35], simultaneously altering their views on ecological and environmental protection.

2.1. Internet Use and the Deviation between Farmers' Willingness and Behavior towards Waste Classification

The Theory of Planned Behavior (TPB) suggests that individual behavior is influenced by subjective attitudes, social norms, and perceived behavioral control [36]. An individual's willingness to act increases when they have a positive subjective attitude toward a particular behavior, as well as strong social norms and perceived control. Stern et al. [37]. introduced the Value Belief Norm (VBN) theoretical model, emphasizing that individual behavior results from a combination of external and subjective factors. According to the VBN theory, individual values can influence personal beliefs and thus affect individual behavior [38]. Guagnano et al. [39] proposed a similar perspective through the Attitudes–Behaviors–Conditions (A-B-C) theory, which suggests that residents' recycling behavior is influenced by both subjective attitudes and external conditions, with external conditions as a key factor in determining the execution of recycling behavior. According to this article, a combination of subjective and external factors determines individual behavior in classifying garbage.

From a subjective perspective, Internet use bolsters farmers' perception of environmental benefits, thereby enhancing their awareness of waste classification. Serving as a medium for information exchange and communication, the Internet exposes farmers to diverse content related to the ecological environment [26]. By conveying and disseminating the importance of environmental protection and the dangers of environmental degradation [28], the Internet amplifies farmers' commitment to waste classification. From one perspective, the Internet can play a positive guiding role. The Internet provides users with a large amount of ecological environmental protection information and publicizes the significance of waste classification for ecological environmental protection, human health, and sustainable development. The wealth of Internet data enlightens farmers about the importance of waste classification, promoting waste classification practice among farmers [31]. From another perspective, the Internet serves a cautionary purpose. By navigating the Internet, farmers can readily grasp the potential risks associated with household waste, fostering a stronger awareness of environmental urgency [40]. This awareness drives them to translate their willingness to classify garbage into tangible actions [41,42]. Evidently, Internet utilization amplifies farmers' awareness around environmental protection, thereby promoting waste classification.

From an external perspective, the use of the Internet serves as a valuable resource for enhancing farmers' knowledge and skills, thereby improving proficiency in waste classification. Firstly, the Internet reduces obstacles associated with information search, acquisition, and sharing. This accessibility empowers farmers with essential knowledge and expertise about waste classification and recycling, like sorting techniques, recycling routes, and site locations [42]. This ease of access allows farmers to bypass challenges related to finances, time, energy, and geographical constraints when dealing with waste classification and management [43]. Secondly, the use of the Internet is also beneficial for expanding farmers' social networks, promoting the exchange of waste classification knowledge and skills among different farmers [7], thus refining their practical skills in the domain. Additionally, the Internet improves and standardizes waste classification practices. The Internet not only guides farmers in executing accurate garbage sorting [44] but also provides insights into tracking waste flows and understanding garbage resource utilization [43]. This standardized approach not only streamlines the waste classification and management process but also deepens farmers' comprehension and confidence in these methods. Evidently, the Internet furnishes farmers with a holistic grasp of the intricacies of waste classification, spurring them towards more efficient practices. Therefore, we propose the following hypotheses:

H1. *Internet use may reduce the deviation between farmers' willingness and behavior towards waste classification.*

H2. *Internet use may enhance farmers' ecological awareness, thereby reducing the deviation between their willingness and behavior in waste classification.*

H3. *Internet use may enhance farmers' knowledge perception, thereby reducing the deviation between their willingness and behavior in waste classification.*

2.2. The Moderating Role of Institutional Constraints

Waste classification is often characterized by administrative mobilization [45]. Therefore, participating in household waste classification is not solely an independent decision made by individuals after weighing various factors. It is also influenced by various social environments, with institutional constraints being a key component of this environment [21]. Specifically, the classification and governance of rural household garbage involves the macroinstitutional environment led by the government and the microinstitutional environment of village and community autonomy [12]. The government or community can strengthen supervision and constraints on farmers' environmental protection behavior by formulating a series of systems and rules, thereby increasing the implicit cost for farmers not engaging in waste classification [5]. Conversely, Internet use is influenced by institutional constraints as well, which moderate the deviation of farmers' willingness and behavior towards waste classification. In the macrocontext, where national efforts are pushing for rural environmental improvements, stringent institutional constraints encourage farmers to leverage the Internet. This enables them to stay informed about governmental and community-based waste classification policies and to acquire relevant skills, thus bolstering their practical competence in waste classification [31]. Additionally, as garbage sorting and disposal predominantly take place within village boundaries, and as villages continuously refine their internal regulations on waste classification, the Internet emerges as a potent tool for village leaders. They use it to advocate for and enforce waste classification and management rules [46], further aiding the transition of farmers' waste classification intent into actionable behavior. The following hypotheses are proposed based on this (Figure 1):

H4. *Institutional constraints may reduce the deviation between farmers' willingness and behavior in waste classification.*

H5. *Institutional constraints have a moderating effect on the path of Internet use on the deviation between farmers' willingness and behavior in waste classification.*

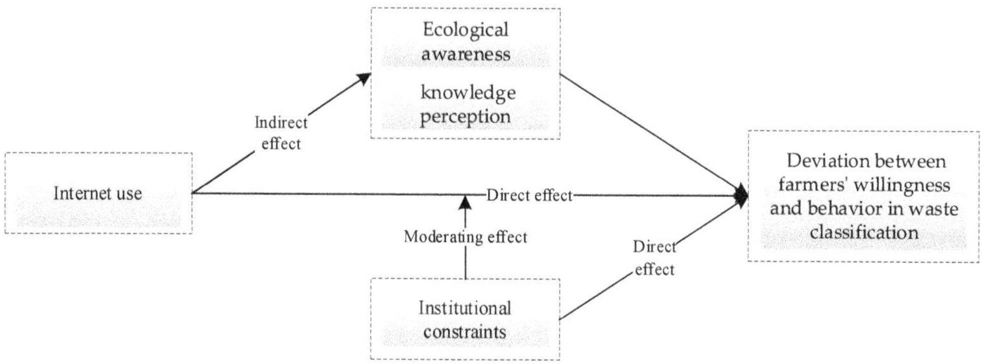

Figure 1. Theoretical mechanism diagram.

3. Data Description and Model Construction

3.1. Data

Data were sourced from the China Land Economic Survey (CLES), conducted by Nanjing Agricultural University in 2021. The survey used the Probability Proportional to Size (PPS) sampling technique, covering 48 villages in 13 prefecture-level cities in Jiangsu Province, and collected a total of 2420 household samples. Tailoring the dataset to fit this study's requirements, a refined sample size of 1933 households was chosen for the analysis. Jiangsu Province, as one of the most developed provinces in China's economy, is also far ahead in terms of network coverage. At present, the fiber optic network in Jiangsu Province will be 100% connected to households, and the 5G network signal coverage rate will reach 100%, achieving full coverage of rural network communication projects. Jiangsu Province not only has a good construction of rural Internet infrastructure, but also has a relatively sound waste classification policy system and a complete waste classification, transportation, and disposal system. Thus, the dataset employed in this article offers substantial representational validity.

3.2. Variables

3.2.1. Dependent Variable

The dependent variable of this article is the deviation between farmers' willingness and behavior in waste classification. Based on the previous analysis, we focus on how to reduce the deviation between farmers' willingness and behavior in waste classification; that is, how to achieve the transformation of farmers from having willingness without behavior to having willingness with behavior. Guided by the methodologies of Guo et al. [47] and Chang et al. [48], this study conceptualizes the disparity between farmers' intent and behavior in waste classification as instances where farmers exhibit a readiness to classify garbage but fail to manifest this intent into tangible action. Notably, we have excluded samples representing farmers devoid of any waste classification intent. The reason for excluding the sample of farmers who have no willingness of waste classification is because there is a clear evolutionary logic from willingness without behavior to willingness with behavior; that is, the core of transforming farmers' waste classification willingness into waste classification behavior is the expression of behavior. The primary issue for farmers with no intention of waste classification lies in cultivating willingness, and the reasoning logic is still moving from unintentional willingness to willingness. Upon an initial assessment of the dataset, it was observed that fewer than 10% of the farmers exhibited no waste classification intent. Consequently, excluding this subset from the analysis resulted in a minimal loss of sample representation. After removing the sample of farmers who have no intention of waste classification, farmers who have actively engaged in waste classification practices were assigned a '0', indicating a harmonious alignment between their intent and actions. In contrast, farmers who, despite their intent, refrained from actual waste classification were designated a '1'. This code highlights a discord between their intent and subsequent behavior. Such a coding structure ensures a clear differentiation between farmers who translate their classification intent into action and those who do not, providing a comprehensive understanding of behavioral dynamics within the context of waste classification.

3.2.2. Key Explanatory Variable

The key explanatory variable of this paper is Internet use. Farmers who use computers, smartphones, or Internet cafes in their daily lives are categorized as Internet users, with a value of 1. Users who do not use the Internet will be assigned a value of 0.

3.2.3. Mediation Variable

The mediating variable of this article is farmers' cognition of waste classification, including ecological awareness and knowledge perception. Ecological awareness evaluates farmers' understanding of the importance of waste classification. Knowledge perception evaluates farmer's grasp of the skills of waste classification. It aims to measure their awareness of the potential repercussions unclassified waste might pose on the rural ecological landscape. Referring to existing research [22,28], we use two questions: (1) Do you think that the non-classification of household waste has a significant impact on the rural ecological environment? (2) Do you understand the relevant knowledge of household waste classification? To respectively characterize farmers' ecological awareness and knowledge perception of waste classification.

3.2.4. Moderation Variable

The chosen moderating variable for this study is institutional constraints. Drawing inspiration from the work of Song et al. [12], this study uses the following question: "Has the government established reward and punishment measures for rural household waste classification?" This question serves to characterize the presence and influence of institutional constraints. By incorporating this question, this study aims to gauge the level of governmental intervention and regulation in the area of rural waste classification and understand its potential impact on the behaviors and practices of farmers.

3.2.5. Control Variable

In previous studies, it has been shown that a farmer's willingness to protect the environment can be influenced by both the characteristics of the village where he lives, as well as his personal and family characteristics [49]. Therefore, this study controls the characteristics of the village where the respondents are located, and the individual and family characteristics of the respondents. The village attributes considered include its classification as a suburban village, the funds allocated for environmental activities, ongoing environmental governance initiatives, and the educational background of the village party secretary. The personal characteristics of the respondents include age, gender, whether they are village cadres, education level, and health status. The household characteristics of the respondents include the size of the household population and the annual household income. The definition and description of the variables are shown in Table 1.

Table 1. Description of variable assignment and descriptive statistics.

Categories	Variables	Variable Meaning and Assignment	Mean	S.D.
Dependent variable	Deviation	Having only willingness without behavior = 1; Having both willingness and behavior = 0	0.399	0.490
Independent variable	Internet use	Whether farmers use the Internet? Yes = 1; No = 0	0.484	0.500
Mediation variable	Ecological awareness	Do you agree that not classifying garbage has a significant negative impact on the ecological environment? Strongly disagree = 1; Strongly agree = 5	4.166	0.840
	Knowledge perception	Do you have any knowledge and skills related to waste classification? Never heard = 1; Well understood = 5	3.361	1.112
Moderation variable	Institutional constraints	Does the government establish reward and punishment measures for waste classification? Yes = 1; No = 0	0.295	0.456
Control variable	Suburban village	Is this village located in the suburbs of the city? Yes = 1; No = 0	0.412	0.492
	Environmental expenditure	The environmental governance expenditure of this village in the previous year (Unit: CNY 10,000)	59.328	87.643
	Environmental governance projects	Is there an environmental governance project in this village? Yes = 1; No = 0	0.304	0.460
	The education level of the village party secretary	Education years of the village party secretary (years)	13.331	2.651
	Gender of the respondents	What is the gender of the respondents? Male = 1; Female = 0	0.734	0.442
	Respondents' age	Age of the respondents (years)	61.430	11.380
	Cadre	Are the respondents village cadres? Yes = 1; No = 0	0.160	0.367
	Respondents' education level	Education years of the respondents (years)	7.422	3.941
	Health of the respondents	The health status of the respondents: Very poor = 1; Very good = 5	4.073	1.043
	Household population of the respondents	Unit: Person	3.109	1.604
	Annual household income	Unit: CNY 10,000	2.507	1.067

3.3. *Model*

3.3.1. Bivariate Logit Model

Considering that the dependent variable, "the deviation between farmers' willingness to classify garbage and their behavior", is a 0–1 discrete variable, a binary Logit model is used for the empirical testing. The model settings are as follows:

$$Deviation_i = \alpha_0 + \alpha_1 Internet_i + \alpha_2 Control_j + \mu_i \quad (1)$$

Among the variables, $Deviation_i$ indicates whether the willingness and behavior of the i-th farmer towards waste classification deviate; $Internet_i$ represents whether the i-th farmer uses the Internet; $Control_j$ is a series of control variables; μ_i is the random error term.

3.3.2. Mediating Effect Model

Referring to the classic testing program for mediating effects [50], we used the stepwise regression method to test the mediating effects of ecological awareness and knowledge perception. The specific estimation formula is as follows:

$$Deviation_i = \alpha_0 + \alpha_1 Internet_i + \alpha_2 Control_j + \mu_i \quad (2)$$

$$M_i = \mu_0 + \mu_1 Internet_i + \mu_2 Control_j + \mu_i{}' \quad (3)$$

$$Deviation_i = \varphi_0 + \varphi_1 Internet_i + \varphi_2 M_{it} + \varphi_3 Control_j + \mu_i{}'' \quad (4)$$

Equation (2) is the same as Equation (1). In Equation (3), M_i represents a mediation variable. Based on the significant coefficient of the key explanatory variable in Equation (2), the significance of the key explanatory variable in Equation (3) and the coefficient of the intermediate variable in Equation (4) are sequentially tested. If both are significant, then the intermediate effect exists. Meanwhile, if the coefficient of the core explanatory variable in Equation (4) is not significant, it indicates the existence of a complete mediating effect. If φ_1 is significant and φ_1 and $\mu_1 \times \varphi_2$ are the same sign, it indicates the existence of partial mediating effects; if the symbols are different, there is a masking effect.

3.3.3. Moderating Effect Model

We used a moderating effect model for testing and constructed the following regression model:

$$Deviation_i = \delta_0 + \delta_1 Internet_i + \delta_2 Mod_i + \delta_3 Internet_i \times Z_i + \delta_4 Control_j + \mu_i{}''' \quad (5)$$

In Equation (5), Mod_i represents the moderating variable "institutional constraints"; $Internet_i \times Mod_i$ represents the interaction term between institutional constraints and Internet use; δ_0 is a constant term; δ_1, δ_2, δ_3, and δ_4 are the coefficients to be estimated; $\mu_i{}'''$ is a random error term.

4. Empirical Analysis Results

4.1. Description of the Statistical Analysis

From the 2021 CLES samples, 2176 households, accounting for 89.9%, were willing to perform waste classification. Farmers from 237 households were unwilling to perform waste classification, accounting for 9.8%. There were also seven samples with missing values. According to the results, rural residents in Jiangsu Province have a high willingness to classify garbage. Based on the previous analysis, we only retained the samples of farmers with waste classification willingness, and the samples with missing or abnormal key variables were further excluded. Ultimately, 1933 farmers with waste classification willingness were retained. As shown in Figure 2, among these 1933 household samples, 1161 households (60.1%) implemented waste classification, and 39.9% of households showed a deviation in their willingness and behavior towards waste classification. This is consistent with the research findings of previous studies [3,22,51]. In terms of Internet use, 56.2% of the sample of farmers who expressed the willingness and behaviors of waste classification use the Internet. Among the sample of farmers who expressed willingness without behavior, only 36.5% use the Internet, which is significantly lower than the former. This suggests that Internet use may promote the conversion of farmers' willingness to classify garbage into waste classification behavior, thereby reducing the deviation between farmers' willingness to classify garbage and their behavior.

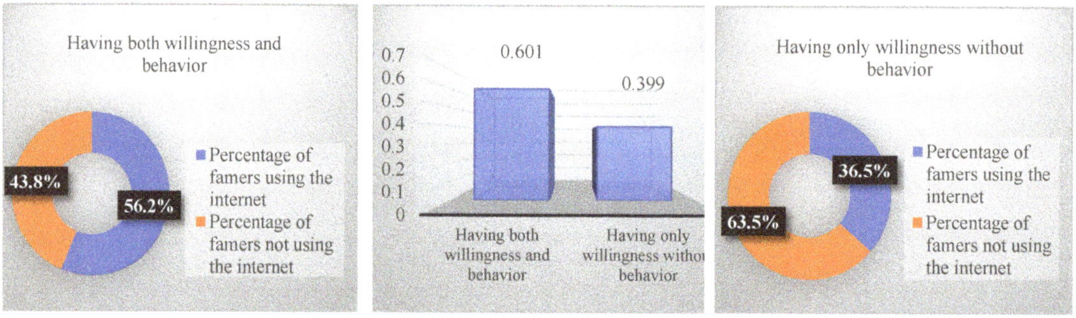

Figure 2. Survey results of farmers' willingness and behavior to classify garbage based on Internet use.

4.2. *Analysis of the Influence of Internet Use on the Deviation between Farmers' Willingness and Behavior towards Waste Classification*

In Table 2, the regression results for the binary Logit models and stepwise regression are shown. There is a significant negative impact of Internet use on farmers' willingness and behavioral deviation in waste classification, regardless of whether control variables are included. Taking columns (5) and (6) as examples, the regression coefficient for Internet usage is −0.492, which is significant at the 1% level. The marginal effect shows that for every 1 unit increase in Internet use, the probability of farmers' willingness to deviate from their behavior in waste classification decreases by 10.4%. Preliminary results show the validity of hypothesis H1.

Table 2. Impact estimation results.

Variable	Explained Variable: Deviation between Willingness and Behavior					
	Coefficient (1)	**Marginal Effect (2)**	**Coefficient (3)**	**Marginal Effect (4)**	**Coefficient (5)**	**Marginal Effect (6)**
Internet use	−0.804 *** (0.095)	−0.185 *** (0.020)	−0.770 *** (0.099)	−0.167 *** (0.020)	−0.492 *** (0.117)	−0.104 *** (0.024)
Suburban village			−0.860 *** (0.107)	−0.187 *** (0.022)	−0.667 *** (0.123)	−0.140 *** (0.026)
Environmental expenditure			−0.002 *** (0.000)	−0.001 *** (0.000)	−0.002 *** (0.000)	−0.000 *** (0.000)
Environmental governance projects			−0.235 ** (0.113)	−0.051 ** (0.025)	−0.281 ** (0.116)	−0.059 ** (0.024)
The education level of the village party secretary			−0.042 ** (0.020)	−0.009 ** (0.004)	−0.027 (0.020)	−0.006 (0.004)
Gender of the respondents					−0.289 ** (0.118)	−0.061 ** (0.025)
Respondents' age					0.020 *** (0.006)	0.004 *** (0.001)
Cadre					−0.391 *** (0.149)	−0.083 *** (0.031)
Respondents' education level					0.005 (0.015)	0.001 (0.003)
Health of the respondents					−0.114 ** (0.051)	−0.024 ** (0.010)
Household population of the respondents					−0.058 * (0.032)	−0.012 * (0.006)
Annual household income					−0.137 ** (0.054)	−0.029 ** (0.011)
Wald chi2	71.03 ***		178.34 ***		223.30 ***	
Pseudo-R2	0.028		0.073		0.093	
N	1933		1933		1933	

Note: Standard errors in parentheses; ***, **, and * indicate significance levels of 1%, 5%, and 10%, respectively.

According to the control variables, the results are consistent with Chen et al. [21] and Kang et al. [52]. Village characteristics like village type, environmental expenditure, and environmental governance projects can significantly reduce farmers' willingness and behavioral deviation for waste classification. The education level of village party secretaries has a negative but not significant impact on farmers' willingness and behavioral deviation for waste classification. Respondents' age, health status, and service as village cadres can significantly reduce the deviation of farmers' willingness and behavior towards waste classification. The impact of respondents' education level is not significant. Gender difference has a significantly negative impact on the deviation between farmers' willingness and behavior towards waste classification. The possible reason is that waste classification is a problem-solving activity, and, compared to women, men often have a more adventurous spirit and problem-solving ability [26]. Therefore, it is likely that men are more willing to learn the relevant knowledge and skills of waste classification through the use of the Internet, and convert their willingness to behavior. Simultaneously, the age of the respondents can significantly reduce the deviation in their willingness and behavior towards waste classification. The possible reason is that implementing waste classification requires a certain amount of mental and physical energy to learn relevant knowledge and carry out relevant actions. For older farmers, it is more difficult and costly for them to accept waste classification knowledge and implement waste classification behavior.

4.3. Robustness Test

4.3.1. Replacing the Key Explanatory Variables

We replaced the key explanatory variable of "Internet use" to "Do you have a computer with Internet access in your home?". The regression results are shown in column (1) of Table 3, and the regression coefficient is significantly negative at the level of 1%. In addition, the marginal effect indicates that for every unit increase in the number of computers that households can access online, the probability of farmers' willingness to deviate from their behavior in waste classification decreases by 5.3%.

Table 3. Results of the robustness test.

Variable	Explained Variable: Deviation between Willingness and Behavior					
	Coef. (1)	Mgn. (2)	Coef(Probit) (3)	Mgn(Probit) (4)	Coef. (5)	Mgn. (6)
Computer with Internet access	−0.253 *** (0.076)	−0.053 *** (0.016)				
Internet use			−0.308 *** (0.071)	−0.107 *** (0.024)	−0.464 *** (0.124)	−0.101 *** (0.027)
Control variables	Yes	Yes	Yes	Yes	Yes	Yes
Wald chi2	165.03 ***		240.25 ***		171.56 ***	
Pseudo-R2	0.088		0.093		0.082	
N	1933		1933		1623	

Note: Standard errors in parentheses; *** indicates significance level of 1%.

4.3.2. Replacing the Regression Model

Compared to the Logit model, the Probit model assumes a stricter probability distribution. Therefore, we used the binary Probit model for retesting, and the regression results are shown in columns (3) and (4) of Table 3. Both the regression coefficient and marginal effect are significantly negative at the 1% level, indicating that Internet use can significantly reduce the discrepancy between farmers' willingness and behavior to classify garbage. The regression results of the two models are consistent.

4.3.3. Subsample Regression

Village cadres are the main implementers of the government's waste classification policy and are responsible for village environmental governance. Thus, they are more likely to carry out waste classification in their daily lives. Therefore, we conducted a retest after excluding the sample of village cadres as the respondents. The regression results are shown in columns (5) and (6) of Table 3, and the impact of Internet use on farmers' willingness and behavior to classify garbage is still significantly negative. Therefore, hypothesis H1 has been further validated.

4.4. Endogeneity Test

Endogeneity issues may arise from various sources, potentially distorting the relationship between variables in a study. In this context, although there is no apparent reverse causation between Internet use and farmers' attitudes and actions regarding waste classification, it is essential to address possible endogeneity concerns. To mitigate these concerns, we employ the instrumental variable method, a powerful technique that can help offer more consistent estimations in the presence of endogeneity. According to Xu et al. [22] and Deng et al. [53], we selected the proportion of rural households using the Internet in villages as the instrumental variable. There are two reasons: (1) A village with a high proportion of Internet users shows that, in such an environment, individuals are surrounded by Internet peers. This prevalence often creates a group effect, encouraging others in the vicinity to also adopt Internet usage, ensuring the relevance of the instrument [54]. (2) The chosen instrumental variable is a village-level indicator, which will not directly influence the willingness and action of rural households to classify garbage. This ensures the instrument's exogeneity, meaning that it will not be correlated with the error term in the regression. Incorporating the percentage of Internet-using rural households within villages as an instrumental variable can, in theory, counteract biases stemming from endogeneity, providing more accurate and reliable findings.

To test the endogeneity, the IV Probit model was utilized. Comparisons have shown consistency between the Probit and Logit model conclusions. In the first stage, the endogenous variable (Internet use) was regressed on the instrumental variable (proportion of village Internet usage), yielding predicted values. Results, presented in column (1) of Table 4, show a significant positive impact of village Internet usage proportion on farmers' Internet use. The F-value is 62.32, and p-values for the Wald endogeneity test and AR test are both below 0.01, indicating no weak instrumental variable problem. In the second stage, the IV Probit method was applied to regress the dependent variable (difference in willingness and behavior) on the fitted values from the first stage. Column (2) of Table 4 displays the results, maintaining a significant negative association at the 1% level, even after addressing potential endogeneity. These findings reinforce the initial conclusions, supporting hypothesis H1.

Table 4. Findings from the endogeneity examination.

Variable	Internet Use		Deviation	
	Coef. (1)	**Std. Err. (2)**	**Coef. (3)**	**Std. Err. (4)**
Internet use rate	0.440 ***	0.079		
Internet use			−1.776 ***	0626
Control variables	Yes		Yes	
Wald chi2			178.96 ***	
The first-stage F-value	62.32 ***			
Wald test			9.76 ***	
Weak IV AR test			8.04 ***	
Adj R2	0.280			
N	1933		1933	

Note: Standard errors in parentheses; *** indicates significance level of 1%.

4.5. Mechanism Analysis

4.5.1. The Mediating Role of Ecological Awareness and Knowledge Perception

To further understand the underlying mechanisms driving this relationship, this study delved into the potential mediating roles of ecological awareness and knowledge perception.

Table 5 displays the results derived from the stepwise regression analyses. According to column (1), Internet use can significantly and positively affect farmers' ecological awareness of waste classification. This implies that Internet use can significantly reduce the deviation between farmers' willingness and behavior to classify garbage, and ecological awareness can also significantly reduce the deviation between farmers' willingness and behavior to classify garbage. Therefore, ecological awareness plays a partial mediating role in the impact of Internet use on farmers' willingness and behavior deviation towards waste classification. Further calculations indicate that the mediating effect of ecological awareness accounts for 12.9% of the total effect. Similarly, as shown in columns (3) and (4), knowledge perception also has a partial mediating effect. The mediating effect of knowledge perception accounts for 52.6% of the total effect. At this point, H2 and H3 have been validated.

Table 5. Mediation effect analysis.

Variable	Internet Use→Ecological Awareness→Deviation		Internet Use→Knowledge Perception→Deviation	
	Ecological Awareness (1)	Deviation (2)	Knowledge Perception (3)	Deviation (4)
Internet use	0.332 *** (0.101)	−0.452 *** (0.118)	0.342 *** (0.096)	−0.396 *** (0.122)
Ecological awareness		−0.157 *** (0.060)		
Knowledge perception				−0.756 *** (0.057)
Control variables	Yes	Yes	Yes	Yes
Wald chi2	103.90 ***	228.28 ***	437.68 ***	325.58 ***
Pseudo-R2	0.024	0.096	0.080	0.175
N	1933	1933	1933	1933

Note: Standard errors in parentheses; *** indicates significance level of 1%.

4.5.2. The Moderating Effect of Institutional Constraints

Earlier theoretical discussion suggested that institutional constraints might play a dual role concerning the deviation of farmers' willingness and behavior towards waste classification. These constraints might not only exert a direct influence but might also modulate the extent to which Internet use affects this deviation. Table 6 elucidates these dynamics. Columns (1) and (2) show that institutional constraints have a significant negative impact on the deviation of farmers' willingness and behavior towards waste classification. Column (3) shows that the interaction coefficient between institutional constraints and Internet use is significantly negative at the 10% level. Also, the interaction coefficient is consistent with the coefficient sign of the key explanatory variable in the benchmark regression results; the impact of Internet use on farmers' willingness and behavior deviation in waste classification is moderated by institutional constraints. Therefore, the hypotheses H4 and H5 were validated.

Table 6. Analysis of regulatory effects.

Variables	Explained Variable: Deviation between Willingness and Behavior		
	(1)	(2)	(3)
Internet use		−0.497 *** (0.120)	−0.413 *** (0.130)
Institutional constraints	−1.288 *** (0.131)	−1.287 *** (0.131)	−1.088 *** (0.175)
Internet use × Institutional constraints			−0.457 * (0.266)
Control variables	Yes	Yes	Yes
Wald chi2	264.37 ***	288.55 ***	274.80 ***
Pseudo-R2	0.127	0.134	0.136
N	1902	1902	1902

Note: Standard errors in parentheses; ***, and * indicate significance levels of 1%, and 10%, respectively.

4.6. Heterogeneity Analysis

From the perspective of village types, as shown in columns (1) and (2) of Table 7, compared to ordinary suburban villages, the negative impact of Internet use in suburban villages is more significant. The possible reason is that the construction of waste classification infrastructure in suburban villages is relatively complete, making it more convenient for farmers to carry out waste classification. An equally enlightening trend is observed when considering the presence or absence of environmental governance projects in the villages. As shown in columns (3) and (4) of Table 7, villages with established environmental governance projects witness a stronger positive influence of Internet use than in villages without such projects. The underlying rationale for this could be the structured, often obligatory nature of these governance projects. Since these initiatives are typically orchestrated by governmental bodies, they possess a certain authoritative weight, pushing residents to act in line with their waste classification intentions.

Table 7. Heterogeneity analysis under different dimensions.

Variables	Explained Variable: Deviation between Willingness and Behavior			
	Suburban Village (1)	Non-Suburban Villages (2)	With Project (3)	No Project (4)
Internet use	−0.740 *** (0.203)	−0.380 *** (0.146)	−0.587 ** (0.241)	−0.467 *** (0.136)
Control variables	Yes	Yes	Yes	Yes
Wald chi2	99.47 ***	90.40 ***	111.03 ***	135.39 ***
Pseudo-R2	0.111	0.061	0.159	0.076
N	797	1136	588	1345

Note: Standard errors in parentheses; ***, and ** indicate significance levels of 1%, and 5%, respectively.

5. Conclusions and Recommendations

In recent years, with a heightened global focus on ecological preservation, the management and categorization of household waste has gained traction as a critical academic and practical concern. This study, anchored in data from the 2021 China Land Economy Survey, offers a deep dive into how Internet usage can shape the alignment between farmers' willingness and behavior regarding waste classification in rural China and analyzes the mechanisms of ecological awareness, knowledge perception, and institutional constraints in this impact path. The results indicate the following: (1) In rural China, there is a significant deviation between farmers' willingness and behavior to classify waste. Survey data show that, although nearly 90% of the sample farmers are willing to carry out waste classification, nearly 40% of the farmers who are willing to carry out waste classification have not implemented waste classification behavior. (2) The use of the Internet has a significant negative impact on the deviation between farmers' willingness and behavior towards waste classification. In other words, the use of the Internet can convert farmers' willingness to waste classification into waste classification behavior. From the perspective of marginal effects, for every increase in Internet use of 1 unit, the probability of farmers' willingness to deviate from their behavior in waste classification will decrease by 10.4%. (3) Mechanism analysis shows that ecological awareness and knowledge perception play a partial mediating role in the impact path of Internet use on the deviation of farmers' willingness and behavior towards waste classification. Specifically, the mediating effects of ecological awareness and knowledge perception account for 12.9% and 52.6% of the total effects, respectively. (4) Further analysis reveals that institutional constraints can not only directly and negatively affect the deviation of farmers' willingness and behavior towards waste classification, but also have a positive moderating effect on the impact path of Internet use on the deviation of farmers' willingness and behavior towards waste classification. The negative effects of Internet use on farmers' willingness and behavior towards waste classification are exacerbated by institutional constraints. (5) A heterogeneity analysis found greater negative impacts of Internet use on farmers' willingness and behavior in waste classification in suburban villages and villages with environmental governance projects.

Based on this study's findings, it is evident that to address the discrepancy between farmers' willingness and behavior regarding waste management in rural China, comprehensive policy measures are required. We suggest the following: (1) Strengthen the construction of rural Internet infrastructure. We need to accelerate the construction of the rural network infrastructure, increase the penetration rate of rural Internet, and provide good conditions for ensuring that farmers use digital tools. Accelerate the development of the Internet and digital platforms for rural residential environment improvement, give attention to digital technology in the classified treatment of rural household garbage, and improve the efficiency of rural household garbage treatment. (2) Strengthen waste classification policies and knowledge dissemination. Online media should appropriately strengthen publicity and education on environmental protection and waste classification so that farmers can understand the importance of household waste classification and treatment through the Internet, and thereby enhance their awareness of environmental responsibility. At the same time, diversified waste classification knowledge and skills training should be provided to farmers through the Internet to improve their practical ability in waste classification. (3) Formulate and implement reward and punishment measures for waste classification. The government should continue to improve the household waste classification system and policy system, improve the constraint and incentive mechanism for household waste classification, and actively create a good atmosphere for everyone to participate. Village cadres should pay more attention to the classification and management of rural household waste, improve the reward and punishment rules for village waste classification by revising village regulations and implementing point system management, and promote the conversion of farmers' willingness to classify garbage into waste classification behavior. (4) Reasonably arrange garbage sorting and collection stations to effectively promote easy and convenient sorting services. Rural communities need to combine the characteristics of village population size, agricultural production, and daily life, as well as the amount and composition of household waste generated to set corresponding garbage classification and recycling standards and collection time plans for different types of waste, and to set up appropriate and easily identifiable garbage collection stations. Moreover, when planning the geographical distribution of garbage collection sites in a reasonable manner, special attention should be paid to the spatial distribution characteristics of rural residential areas in order to reduce the time and energy consumption of rural residents for household waste classification and recycling, enhance the convenience of garbage classification services, and truly achieve the reduction, resource utilization, and harmless treatment of rural household waste.

While offering valuable insights into the relationship between Internet use and farmers' behaviors, there are certain limitations. The binary distinction of farmers either using or not using the Internet might oversimplify the reality. Variations in Internet usage, such as frequency, purpose, or the nature of content consumed, might all influence farmers' behaviors differently. Some might use the Internet primarily for entertainment, while others might engage in educational or agricultural forums, which could provide varying levels of exposure to waste management practices. Moreover, this study's focus on farmers already inclined to classify waste potentially overlooks a significant portion of the population. Those farmers without an initial willingness to classify garbage represent a crucial demographic to study, as they might need different interventions or strategies to shift their mindset. Convincing this group might present unique challenges, and understanding the barriers to their engagement could be instrumental in devising more comprehensive waste management strategies. Future research might benefit from a more nuanced understanding of farmers' Internet use and from expanding this study's scope to include those without an initial inclination towards waste classification.

Author Contributions: Conceptualization, W.C. and H.Z.; methodology, W.C. and Y.C.; software, W.C.; validation, W.C. and Y.C.; formal analysis, W.C.; investigation, H.Z. and Y.C.; resources, W.C.; data curation, W.C.; writing—original draft preparation, W.C.; writing—review and editing, H.Z. and Y.C.; visualization, W.C. and Y.C.; supervision, H.Z.; funding acquisition, W.C. All authors have read and agreed to the published version of the manuscript.

Funding: This research was funded by the 2023 Jiangsu Province Graduate Research and Practice Innovation Program Project (grant number KYCX23_1104).

Institutional Review Board Statement: Not applicable.

Informed Consent Statement: Not applicable.

Data Availability Statement: If necessary, we can provide the raw data.

Acknowledgments: We would like to thank Annalise Guthrie at the University of Kentucky for his assistance with the English language and grammatical editing of the manuscript.

Conflicts of Interest: The authors declare no conflicts of interest.

References

1. Wang, Y.; Li, H.; Wu, Z. Attitude of the Chinese public toward off-site construction: A text mining study. *J. Clean. Prod.* **2019**, *238*, 117926. [CrossRef]
2. Liu, M.; Feng, Q. Does Social Capital Promote Garbage Classification? Evidence From China. *SAGE Open* **2021**, *11*, 749–765. [CrossRef]
3. Wang, S.; Wang, J.; Shu, Y.; Li, J.; Zhou, K. From intention to behavior: Comprehending residents' waste sorting intention and behavior formation process. *Waste Manag.* **2020**, *113*, 41–50. [CrossRef] [PubMed]
4. Bai, R.; Lin, B. Are residents willing to pay for garbage recycling: Evidence from a survey in Chinese first-tier cities. *Environ. Impact Assess. Rev.* **2022**, *95*, 106789. [CrossRef]
5. Tang, L.; Luo, X.; Zhang, J. Social supervision, group identity and farmers' domestic waste centralized disposal behavior: An analysis based on mediation effect and regulation effect of the face concept. *China Rural Surv.* **2019**, *40*, 18–33.
6. Zhang, B.; Lai, K.; Wang, B.; Wang, Z. From intention to action: How do personal attitudes, facilities accessibility, and government stimulus matter for household waste sorting? *J. Environ. Manag.* **2018**, *233*, 447–458. [CrossRef] [PubMed]
7. Jia, Y.; Zhao, M.J. The impact of domestic waste pollution perception and social capital on the farming households' sorting of waste: Based on the survey of 1374 farming households in Shaanxi Province. *Resour. Sci.* **2020**, *42*, 2370–2381. [CrossRef]
8. He, Y.; Huang, S.; Chen, S.; Huang, K. How Environmental Policies Affect Farmers' Willingness to sort Household Garbage: An Analysis of the mediating effect based on social norms and value cognition. *World Agric.* **2022**, *5*, 95–107.
9. Jiang, L.; Zhao, X. Classified Treatment Management of Rural Domestic Waste: Model Comparison and Policy Enlightenment Based on a Case Study of Four Ecological Conservation Areas in Beijing. *China Rural Surv.* **2020**, *41*, 16–33.
10. Lu, S.; Li, X.; Du, H. Comparative analysis and some suggestions of typical governance modes of rural household waste. *World Agric.* **2018**, *2*, 4–10+210.
11. Li, X.; Bi, F.; Han, Z.; Qin, Y.; Wang, H.; Wu, W. Garbage source classification performance, impact factor, and management strategy in rural areas of China: A case study in Hangzhou. *Waste Manag.* **2019**, *89*, 313–321. [CrossRef] [PubMed]

12. Song, C.; Zhu, H.; Zhang, L. Research on the deviation between rural residents' willingness and behavior of domestic waste classification from the perspective of institutional constraints and social norms. *J. Arid. Land Resour. Environ.* **2023**, *37*, 73–80.
13. Zhang, Z.; Zhao, X. Supervision intensity, reward-punishment system, and urban residents' household waste classification behavior: A case study of Beijing. *Urban Probl.* **2023**, *7*, 84–92+103.
14. Jia, Y.; Zhao, M. The influence of environmental concern and institutional trust on farmers' willingness to participate in rural domestic waste treatment. *Resour. Sci.* **2019**, *41*, 1500–1512. [CrossRef]
15. Zhu, K.; Gao, Q.; Jin, L. The Influence of Rural Household Income and Trust in Village Leaders on Households' Willingness to Treat the Domestic Solid Wastes: Based on Survey Data of 465 Households in Yunnan Province. *Resour. Environ. Yangtze Basin* **2021**, *30*, 2512–2520.
16. Wang, B.; Yu, F. The Strategic Task of Improving Rural Residential Environment Renovation in the 14th Five-Year Plan Period. *Reform* **2021**, *3*, 111–120.
17. Tian, J.; Gong, Y.; Li, Y.; Chen, X.; Zhang, L.; Sun, Y. Can policy implementation increase public waste sorting behavior? The comparison between regions with and without waste sorting policy implementation in China. *J. Clean. Prod.* **2022**, *363*, 132401. [CrossRef]
18. Han, H.; Zhang, Z.; Peng, W. An Analysis of the Influence Mechanism of Social Capital on Households' Waste Separation. *J. Zhejiang Univ.* **2016**, *46*, 164–179.
19. Qiu, C. Research on the performance of rural household garbage management from the perspective of farmer' participation. *J. Arid Land Resour. Environ.* **2020**, *34*, 37–43.
20. Li, J.; Vatsa, P.; Ma, W. Small Acts with Big Impacts: Does Garbage Classification Improve Subjective Well-Being in Rural China? *Appl. Res. Qual. Life* **2023**, *18*, 1337–1363. [CrossRef]
21. Chen, S.; Li, R.; Ma, Y. Paradox Between Willingness and Behavior: Classification Mechanism of Urban Residents on Household Waste. China Population. *Resour. Environ.* **2015**, *25*, 168–176.
22. Xu, D.; Qing, C.; Chen, Y.; He, J.; Zhang, F. Sustainable Development of Rural Human Settlements in the Information Age: Can Internet Use Drive Farmers to Participate in Garbage Classification? *Agriculture* **2023**, *13*, 846. [CrossRef]
23. Ma, W.; Zhu, Z. Internet use and willingness to participate in garbage classification: An investigation of Chinese residents. *Appl. Econ. Lett.* **2021**, *28*, 788–793. [CrossRef]
24. Ma, J.; Ning, G. Internet and Non-farm employment of rural labor in China. *Financ. Econ.* **2017**, *7*, 50–63.
25. Ogutu, S.; Okello, J.; Otieno, D. Impact of Information and Communication Technology-Based Market Information Services on Smallholder Farm Input Use and Productivity: The Case of Kenya. *World Dev.* **2014**, *64*, 311–321. [CrossRef]
26. Zuo, X.; Kang, M.; Lu, J. The impact of social interaction and Internet use on rural residents' willingness to sort domestic waste. *Resour. Sci.* **2022**, *44*, 47–58. [CrossRef]
27. Zhang, F.; Bao, X.; Guo, S.; Deng, X.; Song, J.; Xu, D. Internet use and land transfer in: Empirical evidence from China's rural panel data. *Environ. Sci. Pollut. Res.* **2022**, *29*, 88288–88301. [CrossRef]
28. Peng, D.; Li, Y.; Li, C. Research on the impact of Internet use on environmental protection attitudes and environmental literacy. *Financ. Econ.* **2019**, *8*, 97–109.
29. Gong, X.; Zhang, J.; Zhang, H.; Cheng, M.; Wang, F.; Yu, N. Internet use encourages pro-environmental behavior: Evidence from China. *J. Clean. Prod.* **2020**, *256*, 120725. [CrossRef]
30. Jia, X.; Chen, Y.; Zhao, Y.; Dong, X. The dilemma and mechanism innovation of rural solid waste control in our country: Based on the investigation of rural solid waste control in Hebei, Zhejiang, Guangxi and other places. *Environ. Sustain. Dev.* **2018**, *43*, 66–68.
31. Liu, H.; Lv, J.; Han, X. Research on the influence of Internet use on the Willingness of farmers to separate and dispose household solid waste: Data analysis from CLDS. *Res. Agric. Mod.* **2021**, *42*, 909–918.
32. Zhou, S.; Qing, C.; Guo, S.; Deng, X.; Song, J.; Xu, D. Why "Say One Thing and Do Another" a Study on the Contradiction between Farmers' Intention and Behavior of Garbage Classification. *Agriculture* **2022**, *12*, 1159. [CrossRef]
33. Shi, H.; Huang, H. A psychological analysis of farmers' behavior of adopting agricultural technology. *Guizhou Agric. Sci.* **2013**, *41*, 209–213.
34. Wang, Q.; Long, X.; Li, L.; Kong, L.; Zhu, X.; Liang, H. Engagement factors for waste sorting in China: The mediating effect of satisfaction. *J. Clean. Prod.* **2020**, *267*, 122046. [CrossRef]
35. Zhang, J.; Cheng, M.; Wei, X.; Gong, X.; Zhang, S. Internet use and the satisfaction with governmental environmental protection: Evidence from China. *J. Clean. Prod.* **2019**, *212*, 1025–1035. [CrossRef]
36. Ajzen, I. The theory of planned behavior. *Organ. Behav. Hum. Decis. Process.* **1991**, *50*, 179–211. [CrossRef]
37. Stern, P.; Kalof, L.; Dietz, T.; Guagnano, G. Values, beliefs, and pro-environmental action: Attitude formation toward emergent attitude objects. *J. Appl. Soc. Psychol.* **1995**, *18*, 1611–1636. [CrossRef]
38. Stern, P. New Environmental Theories: Toward a Coherent Theory of Environmentally Significant Behavior. *J. Soc. Issues* **2000**, *56*, 407–424. [CrossRef]
39. Guagnano, G.; Stern, P.; Dietz, T. Influences on attitude-behavior relationships: A natural experiment with curbside recycling. *Environ. Behav.* **1995**, *5*, 699–718. [CrossRef]
40. Hajcak, G.; Olvet, D. The persistence of attention to emotion: Brain potentials during and after picture presentation. *Emotion* **2008**, *8*, 250–255. [CrossRef]

41. Qin, M.; Yang, L.; Pan, L. Application of persuasive knowledge model in health communication: A case study of public service advertising of waste sorting in Hefei. *Mod. Commun.* **2019**, *41*, 77–84.
42. Meng, X. Analysis on residents' behavior of domestic solid waste source separation based on structural equation. *Resour. Sci.* **2019**, *41*, 1111–1119.
43. Zhu, H.; Shan, H.; Shen, Y.; Zhang, L.; Zhou, Y. Research on the influence of digital literacy on farmers' willingness and behavior of household garbage classification. *J. Jiangsu Univ.* **2022**, *24*, 35–53.
44. Xue, C.; Li, H. Environmental Knowledge and Pro-environmental Behaviors of Households: An Analysis Based on Mediation Effect of Environmental Capability and Moderation Effect of Social Norms. *Sci. Technol. Manag. Res.* **2021**, *22*, 231–240.
45. Gu, L.; Li, H. Administrative mobilization and multiple participation: The way to realize participatory governance of garbage classification—Based on Shanghai's practice. *J. Public Manag.* **2021**, *18*, 83–94.
46. Shen, F. Investigation on the Path of Realizing Participatory Governance in Rural Environment: Case Study Based on Digang Village in North Zhejiang. *Issues Agric. Econ.* **2019**, *8*, 30–39.
47. Guo, Q.; Li, H.; Li, S.; Nan, L. Research on Farmers' Pro-Environmental Behavior from the Perspective of Paradox Existing Between Behavior and Willingness: Taking the Organic Fertilizers Application as an Example. *Resour. Environ. Yangtze Basin* **2021**, *30*, 212–224.
48. Chang, Q.; Yan, Y.; Li, X.; Zhang, C.; Zhao, M. Why Say One Thing and Do Another?: A Study on the Deviation of Willingness and Behavior of Farmers' Ecological Production. *J. Agrotech. Econ.* **2021**, *4*, 85–97.
49. Govindan, K.; Zhuang, Y.; Chen, G. Analysis of factors influencing residents' waste sorting behavior: A case study of Shanghai. *J. Clean. Prod.* **2022**, *349*, 131126. [CrossRef]
50. Wen, Z.; Ye, B. Mediating effect analysis: Method and model development. *Adv. Psychol. Sci.* **2014**, *22*, 731–745. [CrossRef]
51. Czajkowski, M.; Kądziela, T.; Hanley, N. We Want to Sort! Assessing Households' Preferences for Sorting Waste. *Resour. Energy Econ.* **2014**, *36*, 290–306. [CrossRef]
52. Kang, J.; Wang, C.; Shen, Z.; Lv, X. Study on farmers' willingness and behavior differences on domestic waste classification disposal: A case study of Zhejiang Province. *Resour. Dev. Mark.* **2018**, *34*, 1726–1730+1755.
53. Deng, X.; Xu, D.; Zeng, M.; Qi, Y. Does Internet use help reduce rural cropland abandonment? Evidence from China. *Land Use Policy* **2019**, *89*, 104243. [CrossRef]
54. Meng, J.; Zhao, L.; Liu, Y.; You, W. Social learning, herd psychology and stock market participation decision. *J. Financ. Res.* **2013**, *7*, 153–165.

sustainability

Article

Enhancing Green University Practices through Effective Waste Management Strategies

Julalak Phrophayak [1], Rapeepat Techarungruengsakul [1], Mathinee Khotdee [2], Sattawat Thuangchon [1], Ratsuda Ngamsert [1], Haris Prasanchum [3], Ounla Sivanpheng [4] and Anongrit Kangrang [1,*]

1 Faculty of Engineering, Mahasarakham University, Kantharawichai District, Maha Sarakham 44150, Thailand; julalakjui03@gmail.com (J.P.); rapeepat.tec@msu.ac.th (R.T.); sattawat.t@msu.ac.th (S.T.); ratsuda.n@msu.ac.th (R.N.)
2 Faculty of Architecture, Urban Design & Creative Arts, Mahasarakham University, Kantharawichai District, Maha Sarakham 44150, Thailand; mathinee.k@msu.ac.th
3 Faculty of Engineering, Rajamangala University of Technology, Isan Khon Kaen Campus, Muang, Khon Kaen 40000, Thailand; haris.pr@rmuti.ac.th
4 Faculty of Water Resources, National University of Laos, Vientiane 01020, Laos; o.sivanpheng@nuol.edu.la
* Correspondence: anongrit.k@msu.ac.th

Abstract: The continuity of activities and projects is important for sustainably developing organizations, especially universities. The purpose of this study is to establish university development guidelines regarding waste management according to the green university ranking criteria for sustainable development by studying and collecting the data for Mahasarakham University covering the last five years (2019–2023). We also analyzed and synthesized lessons learned from the details of the operations and the factors involved in the university's successes while comparing and providing suggestions for its operations in the coming year. This study found that effective waste management led to continuous improvements that achieved the goals of Mahasarakham University. These consisted of five guidelines, as follows: (1) making green university policies and using them as a KPI (key performance index) of the administrative divisions and also announcing the university's waste management policies by applying the principles of 3R (Reuse, Reduce, Recycle); (2) allocating budgets supporting the projects' activities according to the evaluation criteria; (3) driving operations through student and personnel participation as well as the mechanisms used by the committees from various administrative divisions, such as the Student Council, the Student Association and clubs; (4) identifying the main responsible individuals and committees for the green university who communicate and drive the operations while collecting, analyzing and preparing the data; and (5) performing follow-ups and evaluations of the project's activities with the Plan-Do-Check-Act (PDCA) processes, facilitating continuous and sustainable developments and improvements leading to a green university. With the above operations, Mahasarakham University's waste management scores for the six indices during the period 2019–2023 increased annually as follows: 900, 900, 1050, 1275 and 1350.

Keywords: green university; sustainability; waste management; UI Green Metric; sustainable waste treatment

1. Significance and Background

Globally, the principle of sustainable development was officially implemented in 2015 in order to meet the current and future demands of humans in the form of the Global Development Agenda for the next 15 years (2016–2030) based on economic sustainability [1]. The Global Development Agenda considered management, job creation and support for tourists and also social sustainability, considering infrastructure development and local cultural conservation. It also addressed environmental sustainability, considering the effects of human actions resulting in waste and environmental pollution [2,3]. In 2017,

Thailand set out its sustainable development goals in its 20-Year National Strategy and formulated the 12th National Economic and Social Development Plan for preparing and creating the foundations for transforming Thailand into a developed, secure, prosperous and sustainable country. The philosophy of sufficiency economy [4] was employed as the main principle for the national and organizational development in 2021 in order to improve educational institutes and universities according to the sustainable development guidelines. Mahasarakham University prepared Mahasarakham University's Education Plan No. 13 (B.E. 2017–2026) for organizational management as a means to become a smart university. One of the sustainable development goals is green university development through improving the safety, appearance and environmentally friendly properties of the university's campus, as well as facilitating education, creation, research studies, innovation and academic services [5] as important parts of the improvement of the university according to sustainable development [6,7].

Universities efforts to achieve sustainability can be measured using the green university ranking system, which was started in 2010 by the University of Indonesia. It was then known as the UI Green Metric, focusing on three foundations, namely the social foundation, economic foundation and environmental foundation, in order to achieve a balance [8–10]. Socially, the involvement of people and quality of life improvements are emphasized. There is a long-term economic focus on people's mutual benefits. Environmentally, resource efficiency is mainly established by considering the environmental effects [11,12]. The green university ranking is a mechanism supporting universities' efforts to maintain a widely accepted environment. To measure a university's efforts to achieve sufficiency, online surveys were conducted in order to present performance data on the university's projects and policies regarding sustainability according to the frameworks relating to environments, cost efficiency and fairness [13–15]. These surveys used six criteria with a total combined score of 10,000. These were Setting and Infrastructure (SI), Energy and Climate Change (EC), Waste Management (WS), Water Usage (WR), Transportation (TR) and Education (ED) (University of Indonesia, UI Green Metric World University Ranking) [16–18].

The evaluation of universities' sustainability efforts, as exemplified by the UI Green Metric ranking system, encompasses various facets, notably waste management [19,20]. Institutions like the German Jordanian University (GJU) have undertaken waste audits to pinpoint waste streams, thereby bolstering waste reduction, recycling, and composting initiatives [21]. Additionally, the Green Metric Index, devised by Indonesia University, scrutinizes sustainability performance across six key criteria, prominently featuring waste management [22]. Noteworthy strides toward sustainability have been taken by the University of Florence (UniFi), particularly in augmenting waste management strategies and conserving water resources [23]. Moreover, Brazilian universities engaged in the UI Green Metric ranking system evince a mounting inclination towards sustainable practices, inclusive of waste management, mirroring broader societal commitments to sustainability. Undoubtedly, waste management stands as a pivotal component within universities' sustainability agendas, underscoring their endeavors to cultivate eco-friendly campuses and foster a sustainable future.

Past efforts at Mahasarakham University to align with the UI Green Metric framework have encountered challenges rooted in the university's administrative structure, characterized by periodic changes in leadership every four years, engendering a lack of operational continuity and disparate developmental trajectories under each administration. Presently, the PDCA (Plan-Do-Check-Act) system has been adopted to oversee UI Green Metric initiatives, facilitating systematic advancement.

The PDCA (Plan-Do-Check-Act) cycle serves as a systematic approach to continuous improvement, proven effective across various industries. Research underscores its efficacy in augmenting quality, efficiency, and overall performance, with applications spanning manufacturing, healthcare, and retail sectors. Notable instances include its implementation in optimizing assembly processes, resulting in a significant reduction in defects and enhanced management efficacy [24], as well as its role in boosting operational efficiency within the soap manufacturing industry [25]. Furthermore, the PDCA cycle has been instrumental in refining healthcare training programs, fostering soft skill competencies and communication among future healthcare professionals, thereby enriching training modules and participant satisfaction [26]. These illustrations underscore the adaptability and advantages of integrating the PDCA cycle for continuous improvement across diverse contexts. Notably, the application of PDCA methodology to enhance and refine the UI Green Metric framework represents a novel approach aimed at systematic development, leveraging the existing merits of PDCA for this purpose.

Mahasarakham University is focusing on environmentally friendly management and sustainable development to become a green university and to create environments facilitating education, as well as being safe and environmentally friendly in the community [27–31]. The activities of students and other personnel produce many forms of waste [32–34]. These factors were considered as indicators of the evaluation criteria for management and sustainable environments to assess the reuse of wastes and treatment of organic and inorganic wastes and wastewater and measure the implementation of the paper and plastic reduction policies on campus [35,36]. As a result, the goals were achieved and the scores increased continuously over time [37]. However, issues and suggestions arose each year that required consideration for planning and continuous developmental improvements [38]. There were also additional details of the evaluation which demonstrated the developments clearly. Mahasarakham University treated the wastes according to the indicators. If analyses and syntheses are conducted, the planning of the budget management guidelines, activities or projects will be influenced and be truly consistent with the evaluation criteria [39–41].

In essence, this research endeavors to carve a path towards a more sustainable future, not only for Mahasarakham University but for the broader landscape of higher education. By elucidating the nexus between waste management practices and sustainable development objectives, it aspires to inspire transformative change and catalyze a paradigm shift towards greener, more resilient universities.

Therefore, this study includes data about waste management (WS) for the last five years, from 2019 to 2023. Details of the evaluation criteria ae studied. Years with increased scores are used to identify the factors of the successes of this green university regarding its waste management leading to sustainable development and to obtain suggestions for the university's operations in the coming year.

2. Methodology

2.1. Research Areas

This research study was conducted in two areas inside Mahasarakham University: the Kham Riang Campus, Sub-District Kham Riang, Kantharawichai District, Mahasarakham Province and the City Campus in Mueang District, Mahasarakham Province, as shown in Figures 1 and 2. It can be seen from the figures that the two areas of Mahasarakham University are about seven kilometers from each other. The first area is in an urban location covering about 0.58 square kilometers. This is where four faculties are situated—the Faculty of Medicine, the Faculty of Veterinary Medicine, the Faculty of Tourism and Hospitality and the Faculty of Education. There is also Sutthawet Hospital, which is included under the Faculty of Medicine. Kham Riang Area covers about 2.08 square meters (km^2). It is where the President's Office and over 16 faculties/units are situated.

Figure 1. Mahasarakham University, the City Campus.

Figure 2. Mahasarakham University, the Kham Riang Campus.

2.2. The Green University Ranking Criteria Regarding Waste Management

The maximum possible score for the six aspects of the UI Green Metric World University Ranking is 10,000 (University of Indonesia, UI Green Metric World University Ranking). This score was used as a criterion for evaluating the university's development. However, this study used it only for the consideration of waste management, and the waste management evaluation criteria comprised six indicators. Each indicator had a maximum score of 300. So, the total maximum score was 1800. All evaluation criteria were weighted at 18%. The details are presented in Table 1. Initially, the details for each indicator over the last five years (2019–2023) were studied, and these were used for the analysis and planning efficient operations and continuous improvements.

Table 1. The waste management evaluation criteria during the period 2019–2023.

No	Criteria	Point
WS1	Recycling program for university's waste	300
WS2	Program to reduce the use of paper and plastic on campus	300
WS3	Organic waste treatment	300
WS4	Inorganic waste treatment	300
WS5	Toxic waste treatment	300
WS6	Sewage disposal	300
	Total (Weighting 18%)	1800

2.3. The Waste Management Operations

Mahasarakham University's waste management operations enacted principles for making plans, implementing practices, performing follow-ups and fostering improvements by using the Plan-Do-Check-Act (PDCA) processes [42] as the guidelines for the operations in order to continuously and sustainably develop the university. The details of each process are as follows.

2.3.1. Plan

To develop Mahasarakham University into a green university, the university has clearly set strategies leading to practices aligned with the goals of progressive strategies for developing the university into a green university and for conserving environments. These include goals for sustainable development and act as KPIs for the organization [31], steering the operations of the green university committee by providing clearly assigned responsibilities. Waste management committees were also appointed to represent the administrative departments according to the concept of involvement, with clubs supporting students' activities and budget allocations for the organization and students and personnel conducting the project's activities according to the evaluation. Committees and stakeholders met in order to plan the operations while driving the project towards achieving its objectives. The amount of each type of produced waste was analyzed in order to plan the project and waste management guidelines. The following features were considered:

1. A recycling program for the university's wastes (WS1): The types of reusable wastes were surveyed, and suitable containers were prepared with clearly visible signs.
2. A program to reduce the use of paper and plastics on the campuses (WS2): The activities using paper and plastic were analyzed in order to conduct activities and use management systems.
3. Organic waste management (WS3): The sources of organic waste were surveyed, and suitable containers were provided with appropriate signage during the preparation of areas and during management.

4. Inorganic waste treatment (WS4): Disposal or collection points appropriate for the actual amounts of waste were prepared, as well as appropriate routes and times being identified.
5. Toxic waste management (WS5): The sources of toxic waste were analyzed. Practical guidelines were prepared among the organizations in order to manage and provide containers for separating other types of wastes. Clear signage was essential because possibly hazardous toxins were involved. Collection and treatment schedules were set.
6. Sewage disposal (WS6): The sources of wastewater were surveyed and the treatment systems were designed in order to treat the wastewater. Committee meetings were held every month in order to follow the progress of operations according to the indicators.

2.3.2. Do

The university complies with the UI Green Metric criteria for ranking green universities regarding waste management by conducting activities and projects according to the standard evaluation criteria of efficient resource management and waste management using six indicators. The operations that complied with the operational plans in order to become a green university included the following:

1. A recycling program for the university's wastes (WS1) that considered recyclable wastes such as water bottles, plastic glasses, aluminum cans, paper boxes and white paper. The students and personnel were encouraged to conduct activities or projects that separated recyclable wastes in order to add value or benefit to them. Budgets were allocated for supporting the project's activities and public relations through channels such as Facebook, Line, pages, LED screens, posters and brochures. These measures facilitated the development of appropriate perceptions and awareness about the preparation and placing of waste bins for the purposes of separating recyclable wastes in various locations. Adding value to or using the wastes was encouraged while recording statistical data to summarize the results of the project.
2. A program to reduce the use of paper and plastics on the campuses (WS2). The policies about paper and plastic reduction were formulated and announced to all departments in the university for them to follow. Technologies were also used for documentation management in order to reduce the amount of paper so that it was also used efficiently. Students and personnel were encouraged to reduce waste by using reusable products instead of single-use products while conducting their activities and projects.
3. Organic waste treatment (WS3) refers to the management of decomposable wastes such as foods and garden wastes. This type of waste can be managed by providing containers for separation and collections in order to reduce the waste at source.
4. Inorganic waste treatment (WS4) refers to the management of undecomposable wastes such as soft plastics, hard plastics, electronic waste and construction materials which must be appropriately managed, used, collected at collection points, discarded according to schedules and treated correctly.
5. Toxic waste management (WS5) refers to the management of wastes with hazardous components or residues that may be toxic or dangerous to life, property or the environment and that must be managed appropriately. These are classified into three categories: (1) laboratory hazardous wastes; (2) office hazardous wastes such as bulbs, batteries and electronic wastes; and (3) hospital hazardous wastes, referring to wastes contaminated by pathogenic secretions such as used syringes, scalpels, gauzes or cotton wool. Appropriate containers must be provided. Places for separating the other types of wastes must be prepared. The wastes must be treated correctly.
6. Sewage disposal (WS6) refers to the disposal of wastewater from activities. To handle these wastes, the university created a treatment system, collected and treated the wastewater, and checked the quality of the treated water before it was released into the environment or used otherwise. The quality of each type of waste was identified

and its components were analyzed, providing representative values for calculating the total quantity of waste compared to the specified criteria.

7. For the investigation into the overall quantity of waste produced at Mahasarakham University, the study was conducted in two parts: a waste quantity study and a waste composition study. Data collection on waste quantity was carried out through weighing using weighing scales. This process was divided into three parts, as follows:
   - The first part involved the daily generated waste, including general and recyclable waste. Analysis was conducted to identify a sample of waste generated over 5 days to determine the average daily waste generation rate (kg/day) for each day, in order to calculate the annual waste generation quantity.
   - The second part focused on organic waste, primarily generated from food establishments. Surveys were conducted to determine the quantity of waste produced per day per establishment, multiplied by the total number of food establishments, to ascertain the average daily waste generation rate (kg/day) for each day and subsequently calculate the annual waste generation quantity.
   - The third part addressed hazardous waste, which is not regularly produced by every unit, and is mainly produced by faculties with laboratories and hospitals. Data collection was conducted through periodic surveys to determine the quantity of waste generated annually and weighed for disposal. The annual surveyed quantity was then extrapolated. Furthermore, the data obtained from all three parts were utilized to calculate waste density and record the composition of waste types, presented as percentages.

2.3.3. Check

In this study, the data for the last five years (2019–2023) were collected and used to analyze the statistical and qualitative data of the operations leading to the development of a green university regarding waste management. The following six indicators were monitored.

1. The recycling program for the university's wastes (WS1). The ratio of the activities or projects related to recycling in the university.
2. The program to reduce the use of paper and plastics on the campuses (WS2). The university had clear and comprehensive policies that provided guidelines to departments regarding systems to be used for managing operations efficiently, conveniently and quickly while saving energy.
3. Organic waste management focused on utilizing organic wastes as much as possible during the development leading to the production of energy.
4. Inorganic waste management focused on upstream management instead of downstream management. Inorganic wastes must be utilized and promoted by developing them into fuels or other products in order to encourage inorganic waste management. Nevertheless, everything should be based on the available resources. The application of management technologies should consider the worthiness of the operations performed.
5. Toxic waste management considered all hazardous wastes at all levels, not just the downstream treatments. The important principles of the green university emphasized internal management that utilized hazardous wastes in order to reduce their quantity.
6. There was a wastewater treatment system for managing the wastewater during sewage disposal. Devices were installed in order to improve the efficiency of wastewater treatment. Importantly, the treated wastewater was rotated in order to use and maximize the benefits from the resources. Nonetheless, the wastewater management needed to be completed within the university without being confined to specific areas because all areas had to be treated and managed efficiently and safely for the students, personnel and environment.

2.3.4. Act

Improvements were made possible after studying and analyzing the activities' and projects' data regarding waste management development during the last five years (2019–2023). Committee meetings generated suggestions for improving the operations in the following year. By studying the data in the last five years, the follow-ups and evaluations of the activities and projects according to the policies of each indicator showed (i) changes and (ii) waste management development inside the university. This was achieved by conducting various activities according to each indicator and covering all locations within the university and with the involvement of the students and personnel of the university, making the projects successful. Instead of prioritizing upstream waste management, attention was refocused instead on downstream waste management. As a result, the development score was increased. For the aspects with scores that did not change, there was room for improvement by planning budgets for conducting activities and projects according to the indicators employed in the successful waste management guidelines of universities or other organizations. This enabled linking projects in order to establish better development guidelines for the next year. However, the green university ranking criteria were adjusted and had more detail. The evaluation criteria and operations must be applied, and the Plan-Do-Check-Act processes must be implemented when checking and making plans, setting clear goals and following up on results in order to obtain continuous improvements.

2.4. Data Preparation and Input

2.4.1. Data Preparation

The data were collected and presented as graphs and tables as required by the main evaluation guidelines according to the UI Green Metric World University Ranking form of the University of Indonesia, as means and percentages. The credit abilities of the operations for the following considerations are shown in Figure 3.

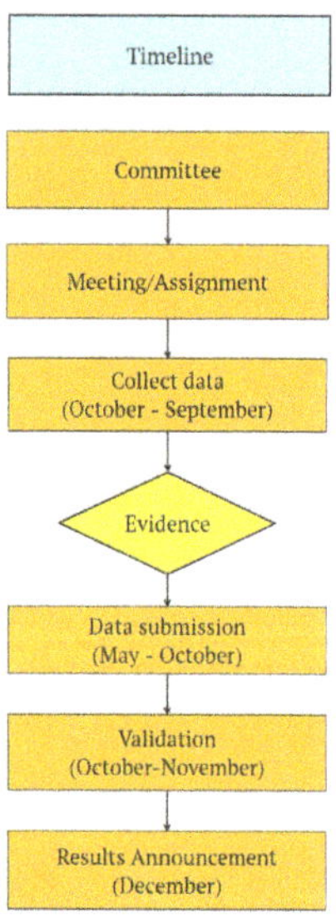

Figure 3. Flowchart of the UI Green Metric workflow at Mahasarakham University.

As illustrated in Figure 3, a crucial step in the UI Green Metric assessment process entails the collection of pertinent data. Particularly concerning waste management, scoring primarily relies on the quantity of specific waste categories. For instance, universities are evaluated based on their ability to manage a notable percentage of recyclable waste, which correlates with achieving higher scores. Therefore, the meticulous collection of raw data to substantiate these claims assumes paramount importance.

To ensure effective data collection, it is imperative to design disposal and waste collection points that facilitate the clear separation of each waste type. This involves implementing methods such as weighing or volume calculation, depending on the nature of the waste. Notably, certain waste types that incur disposal fees are typically measured by weight. In our endeavor, we structured the PDCA framework to enable the segregation of waste types and enhance the clarity of data collection processes. This refined approach was instrumental in accurately measuring the quantity of each waste type, thereby influencing the scores attributed to universities based on the UI Green Metric's predefined criteria for waste management, as outlined in the preceding section.

2.4.2. Inputs

The data were prepared according to the UI Green Metric World University Ranking's form and categorized according to each indicator. The files were recorded in PDF format. The data were inputted into the system by the October of each year. The data were inputted into the system one week before the system was closed each year in order to have time to verify or amend the data. Once the system was closed, documents could not be amended or added.

3. Results

3.1. Results with Indicators

Mahasarakham University continuously operates and makes improvements leading to the development of a sustainable green university according to the indicators of the UI Green Metric World University Ranking criteria. One of the indicators used was waste management, which must be appropriate and focus on sustainable environmental management, facilitating education safety and being environmentally friendly. The operations of a green university must involve cooperation across all sectors, set clear policies and allocate budgets for development and improvement. Waste management inside the university must focus on upstream management starting from the separation process. Waste collection and management apply the principles of 3R (Reuse, Reduce, Recycle), which emphasize reducing and utilizing wastes. By collecting and analyzing the data for 2019–2023, activities and projects were evaluated for each year, reflecting the changes and successes of the waste management practices over time. Continuous developments were recorded for the six indicators, and their scores are shown in Figure 4, in which it can be seen that the maximum possible score was 1800. Mahasarakham University's score was 900 during the period 2019–2020, and this gradually increased to 1350 in 2023. However, this score was not the maximum possible score of 1800, with a further increase of 450 required to reach the maximum. The details of the results for each indicator are presented below.

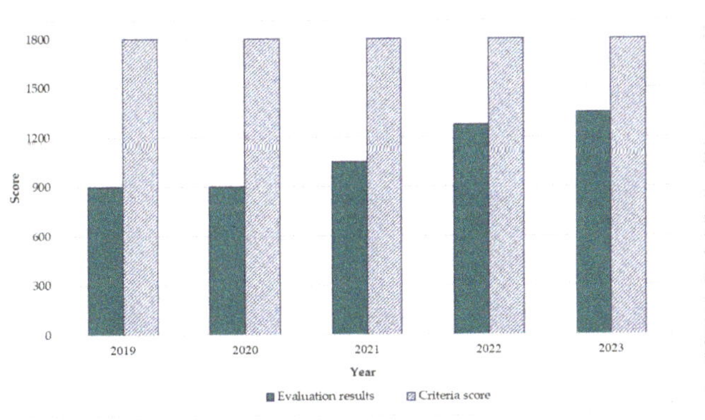

Figure 4. The waste management ranking results during the period 2019–2023.

3.1.1. The Recycling Program for the University's Wastes

The university supported and encouraged students and personnel to separate and add value to or utilize recyclable wastes. Initially, waste separation bins were provided and distributed in order to generate income. Then, the students were encouraged to separate recyclable wastes by engaging in a recycling contest. However, this contest only had a short duration. To promote long-term recyclable waste separation, the university developed other projects, including a seemingly successful project involving a recycling bank and a facility to exchange recyclable wastes for eggs. There were problems with the limitations of the university's regulations, which caused many to withdraw and resulted in delays. Consequently, the number of participants decreased, after which the projects were cancelled. However, the university tried to adjust these activities in order to encourage students and personnel to participate in their operations and to achieve sustainability by conducting voluntary point collection activities with recyclable wastes under the green dormitory project run by the student clubs. Voluntary point collection motivated the students to participate in waste separation in order to collect voluntary points that were issued according to the university's programs, such as the Student Loan Fund (SLF) and fund selection. The point collection criteria were set clearly. As a result, the students were very interested in this project. The waste recycling project was innovatively improved by adding value to the wastes and the project was implemented only in the students' dormitories. To create a network for conducting the project and generalizing the results to other faculties and departments in the university, cooperation with the private sector was implemented in order to actually enter the recycling processes. The project entitled 'Green University Leaves a Turn for the World to Remember with GC YOU Turn' was implemented through the students' organizations and clubs. The application of this project clearly showed students' participation. The project was implemented continuously. The results of the waste recycling project in the university are presented in Table 2.

Table 2. The results of the waste recycling project during the period 2019–2023.

<table>
<tr><th>Year</th><th>Score</th><th>Project Results</th><th>Project Details</th></tr>
<tr><td>2019</td><td>150</td><td>Project 1, 2</td><td rowspan="5">1. Waste sorting bins to create value
2. Organize a contest from recycled waste
3. Conduct "Recycle Waste Bank Project
4. Offer "Egg Exchange Waste Recycling Project"
5. Green Dormitory (recycled waste collects volunteer points)
6. Recycle waste to create inventions
7. Green University Leaves a Turn for the World to Remember with GC YOU Turn.</td></tr>
<tr><td>2020</td><td>150</td><td>Project 1, 2, 3</td></tr>
<tr><td>2021</td><td>225</td><td>Project 1, 4, 5</td></tr>
<tr><td>2022</td><td>225</td><td>Project 1, 5, 6</td></tr>
<tr><td>2023</td><td>225</td><td>Project 1, 5, 6, 7</td></tr>
</table>

According to Table 2, the score increased because of the increased number of projects and was not affected by the evaluations. The form was changed to cover more areas and there was increased participation and increased amounts of recycled and managed wastes in the projects. As a result, the evaluation results were improved. The amounts of waste affected the evaluation criteria. The activities with the greatest effects on the evaluations should be considered in order to continue the operations according to the evaluation criteria and improve the operations. Nonetheless, this indicator could be improved by further studying the details of the amended evaluation criteria and with support for budgets covering the whole university. This would enable continuity, sustained operations and the reviewing of project results, which had the most significant effects on the evaluations in order to generalize the results to cover and comply with the evaluation frameworks according to the factors of the successes of green universities presented by Issaree Rodthatsana (2015). Factors indicative of these successes, including the integration of environmental projects and activities according to the missions of each university regarding each aspect, must involve education, research, academic services, student development, student and internal personnel participation and the areas surrounding educational institutions.

3.1.2. The Program to Reduce the Use of Paper and Plastics on the Campuses

To reduce paper and plastic use, the university encouraged students and personnel to use their own cups by cooperating with the shop owners who joined the project. If they brought their own cups, they would receive discounts. They were also encouraged to use fabric bags in order to reduce their use of plastic bags. Initially, only two activities were implemented in order to comply with the evaluation criteria as consistently as possible. The university implemented and announced these paper and plastic reduction policies for all departments by encouraging them to use reusable products instead of single-use products. Fabric bags replaced plastic bags and refillable cups were promoted continuously. Technology was used in the form of a documentation management system, an electronic document delivery system, an online repair appointment system, an online document submission system and other systems. Conscientious use of paper was also promoted by using both sides of each sheet. This showed the appropriate use of resources during the activities of the projects. Meetings and seminars in the university also implemented waste minimization policies for foods and drinks. For example, foods in pots or trays were ordered. Lunch boxes were ordered in order to reduce waste. A plastic bag reduction project was also implemented are shown in Table 3.

Table 3. The results of the projects.

Year	Score	Project Results	Project Details
2019	150	Total of 3 Projects	1. Ecolife program 2. Say no to plastic program 3. Order refills for food and beverages 4. Electronic Document System 5. E-services system 6. Online document submission system 7. Two-sided paper policy 8. Use environmentally friendly products. 9. Project to reduce receiving, reduce giving, reduce use of plastic bags (MSU No Plastic) 10. Say no to plastic bags project 11. Green Meeting
2020	150	Total of 3 Projects	
2021	225	Total of 5 Projects	
2022	300	Total of 8 Projects	
2023	300	Total of 11 Projects	

According to Figure 5, the scores for the criteria increased throughout the study period, and in later years, a full score of 300 was achieved according to the ranking criteria because the number of projects increased. As a result, the indicators' score was the highest possible score. To continue some projects, they must be conducted continuously according to the evaluation criteria and affect the university's development, especially using management technologies for the purpose of facilitation and efficiency, as well as reducing wasted resources. However, some projects might be halted if their number meets the required level; this is in order to effectively use the budget for managing and developing other things. Nevertheless, other universities' guidelines for conducting projects were also studied and applied to continuously develop Mahasarakham University and to improve the management systems, since the number of competitor institutions is increasing. Some projects such as the waste management and waste reduction projects in the university and the 'No Single-Use Plastic' project could be conducted immediately without budgets. The plastic reduction frameworks were as follows: (1) reject plastic bags from shops; (2) use food carriers and fabric bags; (3) carry cups or mugs; (4) reuse plastic bags; and (5) ask shop owners to change their traditional single-use notes to the coated notes from Kasetsart University, which was ranked nationally as the second best university regarding waste management in 2022 [43].

3.1.3. Organic Waste Treatment

The university separated and managed organic wastes from gardens, including gras and leaves, in order to gain benefits by composting them to form fertilizers. The sources were managed by fermenting the fertilizers. Organic wastes were also used to improve the soil qualities and growth of ornamental plants in the university to reduce chemical usage and residues in the environment. These organic wastes were mostly from gardens.

The university surveyed the sources of organic wastes, finding that some of the organic wastes were produced on a daily basis by the canteen and contained high amounts of food scraps. These wastes needed to be managed appropriately. The university implemented an earthworm fermented fertilizer project in order to treat food scraps produced by the canteen. This project was able to reduce the amounts of waste requiring disposal, and the products were used for nourishing ornamental plants and also generated income for the university. In addition to the canteen, each faculty also produces waste from its internal activities. The Public Duties Certification (PDC) committees set these faculties a KPI. This was an indicator that required waste management and all other operations to reduce and separate the produced wastes at the source. Environmental conservation bins were installed in order to treat the produced food scraps. Organic waste management was also a source of learning management, with wastes being utilized by students, personnel and external organizations. The results of the organic waste management projects are summarized in Table 4.

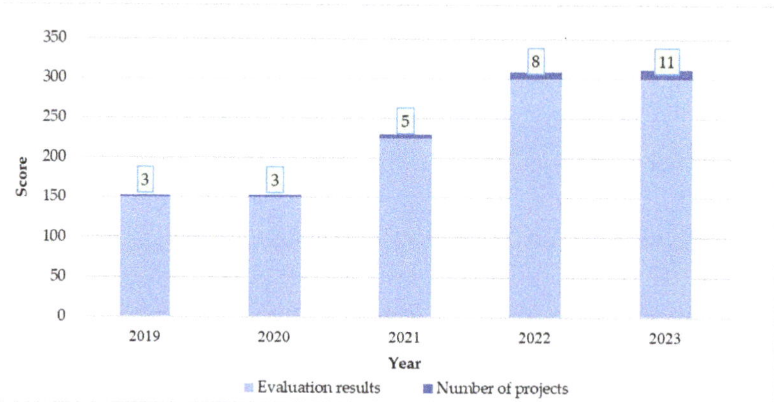

Figure 5. Conducting the projects and the evaluation results.

Table 4. Summary of the results of the organic waste management projects.

<table>
<tr><th>Year</th><th>Score</th><th>Project Results</th><th>Project Details</th></tr>
<tr><td>2019</td><td>150</td><td>Project 1</td><td rowspan="5">1. Bio-fermented water production project (EM)
2. Compost production project
3. Vermicompost production project
4. Environmentally Friendly Fermentation Tank project
5. Pork Rice Bucket Project
6. Project to produce compost in beautiful gardens
7. Wood fuel production project (Charcoal)</td></tr>
<tr><td>2020</td><td>150</td><td>Project 1, 2, 3</td></tr>
<tr><td>2021</td><td>150</td><td>Project 1, 2, 3</td></tr>
<tr><td>2022</td><td>225</td><td>Project 2, 3, 4, 5</td></tr>
<tr><td>2023</td><td>225</td><td>Project 2, 3, 4, 5, 6, 7</td></tr>
</table>

It can be seen in Table 4 that the scores were obviously higher due to the application of the organic waste management projects. However, the number of projects did not affect the evaluation. The projects were still necessary for continuing the organic waste management at the sources, the reduction in the amount of waste for disposal and management budget reduction for improvements. Moreover, this indicator could be improved. The guidelines were studied, and budgets were provided in order to make plans and adjust the projects according to the evaluation criteria, encouraging sustainable management. The conversion of the wastes into energy was considered in order to obtain guidelines for completely managing all of the organic wastes in the university according to the guidelines of Mahidol University and Kasetsart University (which was ranked nationally as the second best university in 2022 regarding waste management). Producing fertilizers and biogas represents another way to effectively manage food scraps. These activities were concrete and appropriate guidelines and forms for improving the projects in the coming year.

3.1.4. Inorganic Waste Treatment

It was found that the amount of inorganic waste in the university was the highest compared to all other types of waste, especially soft and stretched plastic waste (plastic bags). The university also had many restaurants, and the number of students was increased during the study period. The university handled these situations by providing containers, black bags or waste bags for collecting the waste, with these bags being tied and placed at the collection points before being collected to help order and enhance the convenience of waste collection. The university provided waste collection points with doors to prevent dogs accessing the bags. The waste disposal schedules were set clearly in order to collect the wastes before the designated times for the central collection of the combined wastes at the university. The university used two compactor garbage trucks for collecting and transporting the waste, with local organizations burying the waste every day in order to reduce residual waste, flies and bad smells. The wastes were disposed of by local organizations who were given monthly payments. Plastic management included hard plastics, especially containers that had contained chemicals from the laboratories, which were included in order to reduce the number of purchased or degraded containers. The waste collection routes and points are shown in Figure 6, in which it can be seen that the garbage trucks started from the parking area of the university to the first point near the waste separation station (the waste disposal point from Talad Noi Market's canteen to the students' dormitories), the parking area of the Faculty of Sciences, the area behind the Faculty of Arts and Culture, the second personnel condo, the third personnel condo, the area behind the President's Office, the Secondary Education Department of the Demonstration School and the ambulance parking area. After collecting the wastes at all points, the wastes were sent to the waste disposal station (the waste disposal site of Mueang Mahasarakham Municipality), which was 35 km away from Mahasarakham University.

Figure 6. The waste collection points.

The university tried to manage all produced inorganic wastes using the correct disposal treatment. For this, downstream waste management focused only on soft plastics. Inorganic waste reduction methods or utilization practices were not stated, and the evaluation results were not changed. The university not only managed soft plastics but also focused on managing the other types of inorganic wastes by utilizing hard plastics. The evaluation results were improved as a consequence of developing and complying with the evaluation criteria as much as possible. The waste management guidelines of the university also had a positive impact. The guidelines of Suranaree University were studied, since they affected the national waste management ranking for two consecutive years, 2022–2023, regarding complete waste management. At Suranaree University, a Biomass Excellency Center was established. Plastic wastes were processed into fuels and mechanical and biological treatment systems, RDF production and other approaches were established in order to establish guidelines for development.

3.1.5. Toxic Waste Treatment

The university was dedicated to managing toxic wastes for the safety of its students and personnel. The wastes were collected and disposed of correctly by sending the wastes to a private company that was legally registered each year. The details of the toxic wastes and the management guidelines are shown in Table 5. In the table, toxic wastes are classified into three types according to the sources, comprising toxic wastes from laboratories, from buildings and from hospitals.

Table 5. The types of toxic wastes produced by the university and the management guidelines.

Toxic Waste Types	Management and Treatment of Toxic Waste from Laboratories
Toxic waste from laboratories	Including chemical waste and chemical containers. The university provides special receptacles that can prevent leaks and created a storage area separate from other types of waste.
Toxic waste from buildings	Including light bulbs, batteries, and flashlight batteries. The university has set up disposal points or bins for hazardous waste separate from other types of waste with clear symbols. These disposal points are situated at various points and publicized to students and personnel.
Toxic waste from hospitals (infectious waste)	Waste that is contaminated with various secretions that can cause disease, such as syringes, surgical blades, gauze or cotton swabs that are contaminated with blood, etc. The university provides red containers for these wastes and display them clearly with symbol. The university created a collection point separate from other types of waste with clear signage showing the storage location.

The university had no operational policies for reusing toxic wastes. Although it treated all of the produced toxic wastes according to academic principles, the evaluation results were not changed. These results reflect the toxic waste management processes starting from the sources to the final treatment. To comply with the evaluation criteria, upstream toxic waste management and reduction were evaluated. Moreover, guidelines for reusing the toxic wastes were considered. Managing electronic wastes that were considered as a type of toxic waste that must be treated correctly was also emphasized.

3.1.6. Wastewater Treatment

The university established plans and guidelines for reusing wastewater by treating it inside the buildings before it was released from them. The wastewater inside the buildings was treated by collecting and transferring it to an air blowing treatment system (Aerated Lagoon, AL, USA), as shown in Figures 7 and 8, which added oxygen into the wastewater treatment system. The air blowers were installed in order to improve the treatment efficiency. The wastewater quality was checked before and after the treatments before circulating and using the treated water by connecting the system to the watering system with sprinklers, and the wastewater was then used for washing the roads. Wastewater collection pipes often became clogged with fats, especially in the canteen. Consequently, the wastewater overflowed, and bad smells were released. To solve these problems, a budget was allocated for installing grease trap tanks to treat the wastewater before transferring it to the treatment system. The wastewater management system in the university reflected the changes and guidelines for solving or managing the problems according to the evaluation criteria. As a result, the scores increased obviously during the study period. The wastewater treatments should cover all areas of the university and be focused on in order to utilize the treated water to increase the score for this indicator. Tools can also be installed to improve the treatment efficiency and circulate the treated water safely without affecting the environment.

Figure 7. The sewage treatment system of Mahasarakham University.

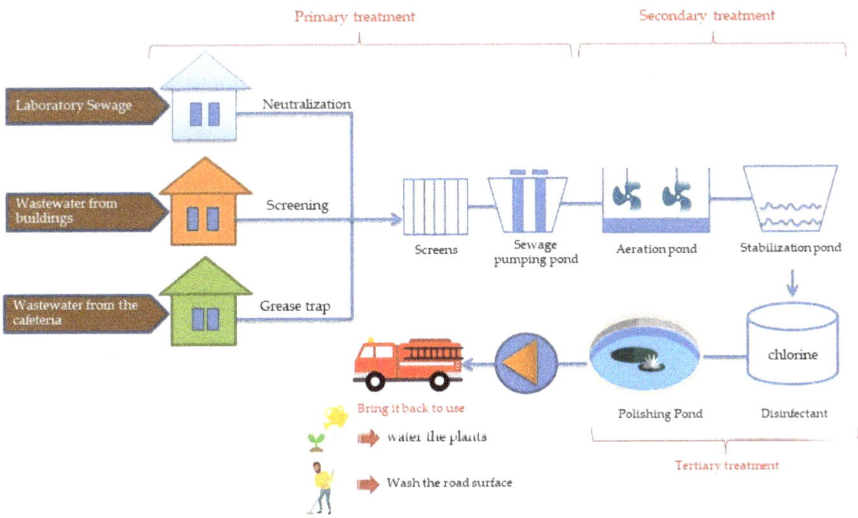

Figure 8. Diagram of the wastewater treatment processes.

As shown in Figure 8, Mahasarakham University designed and built its wastewater treatment system for the initial treatment of wastewater produced from all parts of the university. Coarse sieves were installed to filter waste and scraps out of the pipes. For the wastewater from the canteen, which usually contained fats and caused clogs, grease trap tanks were installed to initially treat the wastewater so as to reduce residues and transfer the water to the wastewater treatment system for treatment before reuse or releasing the treated water to water sources. To ensure that the circulation quality was good, the wastewater quality was checked according to the standard criteria.

3.1.7. The Results of the Waste Component Analysis

The waste components of the faculties and departments of Mahasarakham University over 5 years (2019–2023) were surveyed and analyzed. The results indicate that the waste consisted of 70% general waste, 24% organic waste, 4% recyclable waste, and 2% hazardous waste, as shown in Figure 9. In addition, the waste components were studied from Monday to Friday (5 days) to calculate the average quantities of the waste components. The amount and type of each form of waste were compared with management plans using the appropriate evaluation criteria. The details are shown in Table 6 and Figure 10.

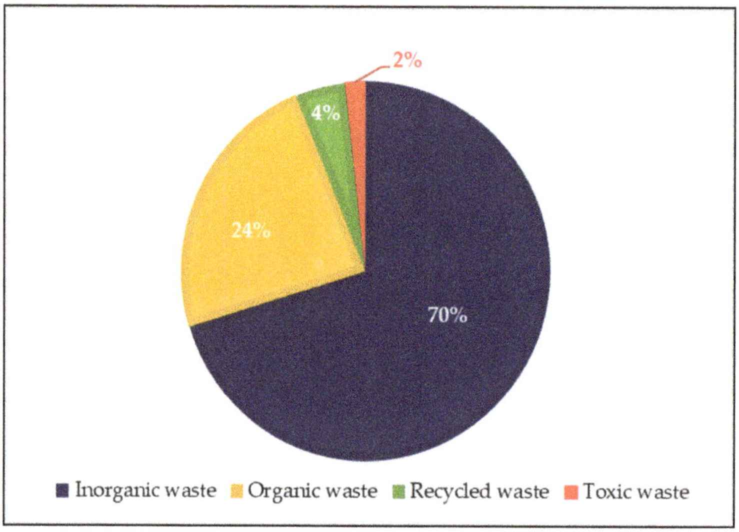

Figure 9. Component amounts of the wastes produced by Mahasarakham University over 5 years (2019–2023).

Table 6. Results from studying the daily amounts of wastes.

Waste Type	Weight (kg)				
	Year 2019	Year 2020	Year 2021	Year 2022	Year 2023
Inorganic waste	4401	3332	990	3544	4238
Organic waste	1050	1232	904	1229	1116
Recycled waste	444	121	158	131	174
Toxic waste	22	56	1	206	142
Total	5917	4741	2053	5109	5670

According to Figure 10, the amount of waste in 2019 was the highest, probably because the number of students was also the highest at that time. Consequently, the amount of waste was increased. The amount of waste decreased in 2021, presumably because the numbers of personnel and students were low due to the COVID-19 pandemic. To slow down the spread of COVID-19, the university applied the "work from home practice" and online learning. As a result, the amount of waste was decreased.

However, the impact of COVID-19 did not significantly affect the statistical scores for the different aspects of waste in the UI Green Metric ranking. This is because the criteria used the total quantity of waste for that year as a starting point and compared proportions based on the managed quantity. The waste components generated during the COVID-19 period consisted mainly of infectious waste and general waste from food delivery services. Meanwhile, the quantity of food scraps within the university itself decreased, leading to a significant reduction in the overall quantity for the year, as food scraps typically contribute substantially to the overall weight of waste.

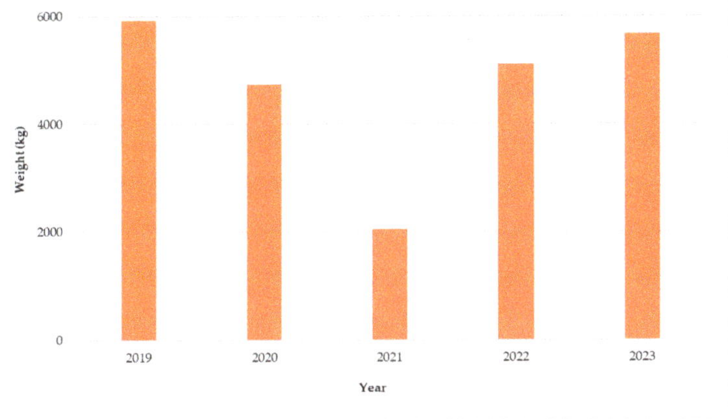

Figure 10. Daily amounts of wastes produced by Mahasarakham University.

Figure 11 shows the results for the amounts of waste produced by all departments in the Mahasarakham University. The amount of each type of waste produced daily is shown. The type of waste with the highest amount was inorganic waste, followed by the organic, recycled and toxic wastes. The amount of each type of waste affected the planning and management of the budgets in line with the evaluation policies and criteria (Division of Buildings and Accommodation, Mahasarakham University).

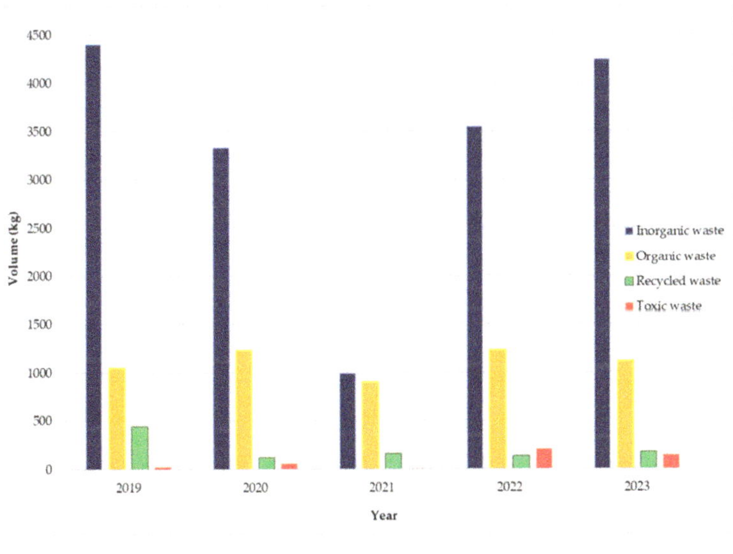

Figure 11. Amounts of wastes produced daily.

3.2. The Results of the Operations under PDCA

To gain success in becoming a green university according to the waste management indicators, the university set clear policies and allocated budgets for operations with respect to the indicators. It also followed, checked and evaluated its activities and projects according to the evaluation criteria. As a result, the overall rank was obviously higher, especially for the program to reduce the use of paper and plastics in the campuses. By setting the policies and using technologies for managing and facilitating the documentation as well as reducing the use of paper, their use was reduced.

The recycling program for the university's waste, the organic waste treatment system and the sewage disposal system reflected the continuous changes to the activities. The ongoing participation of students and personnel as well as the management guidelines covering different areas and focusing on upstream waste management and waste utilization were able to influence the activities of the Students' Association and student clubs. As a result, the evaluation results were improved regarding inorganic waste treatment. These results reflect the activities that affected the two phases of evaluation. During the first four years (2019–2022), soft plastic waste management was the sole focus of attention. Although the methods were correct involving collection and burying the wastes, the evaluation results were not influenced. In the next phase, the focus was on both hard plastic and construction waste management. As a result, the evaluation results were improved. To improve the university's practices so that they comply with the evaluation criteria, guidelines for reducing and utilizing the inorganic wastes should be considered, such as converting waste into electrical energy and producing fuels or refuse-derived fuels (RDFs). To address the unchanged evaluation results regarding toxic waste treatment that only focused on downstream management, reductions and utilizations should be focused on. However, the operations regarding the waste management indicator could be improved. The activities or projects of other universities should be studied in order to establish development guidelines. The amended evaluation criteria should also be studied more carefully in order to make budget allocation plans for conducting activities in compliance with policies in the coming year.

3.3. Evaluation Results

The UI Green Metric World University Rankings during the period 2019–2023 are shown in Figure 12, where it can be seen that the scores for each year indicated continuous changes undertaken in order to become a green university. Mahasarakham University was ranked first among the top 10 universities in the country in the UI Green Metric World University Rankings. Waste management is an indicator of the evaluation criteria that demonstrated the university's dedication to managing waste and developing continuously. Each evaluation indicator had a maximum score of 300. The maximum possible total score across all indicators was 1800. Each indicator was weighted at 18%. By setting the policies for the activities and projects during the period 2019–2023, these affected the six indicators of waste management evaluation. Firstly, the scores for the recycling program for the university's wastes were 150, 150, 225, 225 and 225, respectively, across the 5 years. Secondly, the scores for the program to reduce the use of paper and plastics on the campuses were 150, 150, 225, 300 and 300 across the 5 years. These scores met the evaluation criteria. These results clearly show the achievement of positive developments. Thirdly, the score for organic waste treatment during the period 2019–2021 was about 150. The score during the period 2022–2023 increased to 225. Fourthly, the score for inorganic waste treatment during the period 2019–2022 was 150 without any change, while it increased to 225 in the following year. Fifthly, the score for toxic waste treatment did not change at all, being 150 for all years. Sixthly, the other indicator was sewage disposal. The score was equal to that of organic waste treatment. Regarding the unchanged score for toxic waste treatment, the university needs to establish policies for reusing or reducing toxic wastes at the sources according to the principles of 3R in order to meet the evaluation criteria of the UI Green Metric World University Rankings.

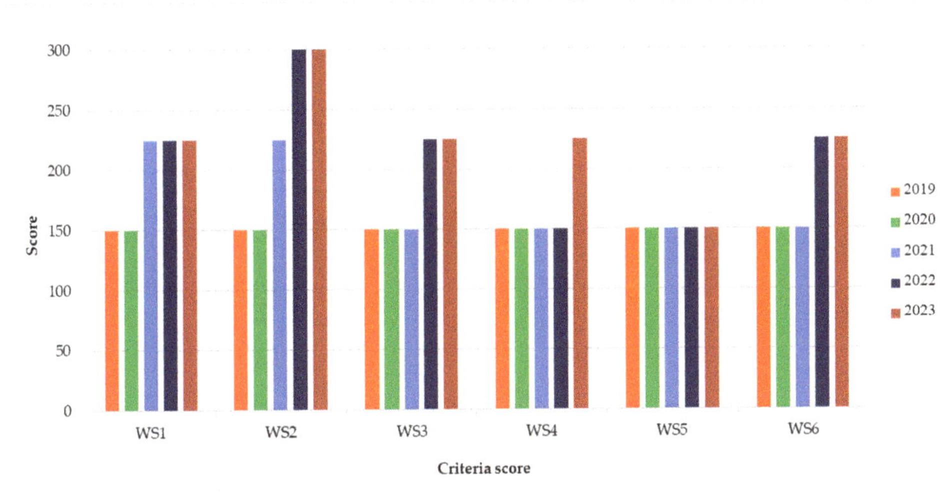

Figure 12. Evaluation results of the aspects of waste management during the period 2019–2023.

3.4. The Waste Management Guidelines for Sustainable Development

Mahasarakham University has improved its operations for sustainable development and made continuous developments. The 17 goals for sustainable development reflect the three pillars of sustainability—society, the economy and the environment. One of the goals for sustainable development is green university development, focusing on organizational development and environmentally friendly management. Waste management is another indicator of the green university ranking evaluation criteria. Wastes must be managed appropriately. Recycling must be promoted. The use of paper and plastics must be reduced. Organic waste treatment, inorganic waste treatment and wastewater treatment according to the principles of 3R must follow the sanitation principles in order to produce environments facilitating education and enhancing the lives of the students and personnel at the university and to safely manage the wastes. Environmental considerations covered the 2nd, 12th and 15th goals, especially the 12th goal, which is responsible consumption and production. Sustainable consumption and production plans must manage and achieve the goals by efficiently managing the use of natural resources, the methods for disposing of toxic wastes and pollutants as well as promoting recycling and reducing wastes in order to implement sustainable consumption plans and to achieve environmental sustainability. The 15th goal is life on land. The land ecosystem was utilized. This included hazardous waste management in a safe and environmentally friendly way are presented in Table 7.

Table 7. Operation results according to the SDGs.

GOALS	Criteria	Project Details
GOAL 2: Zero Hunger	Campus food waste	1. Survey of the amount of food waste that occurs in universities, including cafeterias and food establishments. 2. Providing a tank to hold or separate food scraps for use. 3. Using food waste to produce vermicompost and then use it in agriculture, such as growing organic vegetables and maintaining flowers and ornamental plants within the university

Table 7. *Cont.*

GOALS	Criteria	Project Details
GOAL 12: Responsible Consumption and Production	1. Policy to reduce plastic use within the university and extend it outside the university. 2. Waste disposal policy to measure the amount of waste sent to landfill and recycling.	1. Plastic reduction project: - Ecolife program - Say no plastic program - Say no to plastic bag project - Stop using foam boxes. - Promote the use of personal drinking glasses. - Organize meetings, seminars, training in the form of Green Meeting. 2. Waste recycling project: - Waste sorting bins to create value - Green Dormitory (Recycled wastes collects volunteer points) - Recycle waste to create innovations 3. Collection and disposal project: - Collect and transport waste for disposal by landfill with the municipality for waste that cannot be used.
GOAL 15: Life on Land	Policy on hazardous waste disposal.	The university separates and collects waste that may cause harm to students and personnel, including chemicals from laboratories, hazardous waste from buildings (light bulbs/batteries/batteries/aerosol cans) and hazardous waste from hospitals (infectious waste), which is sent for proper disposal according to academic principles. The university hires a registered company to carry out the disposal once a year.

3.5. The Guidelines for Achieving Sustainability

Mahasarakham University has developed many different projects and activities in order to become a green university. Development commenced in 2011 and continues to this day. As a result, the university is ranked nationally as one of the top 10 universities for waste management. Additionally, the scores used to evaluate its progress have increased every year, as shown in Figures 13 and 14. In 2023, Mahasarakham University attained a cumulative score of 8335 out of 10,000, signifying its commendable performance and satisfactory standing. This achievement is juxtaposed with Kasetsart University, ranked first in Thailand, which secured a total score of 8775 for the same period, and with Wageningen University & Research, leading globally with a total score of 9500 points in 2023 [35]. The university's operations reflected changes and continuous developments. In summary, the factors contributing to the successes of a sustainable green university include the following:

1. Setting and announcing clear policies in order to make the students and personnel aware of and strictly follow the policies.
2. Providing official duties certificates (ODC) as the KPI of each department showing the responsibilities and participation of each department.
3. Creating student and personnel participation networks in all sectors in the form of clubs, the Students' Association and committees. This will drive the activities and projects as well as supporting the reporting of data for preparing the data and achieving the goals.
4. Assigning individuals with responsibilities and clear duties for driving links and integration.
5. Allocating the budgets according to the policies of the activities and projects, thus meeting the green university evaluation criteria.
6. Driving and following activities including Plan, Do, Check and Act processes to facilitate continuous development and improvements.

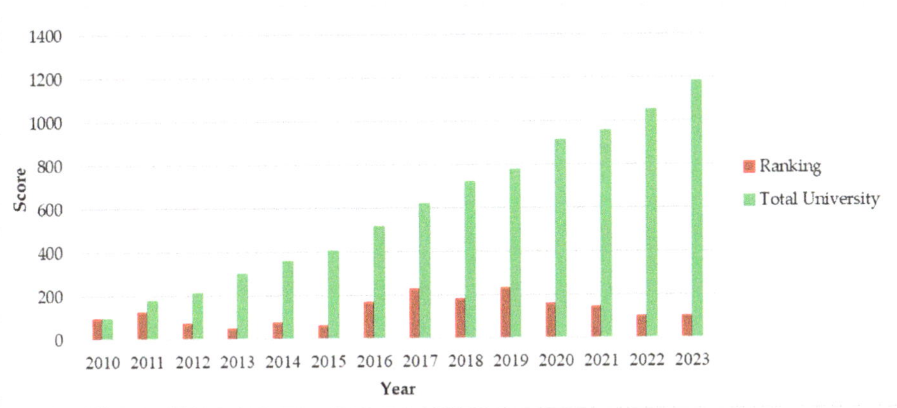

Figure 13. World rankings history diagram [43].

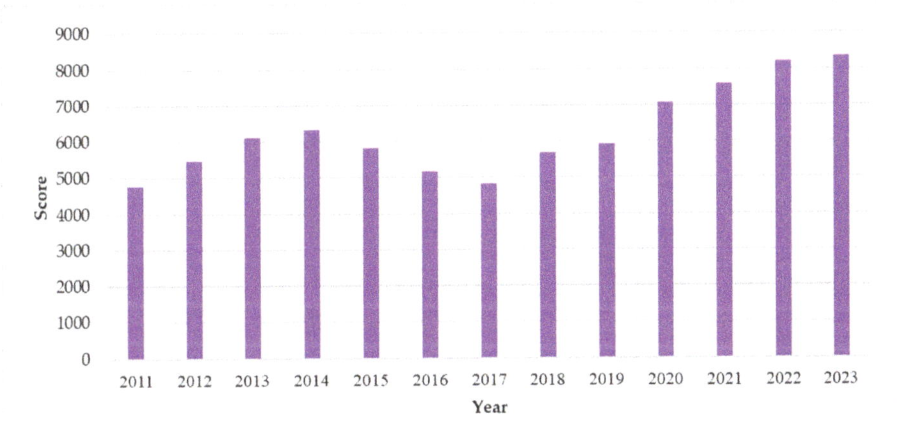

Figure 14. World rankings history diagram (scores) [43].

4. Conclusions

Mahasarakham University has adhered to sustainable waste management guidelines based on the UI Green Metric World University Ranking criteria since 2013, with the Division of Buildings and Accommodation spearheading these efforts. This study identified operational guidelines that drove continuous development from 2019 to 2023, leading to annual enhancements in the university's waste management practices.

According to the above operations, Mahasarakham University's waste management scores regarding the six indicators increased from 2019 to 2023 were established. Specifically, (1) the recycling program's scores were 150, 150, 225, 225, and 225 over consecutive years. (2) The scores of the program to reduce the use of paper and plastics on the campuses were 150, 150, 225, 300, and 300 over consecutive years. (3) The organic waste treatment system's scores were 150, 150, 150, 225, and 225 over consecutive years. (4) The inorganic waste treatment system's scores were 150, 150, 150, 150, and 225 over consecutive years. (5) The toxic waste treatment program's score was 150 every year. (6) The sewage disposal system's scores were equal to those of the organic waste treatment system. As a result, the evaluation scores continuously increased according to the green university ranking criteria.

These findings underscore the effectiveness of the university's waste management strategies over the past five years, as reflected in the increasing evaluation scores aligned with the green university ranking criteria. The practical implications include the establishment of efficient waste management committees, the promotion of recycling initiatives, and the engagement of students and faculty members in sustainability endeavors. Recognizing the imperative need for ongoing improvement, future research avenues could explore advanced waste treatment technologies and strategies for community involvement.

Furthermore, recommendations include the implementation of policies aimed at reusing or reducing waste at its source, following the principles of the 3R approach outlined in the UI Green Metric World University Rankings. To further modernize its operations, Mahasarakham University should examine guidelines and innovations from both domestic and foreign universities to continually enhance its journey towards becoming a sustainable green institution. Additionally, the prospective integration of artificial intelligence may offer avenues for managing environmentally sustainable universities [44,45].

Author Contributions: Conceptualization, J.P., R.T., M.K., S.T., R.N., H.P., O.S. and A.K.; methodology, J.P., R.T., M.K., S.T., R.N., H.P., O.S. and A.K.; validation, J.P. and A.K.; formal analysis, J.P. and A.K.; investigation J.P. and A.K.; writing—original draft preparation, J.P. and A.K.; supervision, J.P. and A.K.; and writing—review and editing, J.P., R.T. and A.K. All authors have read and agreed to the published version of the manuscript.

Funding: This research was financially supported by Mahasarakham University.

Institutional Review Board Statement: Not applicable.

Informed Consent Statement: Not applicable.

Data Availability Statement: This study did not report any data.

Acknowledgments: We would like to thank Adrian Plant for their valuable comments and English language revision of the manuscript.

Conflicts of Interest: The authors declare no conflicts of interest.

References

1. Filho, W.; Tripathi, S.; Guerra, J.; Giné-Garriga, R.; Lovren, V.; Willats, J. Using the sustainable development goals towards a better understanding of sustainability challenges. *Int. J. Sustain. Dev. World Ecol.* **2018**, *26*, 179–190. [CrossRef]
2. Nepal, R.; Irsyad, M.; Nepal, S. Tourist arrivals, energy consumption and pollutant emissions in a developing economy–implications for sustainable tourism. *Tour. Manag.* **2019**, *72*, 145–154. [CrossRef]
3. Kangrang, A.; Prasanchum, H.; Sriworamas, K.; Ashrafi, S.M.; Hormwichian, R.; Techarungruengsakul, R.; Ngamsert, R. Application of Optimization Techniques for Searching Optimal Reservoir Rule Curves: A Review. *Water* **2023**, *15*, 1669. [CrossRef]
4. Arif, A.; Fleischer, L.; Bogiatzis, A.; Asada, H.; Colombo, A.; Zougbédé, K. *Enhancing Governance in Thailand*; OECD Publishing: Paris, France, 2018. [CrossRef]
5. Ramísio, P.; Pinto, L.; Gouveia, N.; Costa, H.; Arezes, D. Sustainability Strategy in Higher Education Institutions: Lessons learned from a nine-year case study. *J. Clean. Prod.* **2019**, *222*, 300–309. [CrossRef]
6. Zhu, B.; Zhu, C.; Dewancker, B. A Study of Development Mode in Green Campus to Realize the Sustainable Development Goals. *Int. J. Sustain. High. Educ.* **2020**, *21*, 799–818. [CrossRef]
7. Nejati, M.; Nejati, M. Assessment of sustainable university factors from the perspective of university students. *J. Clean. Prod.* **2013**, *48*, 101–107. [CrossRef]
8. Puertas, R.; Marti, L. Sustainability in Universities: DEA-GreenMetric. *Sustainability* **2019**, *11*, 3766. [CrossRef]
9. Dagiliūtė, R.; Liobikienė, G.; Minelgaitė, A. Sustainability at universities: Students' perceptions from Green and Non-Green universities. *J. Clean. Prod.* **2018**, *181*, 473–482. [CrossRef]
10. Perchinunno, P.; Cazzolle, M. A clustering approach for classifying universities in a world sustainability ranking. *Environ. Impact Assess. Rev.* **2020**, *85*, 106471. [CrossRef]
11. Ramkissoon, H.; Mavondo, F.; Uysal, M. Social involvement and park citizenship as moderators for quality-of-life in a national park. *J. Sustain. Tour.* **2018**, *26*, 341–361. [CrossRef]
12. McCabe, S.; Joldersma, T.; Li, C. Understanding the benefits of social tourism: Linking participation to subjective well-being and quality of life. *Int. J. Tour. Res.* **2010**, *12*, 761–773. [CrossRef]
13. Amaral, A.; Rodrigues, E.; Gaspar, A.; Gomes, Á. A review of empirical data of sustainability initiatives in university campus operations. *J. Clean. Prod.* **2020**, *250*, 119558. [CrossRef]
14. Ralph, M.; Stubbs, W. Integrating environmental sustainability into universities. *High. Educ.* **2013**, *67*, 71–90. [CrossRef]
15. Pajardo, E.; Kang, D. Sustainable Development Strategies on Campus: Reduce Water Consumption. In Proceedings of the World Environmental and Water Resources Congress, Atlanta, GA, USA, 5–8 June 2022. [CrossRef]

16. Boiocchi, R.; Ragazzi, M.; Torretta, V.; Rada, E. Critical Analysis of the GreenMetric World University Ranking System: The Issue of Comparability. *Sustainability* **2023**, *15*, 1343. [CrossRef]
17. Rakhmetullina, S.; Shaimardanov, Z.; Petrova, O.; Idrisheva, Z.; Kolpakova, V.; Apseitova, A. Green Metrics Questionnaire as the basis of Green University strategy. *IOP Conf. Ser. Earth Environ. Sci.* **2023**, *1194*, 012011. [CrossRef]
18. Hakiki, R. Identification of Green and Sustainable Campus Indicators in Its Implementation at President University. In Proceedings of the 2019 International Conference on Sustainable Engineering and Creative Computing (ICSECC), Bandung, Indonesia, 20–22 August 2019; pp. 287–292. [CrossRef]
19. Muna, H.; Mustafa, J.; Aiman, A.; Batool, A.; Yazan, A.-M.; Bashar, K.H. Sustainable Green University: Waste Auditing, German Jordanian University as a Case Study. *Front. Built Environ.* **2022**, *8*, 884656. [CrossRef]
20. Falsini, S.; Papini, A.; Gentilini, G.; Bagnoli, F.; Pacini, G.C.; Giovannetti, G.; Pierini, M.F. University and Environmental Health: Green Advancement at the University of Florence Revealed by UI GreenMetric Ranking. *IOP Conf. Ser. Earth Environ. Sci.* **2023**, *1194*, 012035. [CrossRef]
21. Gültekin, P. Assessment of Green Campuses for Sustainable Cities and Society With UI Green Metrics: A Case Study of Duzce University. In Proceedings of the 16th International Conference on Environmental Science and Technology, Rhodes, Greece, 4–7 September 2019. [CrossRef]
22. Kusumaningtyas, K.; Fithratullah, R.; Meluk, C. The Academic Community Perception About Implementation of UI GreenMetric-Waste Management Criteria at President University. *J. Environ. Eng. Waste Manag.* **2019**, *4*, 28. [CrossRef]
23. Roberto, R.C.S.J.; Bilotta, P.; de Miranda, M.G.; Janissek, P.R. Análise de Dados da Plataforma UI Green Metric sob a Perspectiva de Índices Socioeconômicos Municipais. *Front. J. Soc. Technol. Environ. Sci.* **2023**. [CrossRef]
24. Nguyen, V.H.; Tran, T.K. PDCA from Theory to Effective Applications: A Case Study of Design for Reducing Human Error in Assembly Process. *Adv. Oper. Res.* **2023**, *2023*, 8007474. [CrossRef]
25. Zhang, Z. *Aplicação da Metodologia PDCA em uma Indústria do Ramo Calçadista*; Universidade Federal de Campina Grande: Campina Grande, Brazil, 2023. [CrossRef]
26. Anonymous. *Aplicação do Ciclo PDCA como Método de Otimização de Processos: Um Caso de uma Indústria de Sabão*; Universidade Federal de Campina Grande: Campina Grande, Brazil, 2023. [CrossRef]
27. Mongkoldhumrongkul, K.; Sukkanta, P. Model of University Development in Thailand 4.0 Era toward "Green Campus Concept". *IOP Conf. Ser. Earth Environ. Sci.* **2022**, *1050*, 012027. [CrossRef]
28. Wang, X.; Li, C.; Lam, C.H.; Subramanian, K.; Qin, Z.-H.; Mou, J.-H.; Jin, M.; Chopra, S.S.; Singh, V.; Ok, Y.S.; et al. Emerging waste valorisation techniques to moderate the hazardous impacts, and their path towards sustainability. *J. Hazard. Mater.* **2022**, *423*, 127023. [CrossRef]
29. Anwar, N.; Mahmood, N.; Yusliza, M.; Ramayah, T.; Faezah, J.; Khalid, W. Green Human Resource Management for organisational citizenship behaviour towards the environment and environmental performance on a university campus. *J. Clean. Prod.* **2020**, *256*, 120401. [CrossRef]
30. Priyadarshini, P.; Abhilash, P. From piecemeal to holistic: Introducing sustainability science in Indian Universities to attain UN-Sustainable Development Goals. *J. Clean. Prod.* **2020**, *247*, 119133. [CrossRef]
31. Ribeiro, J.; Hoeckesfeld, L.; Magro, C.; Favretto, J.; Barichello, R.; Lenzi, F.; Secchi, L.; Lima, C.; Guerra, J. Green Campus Initiatives as sustainable development dissemination at higher education institutions: Students' perceptions. *J. Clean. Prod.* **2021**, *312*, 127671. [CrossRef]
32. Ozanne, L.; Ballantine, P.; McMaster, A. Understanding Food Waste Produced by University Students: A Social Practice Approach. *Sustainability* **2022**, *14*, 10653. [CrossRef]
33. Owojori, O.; Mulaudzi, R.; Edokpayi, J. Student's Knowledge, Attitude, and Perception (KAP) to Solid Waste Management: A Survey towards a More Circular Economy from a Rural-Based Tertiary Institution in South Africa. *Sustainability* **2022**, *14*, 1310. [CrossRef]
34. Dondapati, S.; Kondagunta, N.; Guthi, V.; Ade, A.; Deekala, R. A comparative study of general waste management practices in a campus of a medical university located in the purview of a municipal corporation of a South Indian state. *Int. J. Community Med. Public Health* **2018**, *5*, 5115. [CrossRef]
35. Pandiyarajan, V.; Neelakantan, T.; Sridharan, S.; Ramrao, N. Three "R" Concept in Waste Management for Sustainable Environment. *J. Sustain. Perspect.* **2022**, *2*, 255–262. [CrossRef]
36. Phdungsilp, A. Waste Management and Its Contribution to the Sustainable Development Goals at Dhurakij Pundit University, Thailand. *J. Sustain. Perspectum.* **2022**, *2*, 65–72. [CrossRef]
37. Das, S.; Lee, S.; Kumar, P.; Kim, K.; Lee, S.; Bhattacharya, S. Solid waste management: Scope and the challenge of sustainability. *J. Clean. Prod.* **2019**, *228*, 658–678. [CrossRef]
38. Sharholy, M.; Ahmad, K.; Mahmood, G.; Trivedi, R. Municipal solid waste management in Indian cities—A review. *Waste Manag.* **2008**, *28*, 459–467. [CrossRef]
39. Sawalkar, R.; Undale, S.; Muluk, S.; Mude, G.; Saxena, V.; Pasumarti, S. Strategic waste management practices for environmental sustainability—A case of Indian university. *Manag. Environ. Qual. Int. J.* **2023**. [CrossRef]
40. Shankar, Y.; Khandelwal, R. Sustainable waste management strategy for a campus: A case study of JUET, Guna. *Manag. Environ. Qual. Int. J.* **2017**, *28*, 610–623. [CrossRef]

41. Jaglan, A.; Cheela, V.; Vinaik, M.; Dubey, B. Environmental Impact Evaluation of University Integrated Waste Management System in India Using Life Cycle Analysis. *Sustainability* **2022**, *14*, 8361. [CrossRef]
42. Kong, X.; Zhu, X.; Zhang, Y.; Wu, J. The application of plan, do, check, act (PDCA) quality management in reducing nosocomial infections in endoscopy rooms: It does work. *Int. J. Clin. Pract.* **2021**, *75*, e14351. [CrossRef]
43. GreenMetric—World University Rankings. (n.d.). Available online: https://greenmetric.ui.ac.id/ (accessed on 7 December 2023).
44. Metwaly, A.A.; Elhenawy, I. Predictive Intelligence Technique for Short-Term Load Forecasting in Sustainable Energy Grids. *Sustain. Mach. Intell.* **2023**, *5*, 1–4. [CrossRef]
45. Abouhawwash, M.; Jameel, M.; Askar, S.S. Machine Intelligence Framework for Predictive Modeling of CO2 Concentration: A Path to Sustainable Environmental Management. *Sustain. Mach. Intell.* **2023**, *2*, 1–6. [CrossRef]

sustainability

MDPI

Article

One Year of Mandatory Reusable Packaging in Germany: Opportunities and Obstacles from the Perspective of Consumers and Companies

Klara Theobald [1], Anne Mich [1], Stefanie Hillesheim [2], Susanne Hartard [1] and Holger Rohn [2,*]

[1] Department Environmental Business/Environmental Law, Trier University of Applied Sciences—Environmental Campus Birkenfeld, D-55761 Birkenfeld, Germany; k.theobald@umwelt-campus.de (K.T.); a.mich@umwelt-campus.de (A.M.); s.hartard@umwelt-campus.de (S.H.)

[2] Department of Industrial Engineering, University of Applied Sciences—Technische Hochschule Mittelhessen, Friedberg, D-61169 Gießen, Germany; stefanie.hillesheim@wi.thm.de

* Correspondence: holger.rohn@wi.thm.de; Tel.: +49-6031-604-5769

Abstract: Since the beginning of 2023, the so-called reusable packaging obligation has been in place in Germany in order to reduce the amount of waste from takeaway packaging. Catering companies will then be obliged to offer reusable packaging as an alternative to disposable plastic takeaway packaging. As part of the pilot project 'Mehrweg Modell Stadt' ('Reusable city model'), an open infrastructure for reusable cups is in trial in Mainz and Wiesbaden. The project was subjected to a scientific monitoring process, which included the implementation of quantitative surveys among various stakeholders within the value chain, namely consumers, catering companies, and other companies engaged in the project. This was conducted over two time periods: August 2023 to September 2023, and February to March 2024. The results show a discrepancy between consumers' attitudes and their actual behavior: Despite a high level of sustainability awareness and perceived positive benefits of reusable packaging, perceived barriers stand in the way of actual use. The biggest challenges for companies are the lack of consumer demand, practical handling and hygiene requirements. The results of this quantitative social research provide important insights for the development of targeted measures to promote the use of reusable packaging in takeaway catering and contribute to the discussion on closing the attitude–behavior gap along the entire value chain.

Keywords: reusable packaging obligation; single-use packaging; waste generation; sustainability awareness; consumer survey; takeaway catering companies; reusable quota; takeaway gastronomy; takeaway

1. Introduction

In recent years, the convenience and time-saving nature of takeaway meals in a fast-paced consumer society has led to a steady increase in waste generated by disposable food and drink packaging. In 2017 alone, packaging waste from meal and snack boxes in Germany amounted to almost 120,000 tons, which corresponds to almost 329 tons of waste per day from takeaway catering [1]. This considerable amount of waste prompted Germany to introduce mandatory reusable packaging from 1 January 2023. According to section 33 of the Packaging Act [2], catering businesses and restaurants in Germany are now obliged to offer reusable packaging as an alternative to disposable plastic packaging for takeaway food and drinks. The reusable option may not be offered at a higher price or under worse conditions than the disposable option. Furthermore, companies are obliged to explicitly indicate the availability of reusable packaging. Exceptions to this rule are permitted for small businesses with a sales area of less than 80 m^2 and fewer than six employees.

The German obligation to offer reusable packaging in the out-of-home catering sector can be derived from a series of national and EU-wide regulations that jointly pursue the goal of reducing the environmental impact of single-use plastics and promoting the circular economy.

The genesis of this development can be traced back to the Single-Use Plastics Directive (Directive (EU) 2019/904) [3], which came into force in 2019. The directive's objective is to reduce the consumption of single-use plastics throughout the EU. A crucial aspect of the directive is the requirement for member states to implement measures to reduce the utilization of single-use packaging and to promote more sustainable alternatives. Furthermore, the directive stipulates that manufacturers will be held responsible for the costs of waste management and cleaning up plastic waste.

Concurrently, the European Green Deal was initiated, a comprehensive strategy to make the EU climate-neutral by 2050 [4]. A number of measures are being implemented as part of this plan, including the promotion of clean energy, the improvement of energy efficiency and the reduction of pollution. Furthermore, the Green Deal supports the development of a circular economy, whereby products and materials are used, repaired, reused and recycled for as long as possible.

At the national level, Germany implements these European requirements through the Packaging Act (VerpackG) [2] and the Circular Economy Act (KrWG) [5], among other legislative instruments. The Packaging Act contains specific provisions for the reduction of packaging waste and the promotion of reusable packaging. This is where the obligation to use reusable packaging comes into play, which came into force on 1 January 2023. On 4 March 2024, the European Parliament and the European Council reached a provisional agreement on the revised rules of an EU packaging regulation to reduce, reuse and recycle packaging, increase safety and promote the circular economy [3].

Targets are set for the use of reusable packaging in the food service industry. By 2023, 10% of all takeaway food and 20% of all drinks are to be served in reusable packaging, with the targets being increased, respectively, to 40% and 80% by 2040 [6]. The fact that these regulatory requirements are necessary is also illustrated by the very low proportion of reusable packaging in Germany in 2022: of 13.7 billion packages for food and drinks, only 0.74% were sold in reusable alternatives [7]. The food service industry was the most relevant sector here, accounting for 97.9% of all single-use packaging waste generated (hotel sector: 0.4%, catering sector: 1.7%) [7].

The introduction and expansion of reusable systems has been met with numerous reservations among restaurateurs and system operators. Studies by the WWF [7] and the Institute for Social-Ecological Research and Education (ECOLOG) [8] cite a lack of information regarding the use of reusable containers. Companies lack the capacity to deal with this issue. Language barriers and cost pressure exacerbate the problem. It is also unclear how many companies are subject to the reusable container obligation. If only the number of employees is taken as a criterion, 38.6% of companies are affected by the obligation. No reliable figures can be found based on sales area [7,8]. Looking deeper into the cost pressures, the costs for companies are simply too high, due to limited storage options, logistical challenges at peak times and the need to book deposits. Catering companies in particular also struggle with providing reusable food containers. Finally, many of the companies surveyed in both studies also report low customer demand for returnable containers.

Despite these challenges, most of the companies surveyed have a fundamentally positive attitude towards the use of reusable containers, both for reasons of environmental protection and waste avoidance and as a way of differentiating themselves from the competition. One year after the introduction of the reusable offer obligation in the German food service industry, reliable and comprehensive figures on reusable use are not yet available. However, according to the results of a comparative study by the WWF after one year of the reusable offer obligation in Germany, an initial success has been achieved [9]. In 2023, a total of 14.6 billion disposable packaging units were sold in the takeaway food service sector, compared to 13.6 billion in 2022. The market share of reusable packaging increased from 0.74% to 1.6% during this period, with the share for food at 0.3% and for beverages at 7.0% [9].

In qualitative interviews with system operators, it was reported that the demand from restaurants for reusable systems increased at the turn of the year but decreased again in the summer. This suggests that, despite the positive changes, there are still significant challenges to overcome in order to establish the sustainable use of reusable packaging in the catering sector [9].

Among consumers, the success of reusable systems is strongly influenced by attractive, convenient offers, acceptance, demographic aspects and education. Current studies show that barriers for end consumers are inconveniences in the use of reusable cups and containers, hygiene concerns, the need to pay a deposit or the need for app registration, and the lack of or invisible offers of reusable options [8,10,11]. Therefore, the technical and organizational design of innovative reusable systems can only be implemented in combination with social and societal factors in order to achieve a significant increase in the reusable rate in takeaway gastronomy [12].

Against this legal background and in light of the obligation to offer reusable packaging, numerous pilot projects have been initiated over the past year with the aim of establishing a joint, cross-system and cross-provider take-back system. This encompasses the development of effective organizational structures for return and rinsing logistics, as well as the establishment of shared digital interfaces [13].

One of these projects, the pilot project 'Mehrweg Modell Stadt' [14], was monitored and evaluated as part of this scientific study. Initiated by the Reusable-to-go initiative, the objective is to increase the reuse rate in the cities of Mainz and Wiesbaden by setting up a comprehensive take-back infrastructure. Reusable takeaway coffee cups can be returned at all outlets and at reverse vending machines in public spaces. This pilot project is being monitored scientifically by the universities of applied sciences Umwelt-Campus Birkenfeld (Hochschule Trier) and Technische Hochschule Mittelhessen on behalf of the Environment Ministries of Hesse and Rhineland-Palatinate over a period from July 2023 to May 2024 in order to gain insights into the usage behavior and acceptance of reusable coffee cups [14].

In this context, the research questions are aimed at gaining insights into the so-called attitude–behavior gap of end consumers regarding the use of reusable packaging on the one hand and the economic and organizational challenges of the participating catering companies and companies involved in the value chain on the other. On the basis of a holistic analysis of the influencing factors within the value chain of the various stakeholders, optimization potentials are to be identified in order to achieve an increase in the reusability rate. At the same time, it is to be examined whether the introduction of an overarching take-back infrastructure can have an impact on acceptance and usage behavior.

2. Materials and Methods

To answer the questions posed and to obtain empirical data, the method of quantitative social research with the implementation of standardized surveys was chosen. Due to the holistic view of the value chain, the stakeholders involved were surveyed: out-of-home catering, consumers and reverse logistics and supply chain management (e.g., logistics companies, cleaning companies, IT service providers and system providers). Consumers were surveyed online and on site in the city centers of Mainz and Wiesbaden at two different measurement points, five months apart, to obtain a comprehensive picture of their usage behavior and acceptance of reusable packaging. The second measurement point was during the test phase of the pilot project's overarching return system. This made it possible to investigate its potential influence on acceptance and usage behavior. The companies were surveyed online to capture their perspectives and experiences with reusable packaging.

Figure 1 illustrates the three stakeholder groups in the value chain of the reusable system under consideration: Stakeholder group 01: Takeaway catering; Stakeholder group 02: Consumer; Stakeholder group 03 = Reverse logistics and supply chain management.

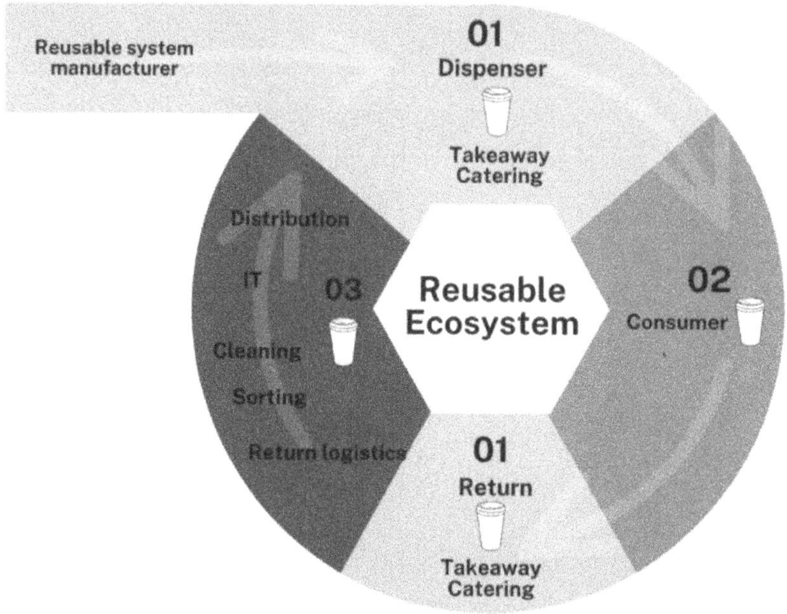

Figure 1. Stakeholders in scientific study 'Mehrweg Modell Stadt'.

2.1. Survey of Companies

The survey of the companies included takeaway catering companies as well as other companies from the areas of logistics, IT service providers, cleaning services and system manufacturers. The questionnaire was developed on the basis of operationalized concepts and in close cooperation with the initiators of the pilot project "Mehrweg Modell Stadt". It contains questions on the current use of reusable containers in the companies surveyed, as well as on obstacles and possible solutions for the use of reusable systems. The data were collected online between mid-January and mid-March 2024. The questionnaire was sent by e-mail to companies in the Mainz/Wiesbaden area and to various associations with the request to forward it. The mailing list consisted of approximately 300 addresses. Despite these efforts, the response rate was very low at 4% for take-away catering companies (12 participating take-away companies) and 4.7% for further companies (14 participating further companies). The data were analyzed using the SPSS statistical program [15]. Due to the limited sample size (12 individuals in the takeaway company group and 14 individuals in the further company group), inferential statistical analyses were not a viable option as they have low statistical power and are susceptible to bias and error probabilities. In contrast, descriptive analyses offered a comprehensive overview of the data, circumventing the methodological limitations previously discussed.

2.2. Consumer Survey

Standardized surveys were used to determine the level of consumer acceptance of reusable food and beverages, identify barriers and challenges to the use of reusable food and beverages, and analyze potential demographic differences. The procedure, implementation and evaluation of the consumer survey were based on the methodology of Mayer [16].

The surveys conducted in the cities of Mainz and Wiesbaden required a random sample of 385 each (simple random sample: n = 385). The measurement model was developed based on the literature review. It assumed that the acceptance of reusable packaging for takeaway food and beverages is influenced by perceived benefits, barriers and context. The theoretical terms were broken down dimensionally and operationalized to map the respective benefit, barrier, and context dimensions. Figure 2 illustrates the operationalization of the three dimensions.

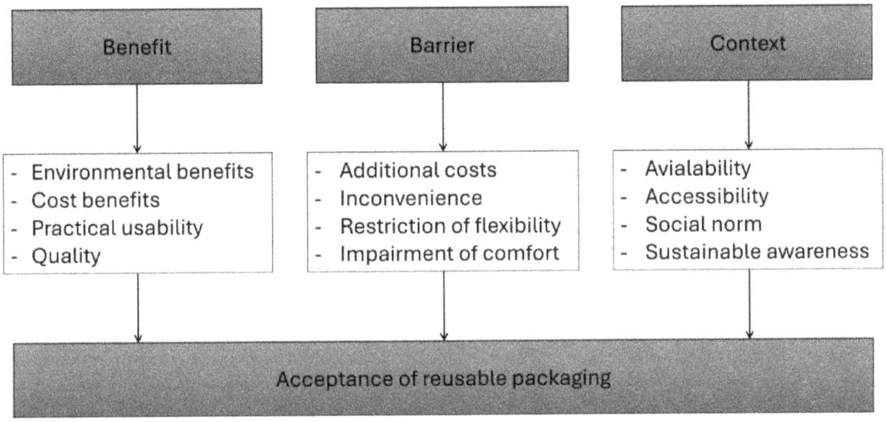

Figure 2. Operationalization of the theoretical concepts of benefit, barrier and context.

As can be seen from the figure, the perceived benefits relate to the environmental benefits, cost benefits and practicality that reusable takeaway can have over disposable. However, there are also barriers to the adoption of reusable takeaway, such as possible additional costs, perceived inconveniences, and restrictions on flexibility and convenience. The context dimension, which includes the availability and accessibility of reusable takeaway, as well as social norms, also plays a role.

The following hypotheses were derived from this assumption:

- H_1: The perceived benefit of reusable packaging has a significant positive influence on the acceptance of reusable packaging in out-of-home gastronomy.
- H_2: The perceived barriers of reusable packaging have a significant negative influence on the acceptance of reusable packaging in out-of-home gastronomy.
- H_3: The higher the approval of the context dimension, the higher the acceptance of reusable packaging in out-of-home catering.
- H_4: The higher the sustainability awareness, the higher the acceptance.

After a pre-test to ensure comprehensibility, a questionnaire was designed based on the operationalized terms and slightly modified. Data collection occurred on two campaign days in August 2023 in the cities of Mainz (14 August 2023) and Wiesbaden (17 August 2023), as well as online from 14 August to 30 September 2023. A second survey of end consumers will be conducted at a second measurement point in the cities of Mainz (13 March 2024) and Wiesbaden (12 March 2024), as well as online from 1 March to 31 March 2024. The first measurement point achieved a sample size of 402 respondents, while the second time point had 99 respondents. The results of these surveys are currently being analyzed using the statistical software SPSS [15].

3. Results

The results of surveys, both of companies and of consumers, are presented below. Only the results of the first measurement point are presented in the consumer survey.

3.1. Survey of Companies

The sample of the business survey comprised a total of 12 takeaway catering companies (TC) and 14 further companies (FC). Figure 3 provides an overview of the sectoral affiliation of the companies; multiple responses were possible for the other companies.

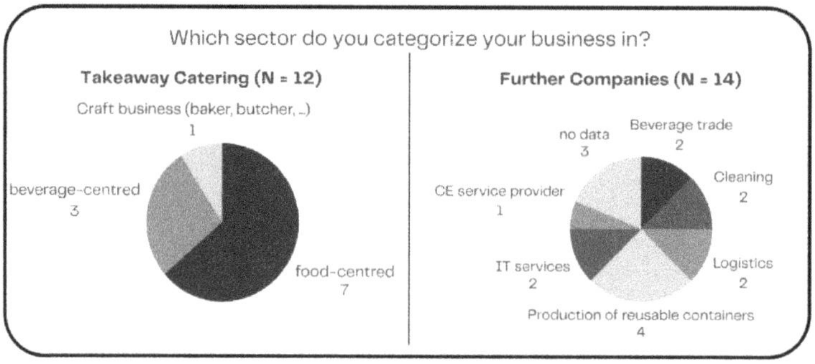

Figure 3. Sector of surveyed companies.

The respondent takeaway catering companies were seven micro-enterprises with fewer than 10 employees, two small enterprises (19–49 employees), one medium enterprise (50 to 250 employees), and one large enterprise with more than 250 employees (Definition based on the European Commission Recommendation (2003/361/EC) according to turnover and employee size classes). One company did not respond to this question. The majority of companies (7) were food-oriented (restaurants, pubs, snack bars, cafés, ice cream parlors [17]), three companies were beverage-oriented (pubs, discotheques and dance halls, bars, entertainment venues, other beverage-oriented gastronomy [17]), and one company was a craft business (bakery or butchery).

The sample of further companies surveyed included two micro, three small, two medium and four large companies. The sectors included beverages (2), cleaning (2), logistics (2), production of reusable containers (4), IT services (2) and circular economy service providers (1). Three companies did not respond.

Not all of the restaurants surveyed were affected by the obligation to offer reusable options or already offer reusable options. Figure 4 shows the distribution of the companies surveyed. The chart also contains information on the development of consumer demand and selected direct statements from the companies surveyed.

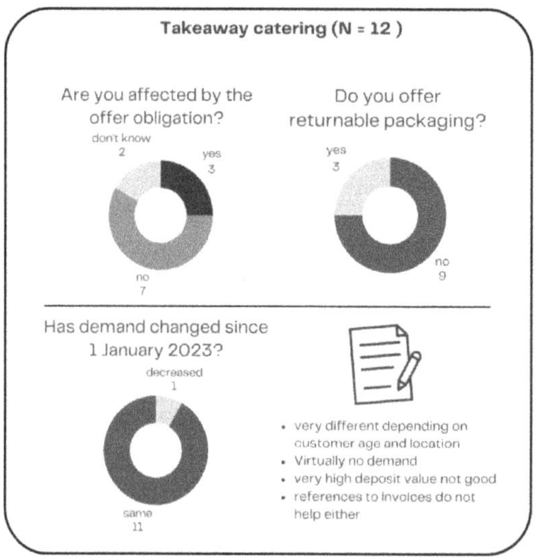

Figure 4. Information on the reusable supply obligation of the companies surveyed and consumer demand.

Six of the restaurants surveyed are legally required to offer reusable options, because they have sufficient square footage and employees. Only three of the respondents were already offering reusable options at the time of the survey in early 2024.

Reasons cited for companies not yet offering reusable packaging included lack of guest acceptance and demand, high cost, not fully developed solutions, complicated implementation, and hygiene concerns.

Consumer demand for reusable packaging is perceived by caterers to be very low and has not changed since the introduction of the reusable packaging obligation on 1 January 2023. One company even reported a decrease in demand. Demand for reusable packaging varies depending on the age of the consumer and the location of the business. However, disposable packaging is preferred by the majority of consumers. The high level of the deposit is also cited as an obstacle

The takeaway restaurants and further companies were asked about the difficulties they perceive in providing reusable options for takeaway food and drinks (see Figure 5).

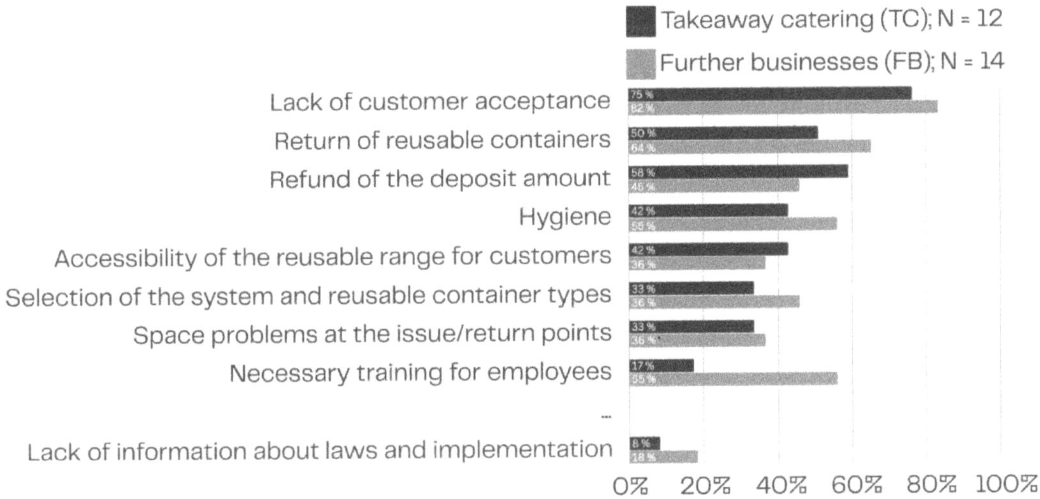

Figure 5. Perceived challenges for companies.

The lack of customer acceptance is the biggest challenge for both stakeholder groups, both for takeaway restaurants and for further companies (TC: 75%; FC: 82%). Returning the deposit (TC: 58%; FC: 45%) and reusable containers (TC: 50%; FC: 64%) are also seen as a significant hurdle. Almost half of the total sample expressed hygiene concerns (TC: 42%; FC: 55%). Around a third of respondents saw challenges in the accessibility of reusable containers for customers (TC: 42%; FC: 36%), the selection of suitable containers for drinks and food (TC: 33%; FC: 45%), as well as the space available at delivery and return points (TC: 33%; FC: 36%). There were slight differences of opinion between the companies surveyed when it came to staff training. The further companies saw a greater need for action here than the catering companies (TC: 17%; FC: 55%). In contrast, a lack of information was perceived as a problem by very few companies (TC: 8%; FC: 18%).

The views of the two stakeholder groups varied considerably with regard to the challenges associated with offering a reusable system to customers (Figure 6).

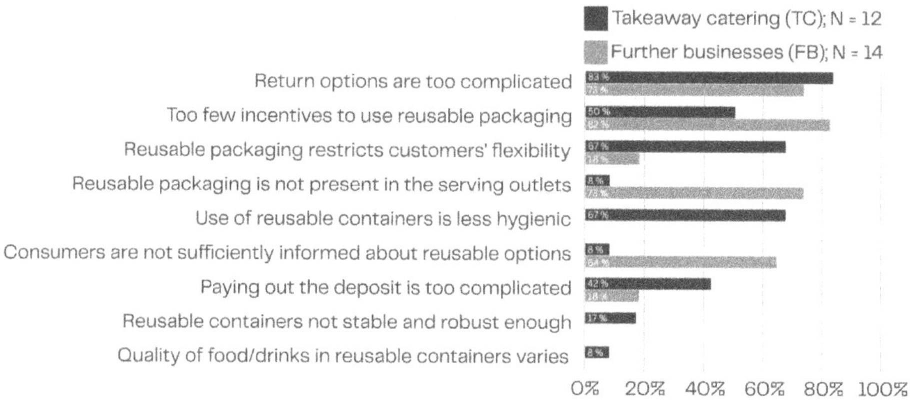

Figure 6. Perceived challenges for customers.

There was an overall consensus that returning reusable containers is very inconvenient. This view was shared by 83% of takeaway restaurants and 73% of other companies.

From this point on, the opinions of the two groups diverged. The takeaway catering companies identified the greatest challenges as being the limited flexibility of customers (67%) and the hygiene of reusable containers (67%). In their opinion, there were too few incentives for the use of reusable containers (50%), e.g., through discounts or bonus points. The selection and deposit systems were also criticized by 42% as being too complicated.

Opinions on the information and presentation of reusable options varied widely. The out-of-home gastronomy sector considered neither the presentation of reusable containers nor the lack of information about them (8%) to be a problem. An 82% share of respondents from further companies, on the other hand, believed that there were too few incentives to use reusable containers. In addition, 73% of respondents believed that the presentation of reusable containers at the point-of-sale was poor, and 64% believed that customers were not sufficiently informed about the reusable offer. Contrary to the takeaway sector, other companies did not see a problem with the hygiene of reusable containers or the stability or possible loss of quality of food and beverages (0%).

The question as to which aspect is particularly important in terms of offering reusable food in takeaway restaurants yielded the following results (see Figure 7).

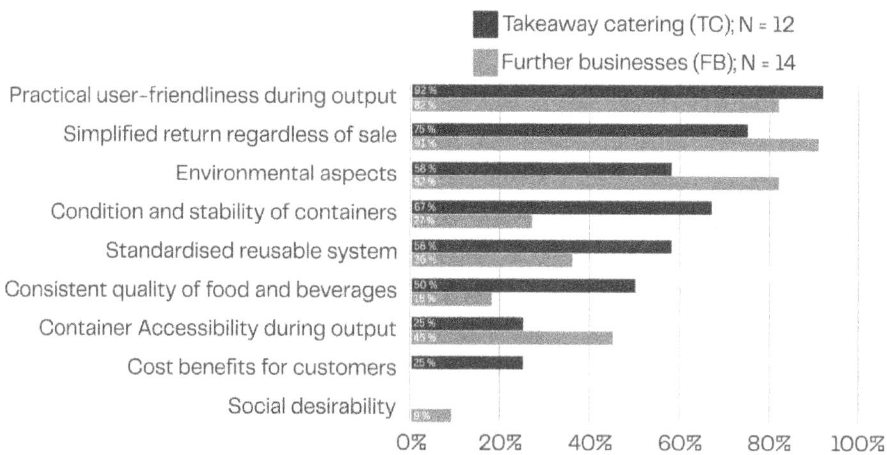

Figure 7. Important aspects when using reusable systems.

The two most important issues for both stakeholder groups were practical usability for serving operations, such as stackability (92% TC and 82% FC), and simplified return regardless of sale (75% for TC and 91% for FC). Environmental aspects such as waste avoidance, climate protection and resource conservation were a high priority for further companies in particular (82%), while only 58% of catering companies considered this aspect to be particularly important. The quality and stability of the containers played an important role for restaurants (67%), while only 27% of further companies emphasized this aspect. A uniform reusable system for all was supported by 58% of restaurants and 36% of further companies. Consistent perceived quality was important for 50% of restaurants and 18% of further companies. Social desirability, on the other hand, played little to no role for the respondents.

3.2. Consumer Survey

Of 402 people surveyed, 59.5% stated that they were female, 32.3% male and 2.2% that they were diverse/non-binary (6% not specified). The largest age group was the 25–34 age group with 20.4%, followed by the 55–64 age group with 18.7%. The group of 14–18-year-olds was the least represented, with 4.7%. Of the respondents, 2.2% did not wish to state their age. In addition to determining the benefit, context and barrier dimensions presented, the correspondence between the stated sustainability awareness and acceptance of reusable packaging among consumers was analyzed. At the first measurement time, 88% of respondents had a high level of sustainability awareness, but at the same time had a low to medium level of acceptance. Just under 25% of respondents had a high level of sustainability awareness and a low level of acceptance, whereas 30% had a high level of sustainability awareness with a medium level of acceptance, and only 32% of respondents had a high level of sustainability awareness with a high level of acceptance.

For consumers, this illustrates the discrepancy between attitude and actual action. The formulated hypotheses H1, H3 and H4 could not be verified, while H2 could be accepted. It can be concluded that although the respondents had a high level of sustainability awareness, a high context dimension, and a positive evaluation of the benefits of reusable packaging, the perceived barriers stood in the way of using reusable packaging or using it more frequently. At measurement point 1, the economic barrier was perceived as an obstacle. According to this, 60% would use reusable packaging more frequently if it were cheaper than the disposable alternative. In addition, 44% felt that returning reusable packaging was inconvenient and limited their flexibility.

The results highlight the need to make reusable packaging more economically advantageous for the end consumer. For example, rebates or a bonus program for the use of reusable packaging could be considered. However, an even greater effect can be expected if disposable packaging is taxed according to the Tübingen model, in which caterers have to pay a packaging tax of EUR 0.50 per item if disposable packaging is used for takeaway food and beverages [18].

The removal of return barriers for the final consumer is another important step towards increasing the reusability rate and thus towards the resource-conserving recycling of packaging. Figure 8 also summarizes the perceived benefits of reusable packaging for consumers.

Although 88% of respondents had a high level of sustainability awareness, they had only a low to medium level of acceptance, once again highlighting the discrepancy between attitudes and actual behavior. Just under half of respondents also saw waste reduction as an advantage of using reusable packaging over disposable packaging. In addition, 24% of respondents considered reusable containers to be more stable than disposable ones, and 16% of consumers surveyed also considered reusable containers to be more leak-proof.

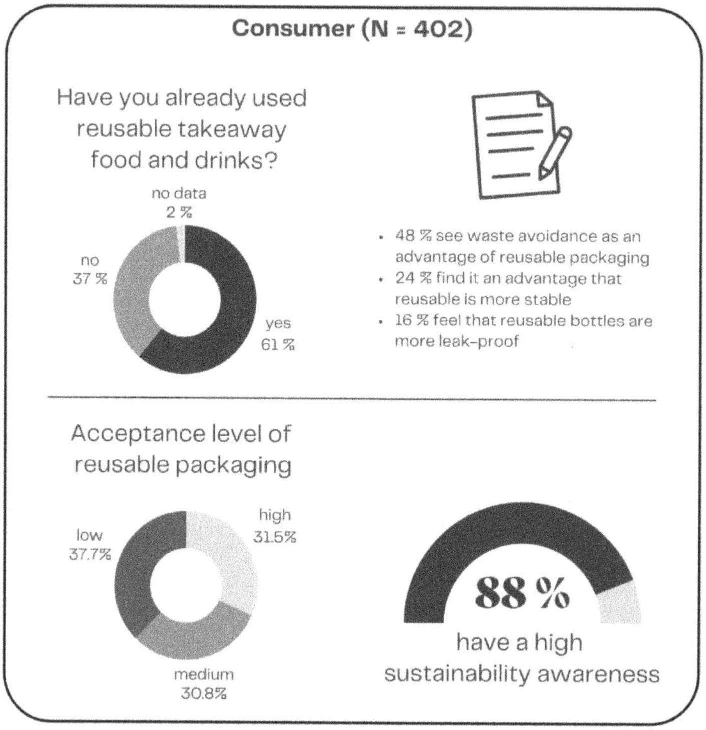

Figure 8. Consumer survey results.

4. Discussion

When reflecting on the quantitative surveys of consumers and companies carried out as part of the Reusable City Model project, it is important to note several critical points that may influence the interpretation of the results. Firstly, the participation of companies in the survey was low despite a large distribution list. This could have affected the representativeness of the results and raises questions about their generalizability. It is worth considering whether a different methodology, such as personal surveys or interviews, could lead to higher participation and more meaningful results. Additionally, while this study focused on urban areas compared with services and user behaviors in rural areas, future studies should aim for broader geographical coverage to obtain a more comprehensive understanding of the situation. It is important to consider these limitations when interpreting the study's results.

Nevertheless, the results are largely in line with the studies cited in the introduction: Companies face significant challenges due to low customer acceptance and demand for reusable packaging, despite providing sufficient information about it. Additionally, return systems for reusables are often considered too cumbersome, leading to logistical challenges for establishments offering them and a lack of flexibility for customers. Many catering companies prefer disposable containers due to hygiene concerns about reusable ones.

According to the latest figures from the German Hotel and Restaurant Association e. V. (DEHOGA), it is unsurprising that companies are reacting cautiously to the new task of introducing and offering reusable packaging. The catering industry has experienced significant sales losses in recent years, with the number of taxable companies in the food-intensive sector decreasing by 11.7% between 2019 and 2021, and by as much as 31.4% in the beverage-intensive sector. In general, businesses are experiencing a decline in guest numbers due to a drop in consumption, with 64.5% reporting subdued expectations. This decline is expected to intensify, particularly due to the increase in VAT from 7% to 19% effective 1 January 2024, a measure introduced in Germany during the pandemic. Consequently, 90.6% of companies are planning to raise their prices, while 69.4% are reducing or eliminating investments, and 28.1% are even laying off employees. Currently, companies are facing significant challenges due to the increasing costs of food and beverages (83.8%), energy (79.5%), personnel (76.0%), rising bureaucracy (75%), and the acute shortage of employees (65.2%) [19].

In summary, companies may lack the personnel and financial capacity to introduce a reusable concept.

The surveys reveal a classic chicken-and-egg problem regarding the use of reusable cups in the to-go sector. On one hand, food services report low demand for reusable options from customers, which has prompted them to continue offering mainly disposable cups. However, customers have expressed a desire for more reusable options and claim that companies do not advertise them enough. This dilemma has resulted in a low reusable cup rate, accounting for less than 10% of the total amount of takeaway food and drink packaging sold.

Companies and customers appear to be in a holding pattern, each waiting for the other to take the initiative. Although both sides are generally willing to use or offer reusable packaging, they are prevented from taking action themselves due to perceived passivity of the other side.

This paradox highlights a discrepancy between consumers' attitudes and their actual behavior when it comes to using reusable packaging, a conclusion that is also reflected by Klöpper and colleagues [9]. The survey results indicate a discrepancy between theory and practice despite positive attitudes towards reusable packaging. Therefore, it is necessary to intensify efforts to promote reusable systems and involve all stakeholders, including politicians, companies, and consumers. Further research is needed to identify specific barriers and opportunities that influence the implementation and use of reusable systems. Examining international projects, such as those in Switzerland [20] or in Aarhus, Denmark [21], where reusable systems are being tested on a large scale, can provide valuable insights. To develop effective strategies for increasing the use of reusable systems and contributing to sustainability, it is essential to have a comprehensive understanding of these factors.

Overall, reusable systems have the potential to be an environmentally friendly solution. However, for efficient and environmentally friendly adaptation, it is crucial to consider and optimize the critical influencing factors and system parameters. According to Kleinhückelkotten and colleagues [8], high circulation rates of reusable systems are decisive, with just 10 to 15 cycles being sufficient for a positive climate balance. However, further research is needed to determine the factors that can encourage all stakeholders to make greater use of reusable systems and increase the number of rounds.

5. Conclusions

A division of responsibility could be established among politicians, companies, and consumers. Politicians should create further clear legal regulations, eliminate information deficits, and introduce controls and sanctions to ensure compliance with the obligation to offer reusable packaging. Targeted information and awareness-raising campaigns aimed at both restaurateurs and customers could help to break up the paradox. These campaigns could highlight the benefits of reusable cups while providing practical tips and guidance on how to use and offer them. Financial support for companies offering reusable options could increase the use of reusable cups.

Restaurateurs should actively promote the reusable packaging to make customers aware of this sustainable alternative. Incentive systems, such as discounts, could encourage customers to use reusable packaging instead of disposable ones. A neutral and organized structure for information and organization could facilitate the joint collection, cleaning, and return of reusable packaging. This, in turn, could increase the rate of reusable packaging in out-of-home catering. By overcoming logistical challenges and promoting the use of reusable packaging, a significant increase in the rate of reusable packaging could be achieved.

Finally, customers have a crucial role in promoting sustainable practices in out-of-home catering by actively demanding reusable solutions.

The results, findings and conclusions of this study will be used in the future to formulate specific requests to policy makers. The main objective is to further promote and support reusable packaging systems. At the same time, these policy demands are intended to adapt and improve the regulatory framework to further facilitate and increase the use of reusable packaging.

Author Contributions: Conceptualization, K.T., A.M. and S.H. (Stefanie Hillesheim); methodology, K.T., A.M. and S.H. (Stefanie Hillesheim); validation, K.T., A.M. and S.H. (Stefanie Hillesheim); investigation, K.T, A.M. and S.H. (Stefanie Hillesheim); resources, K.T, A.M. and S.H. (Stefanie Hillesheim); writing—original draft preparation, K.T., A.M. and S.H. (Stefanie Hillesheim); writing—review and editing, H.R. and S.H. (Susanne Hartard); visualization, K.T., A.M. and S.H. (Stefanie Hillesheim); project administration, K.T., A.M., S.H. (Stefanie Hillesheim), H.R. and S.H. (Susanne Hartard); funding acquisition, H.R. and S.H. (Susanne Hartard). All authors have read and agreed to the published version of the manuscript.

Funding: This study was funded by the Rhineland-Palatinate Ministry for Climate Protection, Environment, Energy and Mobility and the Hessian Ministry for Agriculture and Environment, Viticulture, Forestry, Hunting and Homeland.

Institutional Review Board Statement: Not applicable.

Informed Consent Statement: Informed consent was obtained from all subjects involved in the study.

Data Availability Statement: The raw data supporting the conclusions of this article will be made available by the authors on request.

Acknowledgments: We would like to thank the initiators of the pilot project 'Mehrweg Modell Stadt', the Reusable to Go initiative, for their support. Information on the pilot project can be found at the following website: www.mehrwegstadt.de (accessed on 11 April 2024) [14].

Conflicts of Interest: The authors declare that they are not aware of any competing financial interests or personal relationships that could influence the work in this paper.

References

1. Istel, K.; Detloff, K. Disposable Tableware and To-Go Packaging: Waste Generation in Germany 1994 to 2017. NABU (Naturschutzbund Deutschland) e.V. 2018. Available online: https://www.nabu.de/imperia/md/content/nabude/abfallpolitik/2018_nabu_broschuere_einweggeschirr_to-go.pdf (accessed on 26 March 2024).
2. German Federal Government, German Packaging Act. Available online: https://www.gesetze-im-internet.de/verpackg/VerpackG.pdf (accessed on 6 April 2024).
3. Council of the European Union, Proposal for a Regulation of The European Parliament and of the Council on Packaging and Packaging Waste, Amending Regulation (EU) 2019/1020 and Directive (EU) 2019/904, and repealing Directive 94/62/EC, 15 03 2024. Available online: https://data.consilium.europa.eu/doc/document/ST-7859-2024-INIT/en/pdf (accessed on 9 April 2024).
4. The European Green Deal. Available online: https://ec.europa.eu/stories/european-green-deal/ (accessed on 3 June 2024).
5. Circular Economy Act [KrWG]. 2012. Available online: https://www.bmuv.de/fileadmin/Daten_BMU/Download_PDF/Abfallwirtschaft/kreislaufwirtschaftsgesetz_en_bf.pdf (accessed on 4 June 2024).
6. The Packaging Act. Available online: https://www.verpackungsgesetz.com/themen/die-neue-europaeische-verpackungs-verordnung-eu-verpackv-2025/ (accessed on 11 April 2024).
7. Schüler, K.; Wedekind, N.; Burger, A. *Reusable Food in German Gastronomy: Status Quo, Challenges and Potentials*, 2023rd ed.; WWF: Berlin, Germany, 2023. Available online: https://nbn-resolving.org/urn:nbn:de:101:1-2023020216540582897799 (accessed on 11 April 2024).
8. Kleinhückelkotten, S.; Behrendt, D.; Neitzke, H.P. Reusable Food in Takeaway Gastronomy: Basic Study on the Project 'Climate Protection is(s)t Reusable'. 2022. Available online: https://esseninmehrweg.de/wp-content/uploads/2022/09/Studie_Mehrweg-Takeaway-Gastronomie_2022.pdf (accessed on 26 March 2024).
9. Klöpper, L.; Nguyen, U.; Schüler, K.; Wedekind, N.; Wilhelm, J.; Griestop, L.; Kraas, C. *Reusable Packaging in German Gastronomy: A Reality Check One Year after the Introduction of the Reusable Packaging Obligation*; WWF: Berlin, Germany, 2024.

10. Elisabeth, S.; Alexander, P.; Klara, W.; Janne, W. (Use of (Reusable) Containers for Takeaway Food. In *Results of a Representative Survey*; Zentrum Technik und Gesellschaft, TU: Berlin, Germany, 2024. Available online: https://pur-precycling.de/wp-content/uploads/2024/05/Bericht_Mehrweg_Repraesentativbefragung_PuR.pdf (accessed on 3 May 2024).
11. Jiang, X.; Dong, M.; He, Y.; Shen, J.; Jing, W.; Yang, N.; Guo, X. Research on the Design of and Preference for Collection Modes of Reusable Takeaway Containers to Promote Sustainable Consumption. *Int. J. Environ. Res. Public Health* **2020**, *17*, 4764. [CrossRef] [PubMed]
12. Mich, A. Between attitude and behavior: Attitude-Behavior-Gap in reusable use in takeaway gastronomy. In Proceedings of the 13th Scientific Congress on Circular and Resource Economy, Wien, Austria, 15–16 February 2024; pp. 355–358.
13. Böckel, A.; Pietzke, A.; Wilts, H.; Seyring, N. Piloting of Infrastructure for the Return of Reusable Packaging. Handout on Impact Measurement. March 2024. Available online: https://www.mehrwegverband.de/wp-content/uploads/2024/05/2304 15-Handreichung-Pilotierung-von-Infrastruktur-zur-Ruecknahme-von-Mehrwegverpackungen_compressed.pdf (accessed on 2 June 2024).
14. Pilot Project 'Reusable City Model'. Available online: www.mehrwegstadt.de (accessed on 26 March 2024).
15. IBM. SPSS Statistics. (Version 29.0.2.0). 2024. Available online: https://www.ibm.com/products/spss-statistics (accessed on 24 March 2024).
16. Mayer, H.O. Interview and written survey. In *Fundamentals and Methods of Empirical Social Research*, 6th ed.; Oldenbourg Wissenschaftsverlag: Munich, Germany, 2018.
17. Packaging Tax of the City of Tübingen. Available online: https://www.tuebingen.de/33361.html (accessed on 10 April 2024).
18. DEHOGA. Hospitality Industry 2023. Balance Sheet & Outlook. Results of the DEHOGA Survey. 2023. Available online: https://www.dehoga-bundesverband.de/fileadmin/Startseite/06_Presse/DEHOGA-Booklet_Gastgewerbe_2023_-_Bilanz_und_Ausblick.pdf (accessed on 2 April 2024).
19. Federal Statistical Office. Classification of Economic Activities: With Explanatory Notes. Federal Statistical Office. 2008. Available online: https://www.destatis.de/DE/Methoden/Klassifikationen/Gueter-Wirtschaftsklassifikationen/Downloads/klassifikation-wz-2008-3100100089004-aktuell.pdf (accessed on 25 March 2024).
20. Kooky. Available online: https://www.kooky2go.com/en/home (accessed on 11 April 2024).
21. De Lorenzo, D. Danish City First To Test Return System For Reusable Coffee Cups. Forbes. Available online: https://www.forbes.com/sites/danieladelorenzo/2024/01/17/danish-city-first-to-tests-return-system-for-disposible-coffee-cups (accessed on 11 April 2024).

sustainability

MDPI

Article

Environmental Impact and Sustainability of Bioplastic Production from Food Waste

Katerina Synani [1], Konstadinos Abeliotis [2,*], Kelly Velonia [3], Angeliki Maragkaki [4], Thrassyvoulos Manios [4] and Katia Lasaridi [1]

[1] Department of Geography, Harokopio University of Athens, 17676 Kallithea, Greece; synani@hua.gr (K.S.); klasaridi@hua.gr (K.L.)
[2] Department of Economics and Sustainable Development, Harokopio University of Athens, 17676 Kallithea, Greece
[3] Department of Materials Science and Engineering, University of Crete, 70013 Heraklion, Greece; velonia@materials.uoc.gr
[4] Department of Agriculture, Hellenic Mediterranean University, 71410 Heraklion, Greece; amaragkaki@hmu.gr (A.M.); tmanios@hmu.gr (T.M.)
* Correspondence: kabeli@hua.gr

Abstract: Plastic generation exacerbates the challenge of solid waste management. Moreover, plastics emit substantial amounts of microplastics, which infiltrate the environment and food chain, posing significant environmental risks. Compounded by their production from fossil fuels, such as crude oil and natural gas, plastics present a formidable environmental concern. As a result, bioplastics are an attractive alternative to fossil-based plastics since they use renewable energy sources, aim to alleviate worries about reliance on fossil fuels, and are biodegradable, further enhancing their environmental appeal. Along similar lines, the utilization of food waste to produce bioplastics is attracting international interest. The current study presents the results of a life cycle assessment conducted on bioplastic production from food waste, carried out in a pilot-scale reactor located in Greece. The objective was to ascertain the comparative sustainability of recovering food waste for bioplastic production versus utilizing cultivable raw materials. To this end, an equivalent amount of polylactic acid was produced from corn. The findings revealed a reduction in climate change, eutrophication, and ecotoxicity as a result of the study process. Despite these environmental benefits, the study highlighted that energy consumption throughout the process poses a significant environmental burden. This aspect calls for attention and modification to enhance the entire sustainability of the process.

Keywords: LCA; sustainability; PLA; circular economy; impacts; energy

1. Introduction

The escalation of concerns surrounding solid waste management correlates with the global surge in population and affluence [1], further fueled by the escalating demand for plastic products and packaging. In 2021, global plastic production increased by 4% to exceed 390.7 million tonnes, underscoring the persistent and robust demand for plastics. During the same period, the EU27 + 3 utilized 5.5 million tonnes of post-consumer recycled plastics, marking a 20% increase over 2018, and constituting around 10% of recycled content in plastics conversion [2]. Forecasts indicate that plastic consumption is expected to quadruple by 2060 [3].

Conversely, conventional plastic waste disposal methods remain insufficient, particularly in developing countries [4,5]. Plastic manufacturing accounts for 4% of total global CO_2 emissions. Non-biodegradable plastics endanger ecosystems and human health [6], with their degradation producing micro- and nano-plastics [7], impacting ecosystem quality, human health [8], and climate change [9].

The linear consumption methods prevalent in industrialized countries, termed 'throw-away' [10] and 'take-make-dispose' [11], have led to the global proliferation of plastic waste [12]. This issue has been exacerbated by supply chain globalization [13], mass consumption, and the COVID-19 epidemic [14].

The European Commission [15] has developed a circular economy plan for plastics to optimize resource utilization and extend product lifetimes through multiple use cycles [16]. This concept combines economic approaches with biological aspects, offering the potential for sustainable change [17]. Bioplastics, also known as biomass-based biopolymers, are a viable alternative to petroplastics since they are biodegradable and compostable [18], effectively closing the carbon cycle [19]. Despite representing only 1.5% of the plastics market, bioplastics are estimated to increase their market share [20]. Their growing market share can contribute to reducing reliance on fossil fuels [21], and facilitate the transition to a circular economy [22]. However, this will require a steady supply of bioplastic raw materials [23], and the implementation of appropriate recycling and recovery technologies [24]. Bio-waste, including food waste, holds potential as a carbon feedstock for bio-based commodities and biomaterials [16], which reduces the cost of bioplastic production [25].

The diversion of food waste from landfills to innovative processing mechanisms for converting waste into valuable products or raw materials has seen significant progress in recent years [25]. Furthermore, with the depletion of fossil resources and the challenges posed by climate change, there is an urgent need for a transition to renewable and low-carbon production methods [26].

This has led to the production of bioplastics like polylactic acid (PLA), polyhydroxyalkanoate (PHA), and bio-polypropylene [27], which are derived from food or biomass resources like starch, corn, sugarcane, and lignocellulosic components. These bioplastics play a vital role in transitioning towards a circular economy [28], reducing mineral extraction and lowering carbon footprints, potentially mitigating end-of-life environmental burdens [16].

The raw materials for bioplastics are categorized into three generations based on their state of development. First-generation feedstocks typically include carbohydrate-rich plants suitable for human or animal consumption, such as corn, which is abundant in starch and readily available [29]. First-generation raw materials are based on intensive agriculture, compete with food and feed, and have detrimental effects on humans and the environment [27]. Consequently, alternative raw materials have been tested. Second-generation raw materials comprise basic ingredients unsuitable for food or feed. These may include non-edible crops (e.g., cellulose) or byproducts of first-generation feedstocks, such as corn "hulls" or sugarcane molasses. Separately collected food waste, i.e., that has not ended up in the municipal waste stream, is a second-generation feedstock for bioplastics production [30]. The third-generation raw materials are the most innovative, still in the early stages of development. It contains both algal biomass and industrial or municipal waste, which are being transformed into raw materials through improved processes [31].

The utilization of alternative biomass sources holds promise for producing biomass-based commodities with reduced environmental and socioeconomic implications without competing with the food industry [32]. While food waste represents a valuable raw material for bioplastic production [33], it requires pre-treatment to enhance or modify its physicochemical and biological properties. Proper conversion of such wastes into value-added products can prevent significant economic [34] and energy losses [35].

Despite advancements, industrial-scale production of bioplastics still faces challenges that need addressing. Factors like techno-economics, profitability, raw material availability [36], and market demands must be considered for large-scale manufacturing. The environmental benefits of bioplastics are driving their adoption and promotion [37], with bioplastic production from food waste emerging as a sustainable alternative [38]. Private undertakings need to accurately quantify and demonstrate these benefits to avoid misleading claims [39].

A range of studies have explored the conversion of food waste into bioplastics, highlighting the potential for sustainable and environmentally friendly production. Sakai et al. [40] and Yu et al. [41] both propose innovative methods for this conversion. Sakai et al. [40] focus on the production of high-quality poly-L-lactate (PLLA) biodegradable plastics from food waste, while Yu et al. [41] investigate the use of various carbohydrates and malt wastes as carbon sources for bioplastic production. Moreover, Perotto et al. [42] contribute to the field by introducing a water-based process for converting vegetable waste into bioplastic films, which are biodegradable and environmentally friendly. However, there are significant knowledge gaps in this area, particularly in the environmental assessment of bioplastic properties and their impact on food packaging. Kakadellis and Harris [43] emphasize the need to consider the entire life cycle of food packaging and the trade-offs involved. The potential of converting food waste into bioplastic is a promising path for addressing both environmental and waste management challenges [44]. Bagnani et al. [45] specifically demonstrate the feasibility of producing bioplastics from soy waste on an industrial scale. Furthermore, the lack of standardized methodologies for assessing the biodegradability of bio-based polymers, a key aspect of bioplastic assessment, is highlighted [44]. These gaps underscore the need for further research to optimize the conversion of food waste into bioplastic and ensure its environmental sustainability.

The present research delves into the environmental impact assessment of converting food waste into bioplastic, conducted within the framework of the A2UFood Project [46]. The project aimed to reduce food waste by developing bioplastic production from food waste, aligning with the principles of the circular economy [47]. Tools such as life cycle assessment (LCA) provide valuable insights into the environmental repercussions of a material or product across its full life cycle, from raw material acquisition to disposal [32]. Recently, Ali et al. [16] reviewed several LCA studies that compare the environmental impacts of specific bioplastics to those of petrochemical plastics. LCA studies on bioplastics demonstrate significant differences in selected environmental impacts [30].

The novelty of the study lies in the fact that we employed life cycle assessment (LCA) to examine the environmental impact of producing bioplastic from food waste in a pilot scale unit for the first time in Greece, in addition to the environmental benefits of diverting the same quantity of food waste from landfills. Following the ISO 14040 family standards, LCA investigates environmental consequences across a product's value chain, including goal and scope definition, inventory analysis, impact assessment, and interpretation. Depending on the assessment's aims, LCA might include all system inputs and outputs, such as materials, resources, energy, and emissions [48].

2. Methodology

2.1. Goal and Scope, System Boundaries

The goal of the study is to assess the environmental impact of converting food waste into bioplastic. This assessment is based on the quantity of PLA produced using food waste as feedstock, specifically from the 11 batch operating cycles carried out during the A2UFood Project. In each operating cycle, 65 kg of sorted food waste were utilized, obtained after manual sorting from 125 kg of untreated food waste, resulting in the production of 9 kg of PLA. Thus, the functional unit is defined as the production of 99 kg of PLA. The graphical representation of the system boundaries is shown in Figure 1.

The scope focuses solely on the environmental costs directly associated with the conversion of food waste into bioplastics and does not consider the environmental impacts of food production. However, the conversion process effectively diverts food waste from landfills. Therefore, in this particular model, the diversion of food waste from landfills is regarded as an environmental benefit. This positive impact is factored into the study by subtracting the associated variables from the overall impacts demonstrated by the model. In addition, the assessment excludes the construction of the pilot plant or the operational energy requirements related to cleaning or maintenance of the facilities.

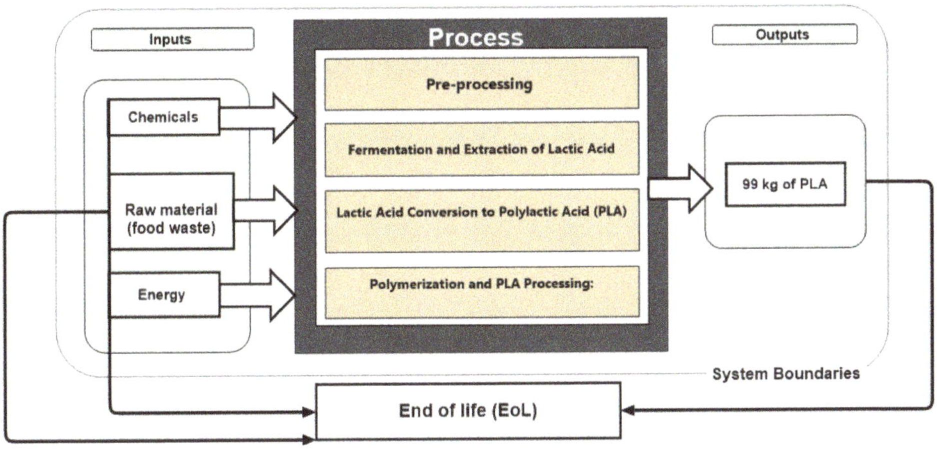

Figure 1. Flow chart with the boundaries of functional units for the baseline scenario.

2.2. Life Cycle Inventory

2.2.1. Raw Material Acquisition

The study assessed the selection, extraction, conversion, and processing steps involved in the PLA production process from food waste [49]. The food waste used in this study was gathered from the Hellenic Mediterranean University's student restaurant. The food waste composition was determined to be 80% raw-fresh foods (vegetables), 10% fruits, and 10% salads (on a wet basis) [50]. The assessment also includes the electricity consumption and the chemicals used in the conversion (Appendix A, Table A1).

2.2.2. Pilot Unit Description

Pretreatment and processing, which encompassed receiving and processing equipment such as conveyor belts, conveyor screws, shredders, pulpers, and monopumps, were thoroughly examined in the present study. Additionally, the energy demands for fermentation processes, including bioreactors, filter systems, heating systems, and low-temperature burners, were recorded. Polymerization, involving polymerization devices, was also covered. Figure 2 depicts the flow chart for the whole process of producing bioplastic in the pilot unit.

Energy consumption was a major consideration in the current assessment. To evaluate the energy requirements for the overall PLA production, a complete inventory of processes requiring power was performed. Energy requirements for the building's fans were also considered, as they were necessary to maintain low ambient temperatures during the production process. The analysis of production energy requirements highlights the significant energy consumption across various operational equipment and systems within an industrial facility. Specifically, the receiving and processing equipment, including conveyor belts, conveyor screws, shredders, pulpers, and monopumps, require 411.8 kWh for 2 h of operation in every 8 h shift across 11 batches with an 85% operational rate. Bioreactor R1 operates continuously for 4 days per batch, consuming 449 kWh, while bioreactor R2, with 4 h of operation over 1 day per batch, uses 18.7 kWh. The filter system, operating 2 h per day for 1 day per batch, consumes 10.3 kWh. The low-temperature cooling system for R1 requires a substantial 3430.5 kWh over 3.5 days per batch. Burners for R1 and R2 consume 3.9 kWh and 1.3 kWh, respectively. Polymerization devices, operating for 6 h per day over 1 day per batch, consume 126.2 kWh. Building air circulation fans, running 8 h daily for 55 days, require 74.8 kWh. Overall, the total energy needs amount to 4451.7 kWh (Appendix A, Table A2).

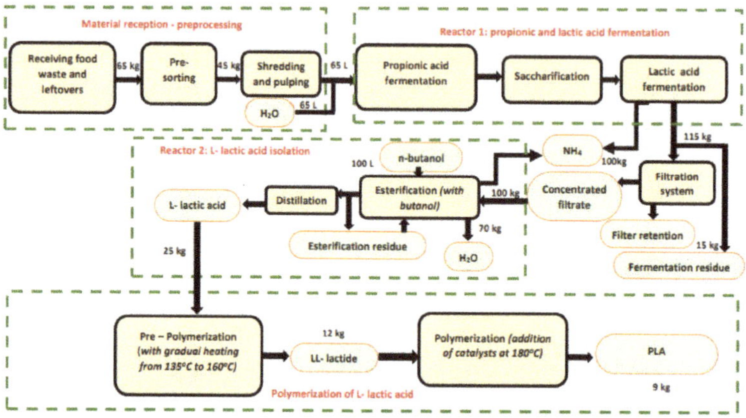

Figure 2. Process flowchart for PLA production in the A2UFood pilot scale unit.

Regarding the energy sources, the electricity mix in Greece in 2020, according to the Hellenic Ministry of Energy and Environment, consisted of oil at 7.22%, natural gas at 43.32%, lignite at 9.48%, renewable energy sources at 28.85%, nuclear energy at 3.21%, and other sources at 7.92%. To ensure a representative evaluation, a custom record for the electricity mix was developed, as no ready-made library specific to Greece was available.

2.2.3. Avoided Landfilling

The present study adopts a gate-to-gate LCA approach [51], focusing solely on the pilot-scale unit for the production of PLA and excluding end-of-life processing and food waste considerations. Biodegradable plastics are an interesting alternative for major questions in materials engineering. Nevertheless, PLA microplastics can pose a threat under specific conditions [52]. End-of-life impacts of bioplastics include the release of microplastics into the natural environment [52]. However, the present study excludes the environmental impacts associated with the disposal or end-of-life processing of the PLA product because it bases its impact assessment on ready-made models, as discussed in the impact assessment Section 2.3 that follows. These models do not include the impacts of microplastics, either in the natural environment or in human health. This is clearly identified by the authors of the present study as a limitation of the applied LCA methodology that requires further research. Furthermore, the production of food that ended up as food waste was not taken into account because it would have been produced regardless of its usage for PLA production.

This study explores the option of diverting food waste disposal to landfills, which is the primary method of waste disposal in Greece, comprising approximately 78% of the total waste generation [53]. Special attention is given to the reduction in methane emissions, with an emphasis on the kilograms of methane not emitted. By diverting 715 kg of food waste from landfills, a reduction of 26.8 kg in methane emissions was achieved. This reduction is based on the understanding that food waste contributes to 58% of methane emissions emitted into the atmosphere in municipal solid waste landfills [54].

2.3. *Life Cycle Impact Assessment*

SimaPro version 9.5 was used to generate the life cycle assessment model, which adhered to the ISO 14040 series of standards [55] The chosen impact assessment method is the Ecological Footprint method 3.0 (adapted) V1.00/EF 3/excluding infrastructure processes, which incorporates sixteen impact categories with their respective units: Climate change (CC)-[kg CO_2 eq.], Ozone depletion (OD)-[kg CFC11 eq.], Ionizing radiation (IR)-[kBq U-235 eq.], Photochemical ozone formation (POF)-[kg NMVOC eq.], Particulate matter (PM)-[disease inc.], Human toxicity, non-cancer (HT-nc)-[CTUh], Human toxicity, cancer (HT-c)-[CTUh], Acidification (A)-[mol H+ eq.], Eutrophication, freshwater (E-f)-[kg P eq.], Eutrophication, marine (E-m)-[kg N eq.], Eutrophication, terrestrial (E-t)-[mol N eq.], Ecotoxicity, freshwater (ECoX-f)-[CTUe], Land use (LU)-[Pt], Water use (WU)-[m^3 depriv.], Resource use, fossils (RU-f)-[MJ], Resource use, minerals and metals (RU-mm)-[kg Sb eq.]. All of the above impact categories are presented in terms of characterization and normalization results. Characterization in life cycle assessment (LCA) research entails modeling direct consequences for human health and other categories [56]. Furthermore, normalization helps in understanding the relative significance of improvements or reductions achieved in specific impact categories, enabling the evaluation of the effectiveness of mitigation efforts or the environmental benefits gained by avoiding certain materials or processes [57].

2.4. Alternative Feedstock Scenario

PLA is a bioplastic primarily derived from starch raw ingredients [58]. Given that starch is the principal feedstock for PLA manufacturing [27], a precise quantification of the starch content within the food waste sample was conducted. Based on this result and upon a review of pertinent literature, corn emerged as a prominent raw material [59]. Consequently, an alternative scenario was explored, utilizing corn for the production of an equivalent mass of bioplastic. Thus, "the production of 304 kg of corn, which served as the alternative raw material, has been avoided by producing bioplastic from food waste".

This illustrates the twofold nature of prevention: one aspect involves diverting food waste from landfills, while the other aspect entails preventing the production of an equivalent amount of corn. An important step was the detailed analysis of the food waste sample composition, with its distribution based on components representing the general variation of food waste composition. This was followed by matching the sample with the starch content, based on the Woodlands Healing Research Center nutritional guide, and one more study [60]. The analytical amounts of these components are listed in Appendix A, Table A3. The next step was to relate the amount of starch contained in the sample to the equivalent amount of corn. Therefore, the use of an equivalent amount of corn, based on its starch content, represents the alternative scenario for the incoming raw material to be used to produce an equal amount of bioplastic.

2.5. Interpretation

The present study on PLA production from food waste provides valuable insights into environmental implications and critical aspects. The findings can aid in decision-making to optimize production processes and reduce environmental effects. At all phases of the life cycle, data collection, measurement, and analysis are crucial for sustainable PLA production. Opportunities for improvement include optimizing the production procedures, reducing energy consumption, reducing garbage output, and developing alternative waste disposal methods. The findings are presented in the results and discussion modules, which include the assessment, limitations, and some recommendations for addressing environmental problems.

2.6. Sensitivity and Uncertainty Analysis

A sensitivity analysis of Greece's electricity mix, considering the environmental impact data presented earlier, indicates the possible repercussions of adjusting the balance of energy sources. Greece has increasingly turned to renewable energy sources [61], comprising over 30% of the electricity mix [62], alongside considerable contributions from natural gas and lignite. However, as sustainability gains prominence and environmental concerns mount, there is an opportunity to rethink this composition. Shifting away from natural gas and lignite in favor of developing renewable energy sources could decrease greenhouse gas emissions [63] as well as other harmful pollutants, thus contributing to global climate change mitigation [64]. Furthermore, prioritizing contributions from nuclear and hydroelectric sources while phasing out carbon-intensive sources could yield lasting benefits regarding environmental protection and energy security [65]. By reevaluating and optimizing the electricity mix, Greece can move towards a more sustainable and resilient energy future. Specifically, according to the Hellenic Ministry of Energy and Environment [66], the energy mix in Greece in 2023 consisted of natural gas at 35.1%, lignite at 9.5%, renewable energy sources at 40.3%, nuclear energy at 3.4%, and hydroelectric energy at 11.7%.

A systematic method for analyzing and minimizing uncertainty in research findings is essential for increasing their reliability [67]. This assessment focuses on the consistency and variability of the data and assumptions used in the LCA process. As indicated in prior research, changes in food waste composition, processing efficiency, and energy utilization have a significant impact on the interpretation of the LCA results. Furthermore, Hong et al. [68] note that meticulously quantifying these uncertainties helps researchers assess the amount of confidence in their results, identify critical areas that require more precise data, and improve the overall dependability of the study's findings. In this regard, an uncertainty analysis was performed using SimaPro software version 9.5 and the Monte Carlo [69] technique (which is incorporated by default in the software) for the total avoidance product in the alternative feedstock scenario.

3. Results and Discussion

3.1. Characterization and Normalization

Table 1 illustrates the outcomes of the study across the various environmental impact categories. The characterization results for each impact category are distributed among three main components, namely: (i) the contribution of chemical additives, (ii) the contribution of electricity, and (iii) the contribution from the avoided landfilling of FW. Figure 3 presents the relative contribution of each of the three components. The electricity component (shown in red in Figure 2) is the major contributor to various impact categories, including climate change, ionizing radiation, freshwater ecotoxicity, and water use. This indicates the crucial environmental impact of electricity use in the bioplastic production process. The chemical addition (shown in blue in Figure 3) greatly affects ozone depletion, photochemical ozone formation, freshwater ecotoxicity, terrestrial eutrophication, land use, resource use, and mineral–metals resource use. This finding emphasizes the importance of careful selection and the potential decrease in chemical additions to reduce their detrimental impact on the environment. Notably, the averted landfilling of FW (indicated in green) has an environmentally favorable impact, particularly in terms of land use and freshwater ecotoxicity. Our findings are in agreement with Nampoothiri et al. [70], who reported that using leftovers or food waste can help lessen the environmental effect of PLA production. Our characterization results reinforce the view that bioplastic production from food waste emerges as a viable option since it has a positive impact on land use and on the reduction of pollution in freshwater ecosystems.

More specifically, reductions in greenhouse gas emissions, with a decrease of 50.9225 kg CO_2 eq., contribute to climate change mitigation. Furthermore, a decrease in ozone-depleting chemicals by -1.6×10^{-6} kg CFC11 eq. helps in preserving the ozone layer. Reduced emissions of non-methane volatile organic chemicals (−0.10398 kg NMVOC eq.) and particle matter (-2.4×10^{-6} illness inc.) result in improved air quality and public health outcomes. Furthermore, reductions in human toxicity, both non-cancer (-9.6×10^{-7} CTUh) and cancer-related (-2.5×10^{-8} CTUh), as well as acidification (−0.34189 mol H+ eq.), indicate favorable effects on human health and ecosystem preservation. Decreased eutrophication in freshwater (−0.0497 kg P eq.), marine (−0.07274 kg N eq.), and terrestrial (−1.16318 mol N eq.) ecosystems helps preserve ecological balance. The reduction in freshwater ecotoxicity (−8161.21 CTUe) and land use (−21,306.69425 Pt) also suggests positive effects on biodiversity conservation. However, please note that some authors have questioned the overall positive influence of PLA manufacturing on climate change and sustainability, particularly regarding land use [71].

Table 1. Characterization impact assessment results for the production of 99 kg of PLA from FW.

Impact Category (Unit)	Total	Chemical for PLA	Electricity, GR_2021	Avoided FW
CC (kg CO_2 eq.)	2492.15	1109.93	1433.14	−50.93
OD (kg CFC11 eq.)	0.0004	0.0003	0.00013	-1.6×10^{-6}
IR (kBq U-235 eq.)	238.23	74.56	165.79	−2.12967
POF (kg NMVOC eq.)	5.99	3.55	2.5425	−0.10398
PM (disease inc.)	8.86×10^{-5}	4.75×10^{-5}	4.35×10^{-5}	-2.4×10^{-6}
HT-nc (CTUh)	1.87×10^{-5}	1.23×10^{-5}	7.38×10^{-6}	-9.6×10^{-7}
HT-c (CTUh)	5.37×10^{-7}	3.03×10^{-7}	2.58×10^{-7}	-2.5×10^{-8}
A (mol H+ eq.)	14.33	5.845239	8.83	−0.34
E-f (kg P eq.)	2.24	0.38	1.96	−0.05
E m (kg N eq.)	2.01	1.02	1.06	−0.07
E-t (mol N eq.)	18.45	12.41	7.25	−1.16
ECoX-f (CTUe)	31,178.8	35,891.19	3448.83	−8161.2
LU (Pt)	7175.5	28,241.53	240.65	−21306.7
WU (m^3 depriv.)	11,037.2	2063.44	9211.42	−237.7
RU-f (MJ)	18,815.9	19,070.01	1.43	−255.4
RU mm (kg Sb eq.)	0.015	0.02	0	−0.0007

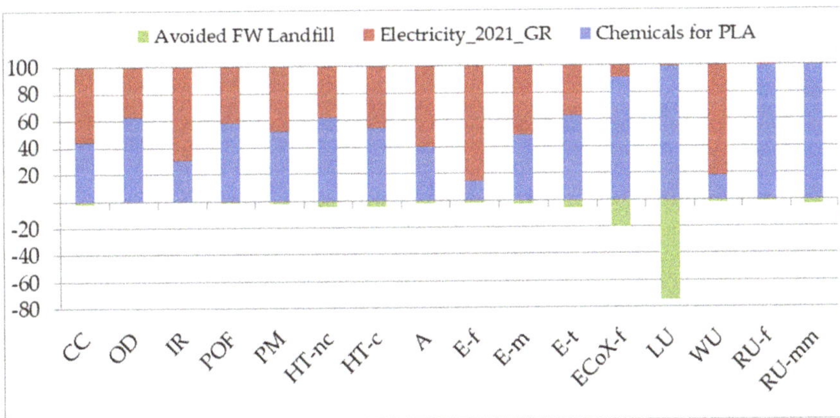

Figure 3. Characterization results for the production of PLA from FW.

Moving on to the normalization phase, presented in Figure 4, our findings indicate that the three most important categories are freshwater eutrophication, water use, and freshwater ecotoxicity. Electricity is the major contributor for the first two categories, while chemicals are mainly responsible for the freshwater ecotoxicity impacts. Electricity generation, particularly from fossil fuels, has a huge environmental impact [72]. To address these challenges, a transition towards cleaner and more sustainable energy sources and improved energy efficiency is of paramount importance. By reducing the reliance on fossil fuels and embracing cleaner alternatives such as wind and solar energy, the negative environmental and health implications may be alleviated. Moreover, in the present pilot-scale case study, several practices and strategies can be implemented to reduce the total energy needs for the specific equipment and systems outlined. For the receiving and processing equipment, replacement of existing motors in conveyor belts, screws, shredders, pulpers, and monopumps with high-efficiency models that use less power is recommended. Then, for bioreactors R1 and R2, improvement of the insulation is recommended to minimize heat loss and maintain optimal temperatures with less energy. Finally, for polymerization devices, replacement of existing equipment with energy-efficient models is recommended, that require less power, and optimizing the polymerization process to reduce the duration of operation without affecting the quality of the final product can reduce energy consumption. By implementing these practices, the pilot-scale unit can reduce its total energy consumption, leading to cost savings and a lower environmental impact.

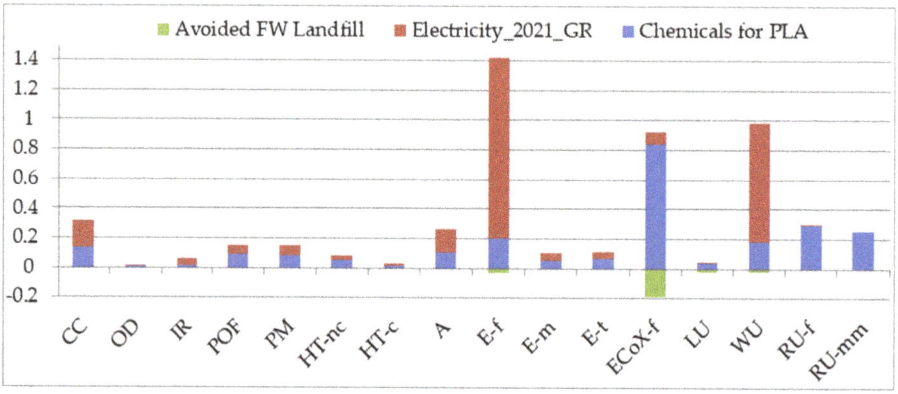

Figure 4. Normalization results for the impact categories for the production of PLA from FW.

3.2. *Alternative Feedstock Scenario*

Again, in this scenario, the characterization results presented in Table 2 are divided into three major components: the contribution of chemical additives, the contribution of electricity, and the contribution from avoiding food waste landfilling and corn not utilized for producing the intended quantity of bioplastic. Figure 5 presents the relative contribution of each of the three components. The electricity component (red in Figure 5) is the major contributor to climate change, ionizing radiation, freshwater ecotoxicity, acidification, and water use. The chemicals (blue color) are the major contributors to ozone depletion, photochemical ozone formation, human toxicity (cancer and non-cancer effects), freshwater ecotoxicity, terrestrial eutrophication, land use, resource use, and mineral–metals resource use. In the present scenario, it is noteworthy the environmentally positive impact, particularly on marine and terrestrial eutrophication, freshwater ecotoxicity, and land and water use, of the prevented landfilling of FW and corn production (green color). First- and second-generation PLA tend to have significant environmental consequences in various areas, including human toxicity, freshwater eutrophication, and marine eutrophication [32], as confirmed by the findings of this study. Additionally, changes in land use can have long-term environmental and ecological consequences [73].

Particularly, the positive savings (as represented by negative values) demonstrated in Figure 5 across multiple categories represent the prevention that contributes to these environmental impacts. The current study's summarized data highlight notable reductions in various environmental impact categories, affirming the benefits of converting food waste into bioplastic feedstock. Specifically, regarding climate change, a total reduction of 182.13 kg CO_2 eq. is observed, primarily attributed to the avoidance of greenhouse gas emissions associated with fossil fuels or their derivatives, amounting to 155.77 kg CO_2 eq. Additionally, biogenic emissions savings of 25.55 kg CO_2 eq. are realized, likely stemming from reduced methane emissions due to diverting food waste from conventional disposal methods. Furthermore, a total reduction of 18,214.4 CTUe in freshwater ecotoxicity is achieved, credited to the avoidance of various organic and inorganic pollutants, as well as metal-based contaminants, thereby safeguarding aquatic habitats. The reduction in acidification by 1.87 mol H+ eq. signifies potential savings from mitigating acidifying agents, which pose threats to ecosystems. Moreover, a decrease in resource usage from fossil fuels by 256.24 MJ indicates reduced reliance on non-renewable energy sources in bioplastic production, contributing to wider sustainability efforts. Finally, the decline in terrestrial eutrophication by 7.24 kg P eq. indicates a favorable impact on nutrient runoff into water bodies, mitigating excessive algae growth and preserving freshwater ecosystems. These findings underscore the environmental benefits of repurposing food waste for bioplastic production.

Table 2. Characterization impact assessment results for the production of 99 kg of PLA for the alternative feedstock scenario for the main impact categories.

Impact Category (Unit)	Total	Chemical for PLA	Electricity GR_2021	Avoided Raw Material
CC (kg CO_2 eq.)	2360.95	1109.94	1433.14	−182.13
OD (kg CFC11 eq.)	0.0004	0.00023	0.00014	-1.4×10^{-5}
IR (kBq U-235 eq.)	225.89	74.56	165.79	−14.463
POF (kg NMVOC eq.)	5.62	3.55	2.54	−0.47
PM (disease inc.)	7.84×10^{-5}	4.75×10^{-5}	4.35×10^{-5}	-1.3×10^{-5}
HT-nc (CTUh)	1.72×10^{-5}	1.23×10^{-5}	7.38×10^{-6}	-2.4×10^{-6}
HT-c (CTUh)	4.97×10^{-7}	3.03×10^{-7}	2.58×10^{-7}	-6.4×10^{-8}
A (mol H+ eq.)	12.81	5.85	8.825883	−1.86
E-f (kg P eq.)	2.20	0.33	1.96	−0.09
E m (kg N eq.)	0.02	1.02	1.06	−2.06
E-t (mol N eq.)	12.42	12.41	7.25	−7.24
ECoX-f (CTUe)	21,125.6	35,891.2	3448.83	−18,214.4
LU (Pt)	6978.47	28,241.5	240.65	−21,503.7
WU (m^3 depriv.)	4327.67	2063.44	9211.42	−6947.19
RU-f (MJ)	18,815.2	19,070.0	1.43	−256.24
RU mm (kg Sb eq.)	0.02	0.02	0	−0.0007

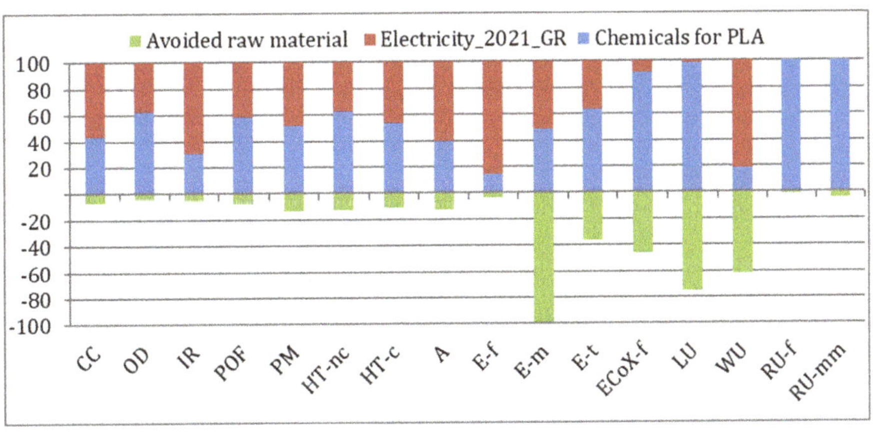

Figure 5. Characterization results for the alternative feedstock scenario, including the total prevention.

Continuing on to the normalization stage, as shown in Figure 6, our findings indicate that the three most significant categories are freshwater eutrophication, water use, and freshwater ecotoxicity. In the same context, electricity is the largest contributor for the first two categories, although at lower values, while chemicals are mostly responsible for the freshwater ecotoxicity impacts.

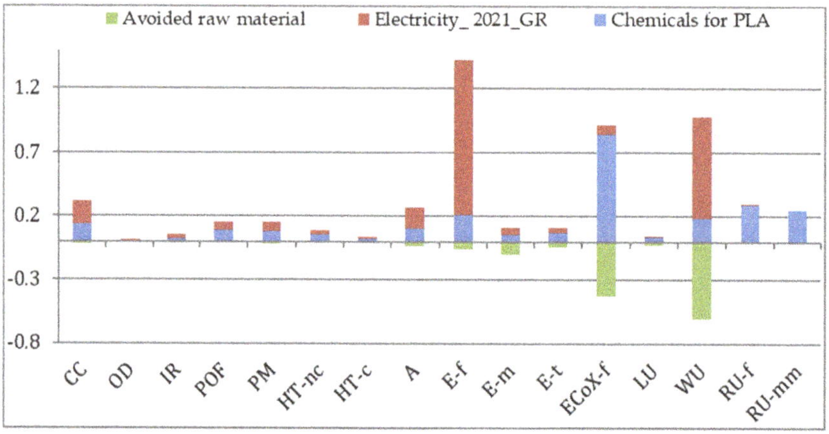

Figure 6. Normalization results for the alternative feedstock scenario, including the total prevention.

The main categories recorded as prevention (Figure 6) at the normalization level include freshwater eutrophication and freshwater ecotoxicity. These categories quantify the savings achieved in reducing nutrient input and the reductions in the release of ecotoxic substances. In life cycle assessments, normalization is a crucial step in assessing environmental impact [74]. It involves comparing the environmental impact of a product or process to a reference value, which is typically based on the average or typical impact of a specific category of products or processes. Overall, normalization and characterization in LCA help in evaluating the environmental impact of various processes and products [75].

3.3. *Sensitivity and Uncertainty Analysis*

Understanding variations in various impact categories is critical in the continuous effort to monitor and reduce the environmental effects of human activities. The following data shows the most current percentage rates of change, emphasizing both gains and areas for concern.

Based on Figure 7 and Table A4 (Appendix A), resuming all operations in 2024 (with the latest energy mix in Greece, which has been implemented since 2023) could potentially lead to a significant 20% reduction in carbon dioxide equivalent, indicating progress in addressing climate change concerns. Additionally, a 21% decrease in chlorofluorocarbon equivalent suggests effectiveness in phasing out ozone-depleting compounds, contributing to the recovery of the ozone layer. Notably, there is a 19% reduction in non-methane volatile organic compound equivalent, indicating progress in reducing air pollution and ozone generation. The significant 47% decrease in illness incidence due to particulate matter indicates successful measures to improve air quality and protect public health. Similarly, the 12% reduction in non-cancerous human toxicity demonstrates attempts to minimize exposure to dangerous chemicals, hence improving human health and well-being. Notable reductions include a 32% decrease in malignant human toxicity, indicating success in reducing carcinogenic dangers, and a 35% reduction in hydrogen ion equivalent, indicating progress in mitigating acidification and maintaining ecological pH equilibrium.

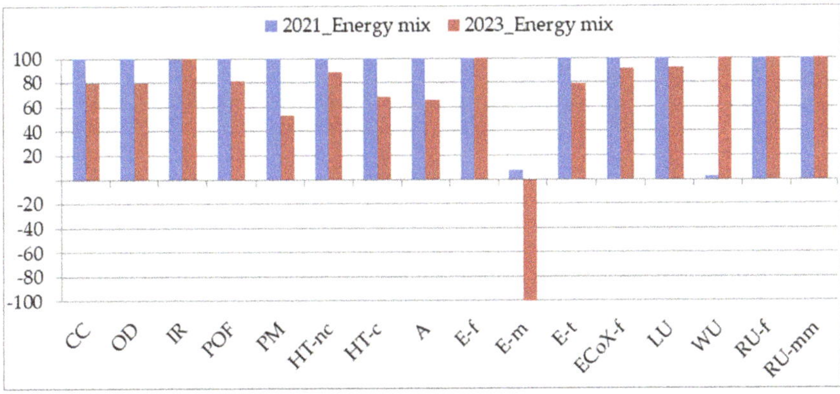

Figure 7. Characterization diagram of sensitivity analysis for the main impact categories.

While there have been considerable advancements in key environmental impact categories, challenges persist that require collective action and innovation. Continued collaboration among governments, industry, and communities is critical to ensuring a sustainable future.

The uncertainty analysis is shown in Figure 8, and the category with a significant difference is human toxicity without cancer effects, which might be attributed to various variables. These include variations in exposure and susceptibility across populations, the complexity of chemical interactions, a lack of comprehensive data on long-term exposure, methodological differences in the assessment of toxicity, and variations in models used to extrapolate data from experimental studies to real-world scenarios. Furthermore, regulatory standards and procedural discrepancies among areas and nations might lead to inaccurate measurements and interpretations of human toxicity data. These characteristics make it challenging to acquire accurate and reliable measures of non-cancer human adverse effects.

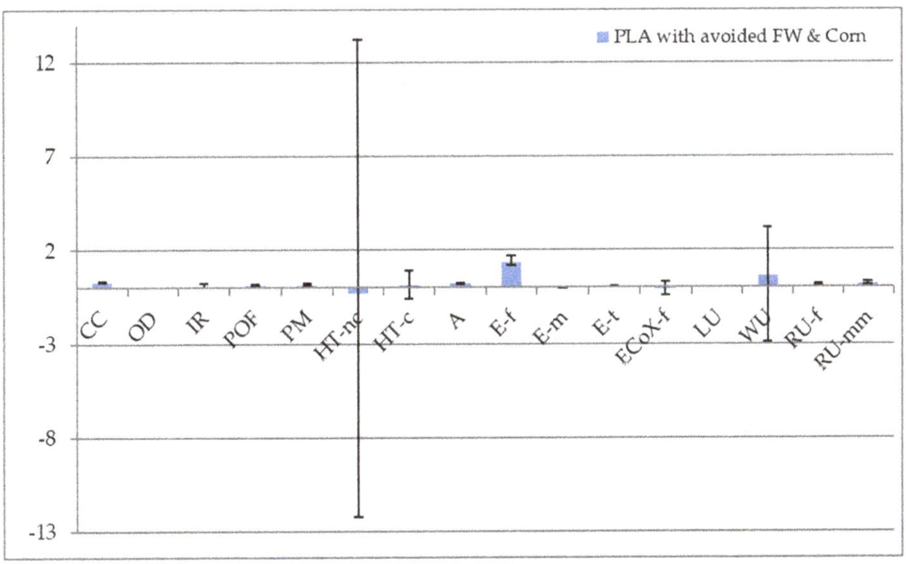

Figure 8. Uncertainty analysis for all avoided products at the normalization level.

The primary advantages of producing PLA from food waste include reduced environmental impact and long-term sustainability [76]. This aligns with environmental aims since converting food waste into bioplastics such as PLA promotes circular economy principles [77] and reduces dependency on polymers produced through fossil fuels [9]. However, significant challenges such as raw material volatility, process optimization, and economic feasibility need to be addressed.

The diversity of raw materials affects the efficiency and consistency of the fermentation process and subsequent PLA synthesis. To produce high-quality PLA, research and development are needed to develop efficient and cost-effective technologies for the pretreatment, fermentation, and purification of lactic acid derived from food waste. The economic feasibility of waste collection, treatment, and conversion procedures relative to standard PLA production methods will determine viability. More numerical details for the uncertainty analysis results are presented in Table A5 in Appendix A.

4. Conclusions and Limitations

Bioplastics produced from organic waste offer a promising alternative for reducing dependence on petroleum-based plastics [78]. This study focuses on evaluating PLA production from food waste, with an emphasis on its potential to promote environmental sustainability. The main benefits are increased resource efficiency, less environmental impact, and long-term sustainability. This strategy applies the principles of the circular economy while also reducing reliance on polymers derived from fossil fuels. Our results indicate that electricity is the major contributor to the most important environmental impacts of PLA production from food waste in the pilot scale unit. Moreover, the electricity required for PLA production still depends on fossil sources. The most important impacts are freshwater eutrophication, water use, and freshwater ecotoxicity. Our findings are in agreement with literature reports that state that switching to second-generation feedstocks, such as separately collected food waste, and improvements in energy efficiency and use of renewable energy are among the technological advances that reduce the environmental impacts of bioplastics [30].

Aside from the environmental benefits, there are certain limitations, such as the economic viability, as waste collection, treatment costs, and conversion process efficiency determine PLA production from food waste. Waste management technology innovation and efficient conversion procedures are crucial to cost-effectiveness. Tsang et al. [35] and Jõgi and Bhat [31] highlighted the potential cost savings resulting from bioplastic production from food waste compared to standard processes while emphasizing the need for more cost-effective solutions. Additional studies underscore the need to examine the economic feasibility of the food-waste-to-bioplastics concept within the context of a circular economy [79]. However, a direct comparative economic analysis is still absent, illustrating the need for more research in this field.

In terms of future research challenges within the use of LCA for the assessment of bioplastics production, we also clearly identify that the plastic littering impacts, even for bioplastics that are marketed as biodegradable, are not included within any current LCA impact category [9]. With the fast-growing accumulation of degradable bioplastics and their resulting microplastics in soils globally, there is an urgent better need to understand their potential effects on soil-microbial-plant systems, which could serve as a starting point for assessing their ecological risk in terrestrial ecosystems [52]. Based on these scientific findings, relative LCA impact models should be developed.

Regarding the scalability of the present process to an industrial one, the use of pilot-scale data is a difficult issue with several challenges. Berndtsson et al. [80] note the necessity for senior management support in scaling up data-driven pilot initiatives. There is also a suggested framework for pilot line scale-up utilizing digital manufacturing, which may improve the accuracy of scale-up models [81]. These studies demonstrate the importance of detailed planning, organizational support, and the technology tools required for scaling up the process.

To summarize, while both food waste and corn may be used as raw materials for PLA production, employing food waste has the potential to reduce waste, increase resource efficiency, and reduce environmental impacts. However, process optimization, scalability, and economic viability issues must be addressed before PLA production from food waste can be more competitive and sustainable than standard corn-based PLA production.

Author Contributions: The collaborative nature of the research outlined in this article involved each author contributing in diverse ways, encompassing unique perspectives, insights, feedback, and guidance that may not be easily quantified or neatly categorized. Despite this complexity, an effort has been made to delineate the authors' contributions, recognizing that: conceptualization. K.S. and K.A.; methodology. K.A. and K.S.; software. K.S. and K.A.; validation. K.A., K.V. and K.L.; formal analysis. K.A. and K.S.; investigation. K.S. and A.M.; resources. A.M., T.M. and K.V.; data curation. A.M., K.V. and T.M.; writing—original draft preparation. K.A. and K.S.; writing—review and editing. K.A. and K.L.; visualization. K.S.; supervision. K.L., K.A. and T.M.; project administration. T.M. and K.L.; funding acquisition. T.M. and K.L. All authors have read and agreed to the published version of the manuscript.

Funding: This study was financially supported by the A2UFood Project (UIA 02-115 "Avoidable and Unavoidable Food Wastes: A Holistic Managing Approach for Urban Environments—A2U Food", Urban Innovative Actions Initiative, EU, (2018–2022)), co-funded by the European Regional and Development Fund through the Urban Innovative Actions Initiative.

Institutional Review Board Statement: Not applicable.

Data Availability Statement: The data presented in this study are contained within this manuscript for reproducibility purposes.

Acknowledgments: The authors would like to acknowledge all who have directly or indirectly helped in carrying out this study.

Conflicts of Interest: The authors declare no conflicts of interest.

Appendix A

Table A1. Chemicals import table for LCI (all the chemicals used were purchased in larger quantities as they are part of the consumables for other laboratory experiments, via on-line ordering).

Chemical	Characteristics/Supply Company	Amount	Unit
Sodium hydroxide	≥98.0–100.5%/Sigma Aldrich (Burlington, MA, USA)	25	kg
Sodium hypochlorite,	without water, in a 15% solution state/market product for cleaning	100	kg
Tetrahydrofuran	Min. 99.5%/Honeywell (Charlotte, NC, USA)	4.5	kg
Acetone,	Pure 100%, Liquid/Manis chemicals (Athens, Greece)	4	kg
Hydrochloric acid from benzene chlorination	Honeywell X31087 hydrochloric acid solution, 6 M HCl	4	kg
1-butanol	p.a. ≥99.5%/Merck (Marousi, Athens, Greece)	176	kg
Tin concentrate (Tin(II) 2-ethylhexanoate)	92.5 100.0%/Sigma Aldrich (Burlington, MA, USA)	1	kg
Enzyme, Glucoamylase, Novozyme, Spirizyme	Amlyloglucosidases Spirizyme Fuel (SPF) from Novozymes	5	kg
Ammonium Hydroxide	liquid 25%/Manis chemicals (Athens, Greece)	25	kg

Table A2. Energy needs import table for LCI.

Production Energy	Hours of Operation (h)		Power of Devices (kW)	Energy Requirements (kWh)
Receiving/processing equipment (conveyor belt, conveyor screws, shredder, pulper, monopump)	2 h in every 8 h shift * 11 batches * 85%	18.7	22.02 kW	411.8
Bioreactor R1	4 days * 24 h/day * 11 batches * 85%	898	0.5 kW	449
Bioreactor R2	1 day * 4 h/day * 11 batches * 85%	37.4	0.5 kW	18.7
Filter system	1 day * 2 h/day * 11 batches * 85%	18.7	0.55 kW	10.3
Low temperature heating system: for R1	3.5 days * 24 h/day * 11 batches * 85%	785	4.37 kW	3430.5
Burner R1	0.5 days * 24 h/day * 11 batches * 85%	112.2	0.035 kW	3.9
Burner R2	1 day * 4 h/day * 11 batches * 85%	37.4	0.035 kW	1.3
Polymerization devices	1 day * 6 h/day * 11 batches * 85%	56.1	2.25 kW	126.2
Building air circulation fans	55 days * 8 h/day * 85%	374	0.20 kW	74.8
Total energy needs				4451.7

Table A3. Starch contained in the case study sample for LCI (https://woodmed.com/index.php/health-information/basic-health-maintenance-diet/135-appendix-3-starch-content-of-vegetables-and-fruits (accessed on 15 February 2024).

Food Category	% Starch	Total Feedstock Quantity (kg)	Quantity of Starch in Feedstock (kg)
Spinach	0.03	25.0	0.75
Broccoli	0.03	28.0	0.84
Cabbage	0.03	29.0	0.87
Brussel Sprouts	0.09	18.0	1.60
White Potato	0.18	100.0	18.0
Pepper	0.06	32.0	1.92
Cucumber	0.03	40.0	1.20
Lettuce	0.03	80.0	2.40
Tomatoes	0.03	70.0	2.10
Carrots	0.09	56.5	5.08
Onions	0.09	93.0	8.30
Beans	0.21	10.0	2.10
Green Peas	0.15	40.0	6.00
Beets	0.09	22.0	1.98
Total amount of vegetables	--	643.5	53.0
Grapefruit	0.09	5.0	0.45
Bananas	0.21	25.0	5.25
Watermelon	0.06	5.0	0.30
Oranges	0.12	20.0	2.40

Table A3. *Cont.*

Food Category	% Starch	Total Feedstock Quantity (kg)	Quantity of Starch in Feedstock (kg)
Peaches	0.12	6.0	0.72
Apples	0.15	10.5	1.58
Total amount of fruits	--	71.5	10.70
	Total Starch		63.9
Corn	0.21	100.0	21.00
Total	--	304	63.9

Table A4. Sensitivity analysis for two electricity mixes and the rate of change.

Impact Category (Unit)	2021 Energy Mix	2023 Energy Mix	% Rate of Change
CC (kg CO_2 eq.)	2360.95	1879.05	−20%
OD (kg CFC11 eq.)	0.00	0.00	−21%
IR (kBq U-235 eq.)	225.90	228.23	1%
POF (kg NMVOC eq.)	5.62	4.56	−19%
PM (disease inc.)	0.00	0.00	−47%
HT-nc (CTUh)	0.00	0.00	−12%
HT-c (CTUh)	0.00	0.00	−32%
A (mol H+ eq.)	12.81	8.38	−35%
E-f (kg P eq.)	2.19	2.20	0%
E m (kg N eq.)	0.02	−0.22	−1360%
E-t (mol N eq.)	12.42	9.78	−21%
ECoX-f (CTUe)	21,125.6	19,234.86	−9%
LU (Pt)	6978.47	6399.44	−8%
WU (m^3 depriv.)	4327.67	186,362.87	4206%
RU-f (MJ)	18,815.2	18,814.65	0%
RU mm (kg Sb eq.)	0.015	0.015	0%

Table A5. Uncertainty analysis for total avoided product at normalization.

Label	PLA with Avoided FW and Corn	Low	High
CC	0.323	0.025	0.036
OD	0.003	0.001	0.001
IR	0.036	0.025	0.205
POF	0.159	0.025	0.042
PM	0.141	0.032	0.09
HT-nc	−0.292	11.9	13.506
HT-c	0.122	0.706	0.802
A	0.225	0.018	0.022
E-f	1.375	0.175	0.32
E-m	0.001	0.01	0.017
E-t	0.069	0.014	0.02
ECoX-f	−0.084	0.358	0.372

Table A5. *Cont.*

Label	PLA with Avoided FW and Corn	Low	High
LU	0.008	0.006	0.004
WU	0.616	3.543	2.544
RU-f	0.108	0.03	0.061
RU-mm	0.198	0.071	0.107

References

1. Ramos, A.; Afonso Teixeira, C.; Rouboa, A. Environmental Analysis of Waste-to-Energy—A Portuguese Case Study. *Energies* **2018**, *11*, 548. [CrossRef]
2. Plastics—The Facts 2022 • Plastics Europe. Available online: https://plasticseurope.org/knowledge-hub/plastics-the-facts-2022/ (accessed on 5 October 2023).
3. OECD. *Global Plastics Outlook: Policy Scenarios to 2060*; OECD: Paris, France, 2022; ISBN 978-92-64-97364-0.
4. Browning, S.; Beymer-Farris, B.; Seay, J.R. Addressing the Challenges Associated with Plastic Waste Disposal and Management in Developing Countries. *Curr. Opin. Chem. Eng.* **2021**, *32*, 100682. [CrossRef]
5. Zheng, J.; Suh, S. Strategies to Reduce the Global Carbon Footprint of Plastics. *Nat. Clim. Chang.* **2019**, *9*, 374–378. [CrossRef]
6. Schmaltz, E.; Melvin, E.C.; Diana, Z.; Gunady, E.F.; Rittschof, D.; Somarelli, J.A.; Virdin, J.; Dunphy-Daly, M.M. Plastic Pollution Solutions: Emerging Technologies to Prevent and Collect Marine Plastic Pollution. *Environ. Int.* **2020**, *144*, 106067. [CrossRef] [PubMed]
7. Rai, P.K.; Lee, J.; Brown, R.J.C.; Kim, K.-H. Micro- and Nano-Plastic Pollution: Behavior, Microbial Ecology, and Remediation Technologies. *J. Clean. Prod.* **2021**, *291*, 125240. [CrossRef]
8. Saley, A.M.; Smart, A.C.; Bezerra, M.F.; Burnham, T.L.U.; Capece, L.R.; Lima, L.F.O.; Carsh, A.C.; Williams, S.L.; Morgan, S.G. Microplastic Accumulation and Biomagnification in a Coastal Marine Reserve Situated in a Sparsely Populated Area. *Mar. Pollut. Bull.* **2019**, *146*, 54–59. [CrossRef]
9. Bishop, G.; Styles, D.; Lens, P.N.L. Environmental Performance Comparison of Bioplastics and Petrochemical Plastics: A Review of Life Cycle Assessment (LCA) Methodological Decisions. *Resour. Conserv. Recycl.* **2021**, *168*, 105451. [CrossRef]
10. Shogren, R.; Wood, D.; Orts, W.; Glenn, G. Plant-Based Materials and Transitioning to a Circular Economy. *Sustain. Prod. Consum.* **2019**, *19*, 194–215. [CrossRef]
11. Blank, L.M.; Narancic, T.; Mampel, J.; Tiso, T.; O'Connor, K. Biotechnological Upcycling of Plastic Waste and Other Non-Conventional Feedstocks in a Circular Economy. *Curr. Opin. Biotechnol.* **2020**, *62*, 212–219. [CrossRef] [PubMed]
12. Tamburini, E.; Costa, S.; Summa, D.; Battistella, L.; Fano, E.A.; Castaldelli, G. Plastic (PET) vs Bioplastic (PLA) or Refillable Aluminium Bottles—What Is the Most Sustainable Choice for Drinking Water? A Life-Cycle (LCA) Analysis. *Environ. Res.* **2021**, *196*, 110974. [CrossRef] [PubMed]
13. Schneiderman, D.K.; Hillmyer, M.A. 50th Anniversary Perspective: There Is a Great Future in Sustainable Polymers. *Macromolecules* **2017**, *50*, 3733–3749. [CrossRef]
14. Yuan, X.; Wang, X.; Sarkar, B.; Ok, Y.S. The COVID-19 Pandemic Necessitates a Shift to a Plastic Circular Economy. *Nat. Rev. Earth Environ.* **2021**, *2*, 659–660. [CrossRef] [PubMed]
15. Document 52018DC0028, Communication from the Commission to the European Parliament, the Council, the European Economic and Social Committee and the Committee of the Regions. A European Strategy for Plastics in a Circular Economy. 2018. Available online: https://eur-lex.europa.eu/legal-content/EN/TXT/?uri=COM:2018:28:FIN (accessed on 10 January 2024).
16. Ali, S.S.; Abdelkarim, E.A.; Elsamahy, T.; Al-Tohamy, R.; Li, F.; Kornaros, M.; Zuorro, A.; Zhu, D.; Sun, J. Bioplastic Production in Terms of Life Cycle Assessment: A State-of-the-Art Review. *Environ. Sci. Ecotechnol.* **2023**, *15*, 100254. [CrossRef] [PubMed]
17. Kircher, M. The Bioeconomy Needs Economic, Ecological and Social Sustainability. *AIMSES* **2022**, *9*, 33–50. [CrossRef]
18. Sidek, I.S.; Draman, S.F.S.; Abdullah, S.R.S.; Anuar, N. Current Development on Bioplastics and Its Future Prospects: An Introductory Review. *INWASCON Technol. Mag.* **2019**, 03–08. [CrossRef]
19. Blum, N.U.; Haupt, M.; Bening, C.R. Why "Circular" Doesn't Always Mean "Sustainable". *Resour. Conserv. Recycl.* **2020**, *162*, 105042. [CrossRef]
20. Bajpai, P. Market for Biobased Packaging Material. In *Biobased Polymers*; Elsevier: Amsterdam, The Netherlands, 2019; pp. 191–196. ISBN 978-0-12-818404-2.
21. Chia, W.Y.; Ying Tang, D.Y.; Khoo, K.S.; Kay Lup, A.N.; Chew, K.W. Nature's Fight against Plastic Pollution: Algae for Plastic Biodegradation and Bioplastics Production. *Environ. Sci. Ecotechnol.* **2020**, *4*, 100065. [CrossRef] [PubMed]
22. Imre, B.; García, L.; Puglia, D.; Vilaplana, F. Reactive Compatibilization of Plant Polysaccharides and Biobased Polymers: Review on Current Strategies, Expectations and Reality. *Carbohydr. Polym.* **2019**, *209*, 20–37. [CrossRef] [PubMed]
23. Zhao, X.; Cornish, K.; Vodovotz, Y. Narrowing the Gap for Bioplastic Use in Food Packaging: An Update. *Environ. Sci. Technol.* **2020**, *54*, 4712–4732. [CrossRef] [PubMed]

24. Muniyasamy, S.; Dada, O.E. Recycling of Plastics and Composites Materials and Degradation Technologies for Bioplastics and Biocomposites. In *Waste Management in the Fashion and Textile Industries*; Elsevier: Amsterdam, The Netherlands, 2021; pp. 311–333. ISBN 978-0-12-818758-6.
25. Sindhu, R.; Gnansounou, E.; Rebello, S.; Binod, P.; Varjani, S.; Thakur, I.S.; Nair, R.B.; Pandey, A. Conversion of Food and Kitchen Waste to Value-Added Products. *J. Environ. Manag.* **2019**, *241*, 619–630. [CrossRef] [PubMed]
26. Rebolledo-Leiva, R.; Ladakis, D.; Ioannidou, S.-M.; Koutinas, A.; Moreira, M.T.; González-García, S. Pursuing Single or Combined Wheat Straw Based Poly(Butylene Succinate) Production Routes: A Life Cycle Approach of First- and Second-Generation Feedstocks. *Sustain. Mater. Technol.* **2023**, *37*, e00683. [CrossRef]
27. Trivedi, A.K.; Gupta, M.K.; Singh, H. PLA Based Biocomposites for Sustainable Products: A Review. *Adv. Ind. Eng. Polym. Res.* **2023**, *6*, 382–395. [CrossRef]
28. Tsiropoulos, I.; Faaij, A.P.C.; Lundquist, L.; Schenker, U.; Briois, J.F.; Patel, M.K. Life Cycle Impact Assessment of Bio-Based Plastics from Sugarcane Ethanol. *J. Clean. Prod.* **2015**, *90*, 114–127. [CrossRef]
29. Vink, E.T.H.; Rábago, K.R.; Glassner, D.A.; Gruber, P.R. Applications of Life Cycle Assessment to NatureWorks^{TM} Polylactide (PLA) Production. *Polym. Degrad. Stab.* **2003**, *80*, 403–419. [CrossRef]
30. Brizga, J.; Hubacek, K.; Feng, K. The Unintended Side Effects of Bioplastics: Carbon, Land, and Water Footprints. *One Earth* **2020**, *3*, 45–53. [CrossRef]
31. Jõgi, K.; Bhat, R. Valorization of Food Processing Wastes and By-Products for Bioplastic Production. *Sustain. Chem. Pharm.* **2020**, *18*, 100326. [CrossRef]
32. Fonseca, A.; Ramalho, E.; Gouveia, A.; Figueiredo, F.; Nunes, J. Life Cycle Assessment of PLA Products: A Systematic Literature Review. *Sustainability* **2023**, *15*, 12470. [CrossRef]
33. Govil, T.; Wang, J.; Samanta, D.; David, A.; Tripathi, A.; Rauniyar, S.; Salem, D.R.; Sani, R.K. Lignocellulosic Feedstock: A Review of a Sustainable Platform for Cleaner Production of Nature's Plastics. *J. Clean. Prod.* **2020**, *270*, 122521. [CrossRef]
34. Swetha, T.A.; Ananthi, V.; Bora, A.; Sengottuvelan, N.; Ponnuchamy, K.; Muthusamy, G.; Arun, A. A Review on Biodegradable Polylactic Acid (PLA) Production from Fermentative Food Waste—Its Applications and Degradation. *Int. J. Biol. Macromol.* **2023**, *234*, 123703. [CrossRef]
35. Tsang, Y.F.; Kumar, V.; Samadar, P.; Yang, Y.; Lee, J.; Ok, Y.S.; Song, H.; Kim, K.-H.; Kwon, E.E.; Jeon, Y.J. Production of Bioplastic through Food Waste Valorization. *Environ. Int.* **2019**, *127*, 625–644. [CrossRef] [PubMed]
36. Elissen, H.J.H.; Kootstra, A.M.J. Production of Short-Chain Volatile Fatty Acids and Lactic Acid during Small-Scale Ensiling of Meadow Grass. Available online: https://edepot.wur.nl/415175 (accessed on 21 November 2023).
37. Woodford, M. Modeling Imprecision in Perception, Valuation, and Choice. *Annu. Rev. Econ.* **2020**, *12*, 579–601. [CrossRef]
38. Gong, L.; Passari, A.K.; Yin, C.; Kumar Thakur, V.; Newbold, J.; Clark, W.; Jiang, Y.; Kumar, S.; Gupta, V.K. Sustainable Utilization of Fruit and Vegetable Waste Bioresources for Bioplastics Production. *Crit. Rev. Biotechnol.* **2024**, *44*, 236–254. [CrossRef] [PubMed]
39. Cecchi, T. Biobased Polymers from Food Waste Feedstock and Their Synthesis. In *Biobased Products from Food Sector Waste*; Springer International Publishing: Cham, Switzerland, 2021; pp. 231–285. ISBN 978-3-030-63435-3.
40. Sakai, K.; Taniguchi, M.; Miura, S.; Ohara, H.; Matsumoto, T.; Shirai, Y. Making Plastics from Garbage: A Novel Process for Poly-L-Lactate Production from Municipal Food Waste. *J. Ind. Ecol.* **2003**, *7*, 63–74. [CrossRef]
41. Yu, P.H.; Chua, H.; Huang, A.L.; Lo, W.; Chen, G.Q. Conversion of Food Industrial Wastes into Bioplastics. *Appl. Biochem. Biotechnol.* **1998**, *70–72*, 603–614. [CrossRef] [PubMed]
42. Perotto, G.; Ceseracciu, L.; Simonutti, R.; Paul, U.C.; Guzman-Puyol, S.; Tran, T.-N.; Bayer, I.S.; Athanassiou, A. Bioplastics from Vegetable Waste via an Eco-Friendly Water-Based Process. *Green Chem.* **2018**, *20*, 894–902. [CrossRef]
43. Kakadellis, S.; Harris, Z.M. Don't Scrap the Waste: The Need for Broader System Boundaries in Bioplastic Food Packaging Life-Cycle Assessment—A Critical Review. *J. Clean. Prod.* **2020**, *274*, 122831. [CrossRef]
44. Ramadhan, M.O.; Handayani, M.N. The Potential of Food Waste as Bioplastic Material to Promote Environmental Sustainability: A Review. *IOP Conf. Ser. Mater. Sci. Eng.* **2020**, *980*, 012082. [CrossRef]
45. Bagnani, M.; Peydayesh, M.; Knapp, T.; Appenzeller, E.; Sutter, D.; Kränzlin, S.; Gong, Y.; Wehrle, A.; Greuter, S.; Bucher, M.; et al. From Soy Waste to Bioplastics: Industrial Proof of Concept. *Biomacromolecules* **2024**, *25*, 2033–2040. [CrossRef] [PubMed]
46. Chroni, C.; Lasaridi, K.; Stylianidis, N.; Velonia, K.; Manios, T.; Daliakopoulos, I.; Tsompanidis, C.; Hafner, G.; Skarvelakis, M.; Drosou, Z. The A2UFood Project—Avoidable and Unavoidable Food Wastes: A Holistic Managing Approach for Urban Environments. *Proceedings* **2019**, *30*, 83. [CrossRef]
47. D'Adamo, I. *Adopting Circular Economy Current Practices and Future Perspectives*; MDPI: Basel, Switzerland, 2020; ISBN 978-3-03928-343-9.
48. Finkbeiner, M.; Inaba, A.; Tan, R.; Christiansen, K.; Klüppel, H.-J. The New International Standards for Life Cycle Assessment: ISO 14040 and ISO 14044. *Int. J. Life Cycle Assess.* **2006**, *11*, 80–85. [CrossRef]
49. Theodorou, A.; Raptis, V.; Baltzaki, C.I.M.; Manios, T.; Harmandaris, V.; Velonia, K. Synthesis and Modeling of Poly(L-Lactic Acid) via Polycondensation of L-Lactic Acid. *Polymers* **2023**, *15*, 4569. [CrossRef] [PubMed]
50. Maragkaki, A.; Malliaros, N.G.; Sampathianakis, I.; Lolos, T.; Tsompanidis, C.; Manios, T. Evaluation of Biodegradability of Polylactic Acid and Compostable Bags from Food Waste under Industrial Composting. *Sustainability* **2023**, *15*, 15963. [CrossRef]

51. Preethi; Kavitha, S.; Rajesh Banu, J.; Arulazhagan, P.; Gunasekaran, M. Environmental Impacts and Sustainability Assessment of Food Loss and Waste Valorization: Value Chain Analysis of Food Consumption. In *Food Waste to Valuable Resources*; Elsevier: Amsterdam, The Netherlands, 2020; pp. 359–388. ISBN 978-0-12-818353-3.
52. Liu, R.; Liang, J.; Yang, Y.; Jiang, H.; Tian, X. Effect of Polylactic Acid Microplastics on Soil Properties, Soil Microbials and Plant Growth. *Chemosphere* **2023**, *329*, 138504. [CrossRef] [PubMed]
53. Perkoulidis, G.; Malamakis, A.; Banias, G.; Moussiopoulos, N. Development of a Methodological Framework for the Evaluation of the Material and Energy Recovery Potential of Municipal Solid Waste Management: Implementation in Five Greek Regions. *Circ. Econ. Sustain.* **2022**, *2*, 313–326. [CrossRef]
54. Powell, J.T.; Townsend, T.G.; Zimmerman, J.B. Estimates of Solid Waste Disposal Rates and Reduction Targets for Landfill Gas Emissions. *Nat. Clim. Chang.* **2016**, *6*, 162–165. [CrossRef]
55. ISO 14040:2006 (En), Environmental Management—Life Cycle Assessment—Principles and Framework. Available online: https://www.iso.org/obp/ui/en/#iso:std:iso:14040:ed-2:v1:en (accessed on 23 June 2024).
56. Boulay, A.-M.; Bulle, C.; Bayart, J.-B.; Deschênes, L.; Margni, M. Regional Characterization of Freshwater Use in LCA: Modeling Direct Impacts on Human Health. *Environ. Sci. Technol.* **2011**, *45*, 8948–8957. [CrossRef] [PubMed]
57. Hélias, A.; Esnouf, A.; Finkbeiner, M. Consistent Normalization Approach for Life Cycle Assessment Based on Inventory Databases. *Sci. Total Environ.* **2020**, *703*, 134583. [CrossRef] [PubMed]
58. Lim, L.-T.; Auras, R.; Rubino, M. Processing Technologies for Poly(Lactic Acid). *Prog. Polym. Sci.* **2008**, *33*, 820–852. [CrossRef]
59. Rezvani Ghomi, E.R.; Khosravi, F.; Saedi Ardahaei, A.S.; Dai, Y.; Neisiany, R.E.; Foroughi, F.; Wu, M.; Das, O.; Ramakrishna, S. The Life Cycle Assessment for Polylactic Acid (PLA) to Make It a Low-Carbon Material. *Polymers* **2021**, *13*, 1854. [CrossRef] [PubMed]
60. Goldschmidt, J.; Sankavaram, K.; Udahogora, M. Advancing Cultural Competencies: Applying the Dietary Exchange List System to Jamaican Foods. *Health Sci. J.* **2018**, *12*. [CrossRef]
61. Karytsas, S.; Vardopoulos, I.; Theodoropoulou, E. Factors Affecting Sustainable Market Acceptance of Residential Microgeneration Technologies. A Two Time Period Comparative Analysis. *Energies* **2019**, *12*, 3298. [CrossRef]
62. Oikonomou, T.I.; Andreosatos, C.; Drosou, V.; Karytsas, C. The Influence of Energy Policy Instruments upon the Promotion of Solar Thermal Technology in Greece. In Proceedings of the BEHAVE 2020—6th European Conference on Behaviour and Energy Efficiency, Copenhagen, Denmark, 21–23 October 2020. Available online: https://c2e2.unepccc.org/wp-content/uploads/sites/3/2021/05/442-the-influence-of-energy-policy-instruments-upon-the-promotion-of-solar-thermal-technology-in-greece.pdf (accessed on 12 January 2024).
63. Rashid Khan, H.U.; Awan, U.; Zaman, K.; Nassani, A.A.; Haffar, M.; Abro, M.M.Q. Assessing Hybrid Solar-Wind Potential for Industrial Decarbonization Strategies: Global Shift to Green Development. *Energies* **2021**, *14*, 7620. [CrossRef]
64. Al Shidi, H.; Sulaiman, H.; Amoatey, P. Shifting to Renewable Energy to Mitigate Carbon Emissions: Initiatives by the States of Gulf Cooperation Council. *Low Carbon Econ.* **2016**, *07*, 123–136. [CrossRef]
65. Sheldon, S.; Hadian, S.; Zik, O. Beyond Carbon: Quantifying Environmental Externalities as Energy for Hydroelectric and Nuclear Power. *Energy* **2015**, *84*, 36–44. [CrossRef]
66. International Energy Agency. *Greece 2023 Energy Policy Review*; IEA Energy Policy Reviews; OECD: Paris, France, 2023; ISBN 978-92-64-32569-2.
67. Plant, A.L.; Becker, C.A.; Hanisch, R.J.; Boisvert, R.F.; Possolo, A.M.; Elliott, J.T. How Measurement Science Can Improve Confidence in Research Results. *PLoS Biol.* **2018**, *16*, e2004299. [CrossRef] [PubMed]
68. Hong, J.; Shaked, S.; Rosenbaum, R.K.; Jolliet, O. Analytical Uncertainty Propagation in Life Cycle Inventory and Impact Assessment: Application to an Automobile Front Panel. *Int. J. Life Cycle Assess.* **2010**, *15*, 499–510. [CrossRef]
69. Adamczak, S.; Świderski, J.; Dobrowolski, T.; Mathia, T.G. Evaluation of Measurement Uncertainty of Profile Parameters Using a Monte Carlo Method. *Mechanik* **2018**, 418–420. [CrossRef]
70. Madhavan Nampoothiri, K.; Nair, N.R.; John, R.P. An Overview of the Recent Developments in Polylactide (PLA) Research. *Bioresour. Technol.* **2010**, *101*, 8493–8501. [CrossRef] [PubMed]
71. Escobar, N.; Haddad, S.; Börner, J.; Britz, W. Land Use Mediated GHG Emissions and Spillovers from Increased Consumption of Bioplastics. *Environ. Res. Lett.* **2018**, *13*, 125005. [CrossRef]
72. Chen, W.; Oldfield, T.L.; Cinelli, P.; Righetti, M.C.; Holden, N.M. Hybrid Life Cycle Assessment of Potato Pulp Valorisation in Biocomposite Production. *J. Clean. Prod.* **2020**, *269*, 122366. [CrossRef]
73. Qiu, H.; Hu, B.; Zhang, Z. Impacts of Land Use Change on Ecosystem Service Value Based on SDGs Report--Taking Guangxi as an Example. *Ecol. Indic.* **2021**, *133*, 108366. [CrossRef]
74. Aymard, V.; Botta-Genoulaz, V. Normalisation in Life-Cycle Assessment: Consequences of New European Factors on Decision-Making. *Supply Chain Forum Int. J.* **2017**, *18*, 76–83. [CrossRef]
75. Lim, S.-R. Application of Life Cycle Assessment to Enhance the Environmental Performance of Process Systems and Products. *Clean Technol.* **2014**, *20*, 339–348. [CrossRef]
76. Fahim, I.S.; Chbib, H.; Mahmoud, H.M. The Synthesis, Production & Economic Feasibility of Manufacturing PLA from Agricultural Waste. *Sustain. Chem. Pharm.* **2019**, *12*, 100142. [CrossRef]
77. Rosenboom, J.-G.; Langer, R.; Traverso, G. Bioplastics for a Circular Economy. *Nat. Rev. Mater.* **2022**, *7*, 117–137. [CrossRef] [PubMed]

78. Samer, M.; Hijazi, O.; Mohamed, B.A.; Abdelsalam, E.M.; Amer, M.A.; Yacoub, I.H.; Attia, Y.A.; Bernhardt, H. Environmental Impact Assessment of Bioplastics Production from Agricultural Crop Residues. *Clean Technol. Environ. Policy* **2022**, *24*, 815–827. [CrossRef]
79. Banu, J.R.; Sharmila, V.G. Review on Food Waste Valorisation for Bioplastic Production towards a Circular Economy: Sustainable Approaches and Biodegradability Assessment. *Sustain. Energy Fuels* **2023**, *7*, 3165–3184. [CrossRef]
80. Berndtsson, M.; Ericsson, A.; Svahn, T. Scaling-Up Data-Driven Pilot Projects. *AI Mag.* **2020**, *41*, 94–102. [CrossRef]
81. Chinnathai, M.K.; Al-Mowafy, Z.; Alkan, B.; Vera, D.; Harrison, R. A Framework for Pilot Line Scale-up Using Digital Manufacturing. *Procedia CIRP* **2019**, *81*, 962–967. [CrossRef]

sustainability

MDPI

Review

Examining the Drivers to Support Improved Construction and Demolition Waste Management for a Circular Economy: A Comprehensive Review Using a Systematic Approach

Mahmoud Alhawamdeh [1,2,*], Juan A. Ferriz-Papi [2] and Angela Lee [3]

[1] Department of Business and Management, Bath Business School, Bath Spa University, Newton St Loe, Bath BA2 9BN, UK
[2] Department of Sustainable Built and Natural Environments, School of Science, Engineering and Environment, University of Salford, 43 Crescent, Manchester M5 4WT, UK
[3] School of the Built Environment, University College of Estate Management (UK) Horizons, 60 Queens Rd, Reading RG1 4BS, UK
* Correspondence: m.alhawamdeh@bathspa.ac.uk

Abstract: With the rapid pace of global urbanisation, construction demolition waste (CDW) constitutes roughly 36% of the total solid waste deposited in landfill sites worldwide, thereby posing a significant challenge to the sustainability of the construction industry. To address this issue, circular economy strategies are proposed as a solution. This paper systematically analyses 55 research articles published in leading peer-reviewed English-language scholarly journals over the past decade. It aims to identify and categorise drivers for enhanced CDW management by synthesising findings from previous research to support the principles of a circular economy. Utilising a PESTLE model for classification and analysis provides valuable insights into disparities and distinctions among categories, regions, and countries. The resulting analysis yields valuable insights into enablers and trends, with the aim of making a substantial contribution to mitigating the impact of construction activities and thus fostering the establishment of an efficient circular economy within the sector.

Keywords: construction waste; construction and demolition waste; waste management; driver; circular economy; sustainability

1. Introduction

The problem of waste is significantly concerning for the construction industry. Construction and demolition waste (CDW) originates from various building and infrastructure activities, including new construction, renovation, demolition, and land clearance. A total of 36% of the overall waste produced is attributed to the sector, which equates to between 2.5 and 3.5 billion tonnes on an annual basis [1].

The existing literature highlights several social, economic, and environmental challenges stemming from CDW. Firstly, there is landfill space depletion. The need for landfill space is rapidly increasing in numerous countries, especially in developing nations. For instance, the UK generates about 138 million tonnes of CDW annually, constituting two-thirds of total landfill waste [2]. EU countries produce over 800 million tonnes yearly, comprising 25–30% of total waste [3], while China exceeds 1.5 billion tonnes [4], and the USA reached 569 million tonnes in 2017 [5]. Gulf Cooperation Council countries generate 66 million tonnes annually, accounting for 55% of the total waste [6]. This growing waste generation occupies bigger spaces to be disposed of. Secondly, there is resource depletion. CDW plays a significant role in depleting global natural resources, including non-renewable energy sources and critical materials such as crushed stone, timber, and metal [7]. Construction operations alone consume approximately 35% of global resources, comprising 12% of water, 25% of steel, and exceeding 50% of sand, crushed rock, and gravel [8–10]. Additionally, the construction industry accounts for approximately 36% of global final energy usage, encompassing embodied energy [8].

Thirdly, there is contamination and pollution. Increasing CDW volumes pose significant environmental challenges, with construction activities significantly polluting soil, water, and air. Globally, approximately 33% of CDW is disposed of without restrictions, rising to 93% in developing nations where open dumping prevails [11,12]. Moreover, the construction industry's energy consumption accounts for 11% of global carbon dioxide emissions [8]. Lastly, financial losses occur due to waste generation in construction projects, leading to increased expenses, with approximately 15% of materials (by value) resulting in waste [13–15]. Estimates suggest that the actual cost of waste in construction activities is about 20 times the cost of disposal [16]. Nationally, CDW poses significant economic challenges, with over 20% of municipal budgets and 50% of local government investments allocated to solid waste management in low- and middle-income nations [12].

As an economic paradigm and approach for resource and waste management, a circular economy is designed to reduce waste, foster sustainability, and optimise the use of resources. In a circular economy, to prolong the lifespan of materials and products, they are reused, refurbished, remanufactured, and recycled [17]. The shift towards a circular economy aims to establish a more regenerative and sustainable economic framework, advantageous for both the economy and the environment. A primary challenge is the significant generation of CDW and the considerable consumption of resources, which presents a major obstacle to the sustainability of the construction sector, the country's economy at large, and the global environment [1]. To tackle this issue and align with circular economy principles, various waste management strategies have been devised to reduce waste throughout the lifecycle of construction projects. These strategies include waste collection, reduction, reuse, recycling (3Rs), low-waste technologies (LWTs), and landfill disposal charges [18–20]. However, despite these efforts, statistics on CDW disposal indicate that the industry has yet to achieve a closed-loop circular economy. Key barriers preventing effective CDW management have been identified by a study conducted by Ferriz-Papi JA et al. [21]. Their paper systematically analysed 54 articles published over the past decade, identifying and classifying a total of 59 challenges for successfully attaining circular economy transition in the construction sector. Understanding the drivers of CDW management is also essential. Comparing both barriers and drivers leads to a more comprehensive understanding of the factors at play, supports the development of balanced and effective strategies, fosters continuous improvement, engages stakeholders and decision-makers, and enhances academic research. Analysing existing drivers is therefore crucial to improving CDW management and achieving zero-waste targets.

Many global publications have attempted to identify key effective CDW management enablers for a closed-loop circular economy. The construction industry must engage with a whole variety of stakeholders including regulators responsible for technical standards, investors, the suppliers of materials and products, and the whole industry at large who are often sceptic about the quality of recycled products. Thus, this review paper aims to facilitate the development of appropriate solutions tailored to regional or national levels for implementing circular economy practices in the construction industry. It categorises these drivers using the PESTLE model (political, economic, social, technological, legal, and environmental), exploring their current trends and impacts on aspects like construction methods, technology access, culture, and country development as a means of identifying research trends and gaps for future development.

2. Materials and Methods

A systematic literature review is a robust, objective, and replicable method used to analyse existing research on a specific topic [22]. It serves as a valuable tool aimed at summarising the extant literature to establish the knowledge boundaries of a given subject and identify areas requiring further research. Given the specific context of CDW management, the PRISMA statement is a practical approach particularly useful for systematically reviewing the literature in social sciences studies and for conducting comprehensive reviews in this field [23]. It provides a comprehensive methodological framework and specific guidance on statistical techniques for meta-analysis and approaches for qualitative synthesis. This entails employing systematic literature searches for peer-reviewed material, screening, critical appraisal, metadata extraction, and content analysis. Figure 1 illustrates a simplified diagram outlining the methodological steps, following the PRISMA checklist.

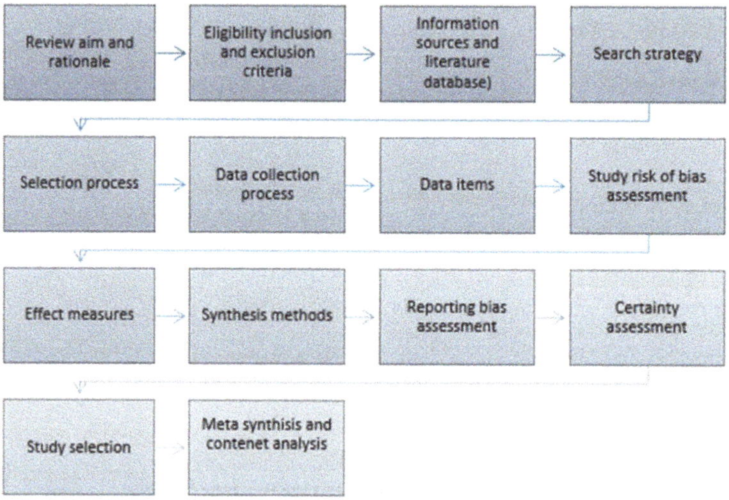

Figure 1. Methodological steps following the PRISMA guidelines for systematic reviews.

As the aim of this paper is to identify drivers for improved CDW management, the selection criteria used was primarily based on the direct relevance to the subject, while also considering studies related to the subject due to their substantial importance. To reduce bias in this process, an objective and transparent approach for research synthesis was adopted, comprising both quantitative and qualitative research papers including journal articles and peer-reviewed conference papers ensuring the quality and reliability of the research. ScienceDirect and Scopus databases, two of the leading citation index organisations, were used for this study. Approximately 99.11% of the journals indexed in Web of Science are also indexed in Scopus, indicating the extensive coverage of Scopus [24]. Consequently, the authors deemed Scopus sufficient for their research due to its significant coverage and its widespread use in previous studies, particularly in systematic reviews across various fields, including construction waste management. The terms 'construction waste management' (CW) and 'construction demolition waste management' (CDW) were used, combined with the terms 'drivers', 'enablers', 'opportunities', and 'practices' to select any papers where they were found in the title, abstract and/or keywords. There was no limit on the country of studies, but only studies written in English were included. This generated 564 papers (as of January 2023).

Thus, in order to limit this wide scope and to focus closely on the drivers of CDW, the search was amended to include those with explicit development on drivers and enablers within the time period of 2013 to 2022. Thus, 196 papers were identified; these were narrowed down following a rapid screening of the titles and abstracts and excluding any review papers as a means to ensure relevance to the timeframe and geographic relevance for the analysis of the metadata. Finally, the full texts of the remaining studies were evaluated based on the inclusion criteria, and any discrepancies were discussed between the authors until a consensus was reached. Notably, some papers could not be accessed, despite the efforts through institutional subscriptions and other available resources, due to restrictions such as a lack of institutional access or unavailability in public repositories.

As a result, 55 papers were identified as suitable for inclusion in this systematic review. A flow diagram illustrating the study selection process is presented in Figure 2.

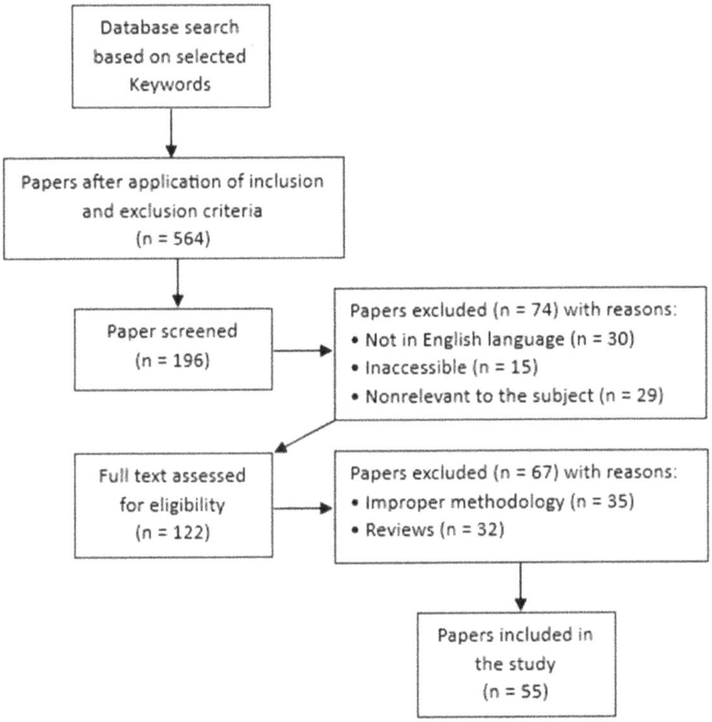

Figure 2. Reporting items diagram for study review process.

3. Results and Discussion

The regression analysis depicted in Figure 3 indicates a low to moderate positive correlation, suggesting an increasing trend in the number of publications related to this topic throughout the review period from 2013 to 2022. Although the significance of this correlation is relatively low, it highlights a growing research interest in this area over time. Notably, the years 2021 and 2022 exhibit the highest number of published papers. A substantial surge is observed in 2021, with a notable increase; 27.3% of the 55 articles reviewed were published in that year alone. In 2020, the European Commission introduced the Construction 2020 strategy and the New Circular Economy Action Plan aimed at stimulating Europe's transition towards a circular economy [25]. Additionally, the second edition of the Global Waste Management Outlook, published by UNEP and the International Solid Waste Association, assesses global waste management trends and strategies for 2020–2024. It has shaped global policies, emphasising the transition to a circular economy and zero-waste strategies [26]. These initiatives collectively created a conducive environment for research and policy development, leading to a marked increase in publications on CDW management by 2021. Another peak is also evident in 2016 and 2017. This surge may be attributed to new waste reduction targets established in the preceding years, such as those set forth in the Paris Agreement of 2015 [27] and the European Waste Framework Directive targets for CDW in 2020, as well as the framework amendment in 2018 [28]. Despite potential disruptions caused by the global COVID-19 pandemic on research output, the overall growth in publications persists.

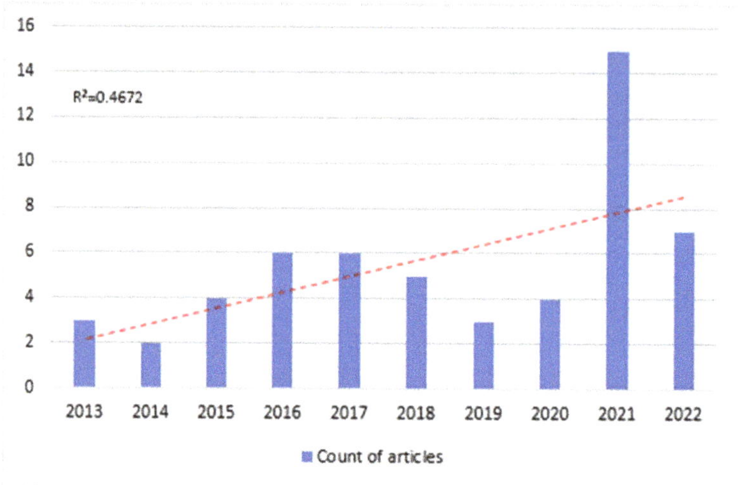

Figure 3. Annual publications trend on CDW from 2013 to 2022.

However, this does not necessarily indicate a universal increase in concern and implementation of improved CDW management practices across the global construction industry. Figure 4 illustrates the distribution of reviewed articles by continent. The geographical location of the papers reviewed in this study is determined by the focus of each paper, specifically the case study or the location where the data were collected. The results revealed that Asia (41%) is the primary source of the reviewed publications. China's contribution to Asian research is remarkable, accounting for 48% of the total articles reviewed. In contrast, no articles were contributed by countries such as India, Indonesia, and Japan despite their robust economies and substantial impact on the Asian economy, especially Japan's highly efficient and innovative construction sector. Consequently, future related research should consider such countries.

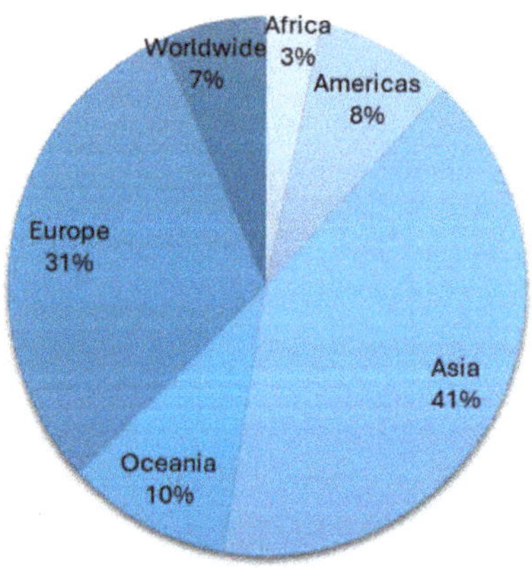

Figure 4. The geographical distribution of the reviewed CDW articles.

Europe is the second primary source of the reviewed publications (31%), with the UK leading with the highest number of reviewed articles, accounting for 35% of European research output in this field. The European Union is recognised for establishing the Waste Framework Directive, which sets the basic concepts and definitions related to waste management, such as recovery and recycling [28]. Meanwhile, Oceania and the Americas show relatively modest research contributions, representing only 10% and 8% of the total, respectively. Four articles were identified from the USA (i.e., [29–32]), despite the enormous CDW volumes generated by the country, which reached 569 million tonnes in 2017 [5], highlighting an urgent need for CDW management. This is in addition to the United States Environmental Protection Agency's initiatives towards CDW management [33]. Finally, Africa has the least reviewed articles, with 3%. This calls for increased research efforts in these underrepresented countries and continents.

3.1. PESTLE Drivers Classification

A frequency count was conducted for each driver, during the review process, and referenced to the source article. In total, 200 references to drivers were identified across the 55 reviewed papers. These references were categorised into 29 drivers, summarised in Table 1. The drivers were then ranked in ascending order of frequency and categorised using the PESTLE analysis approach (political, economic, social, technological, legal, and environmental). Professor Francis Aguilar is credited with creating the ETPS analysis in 1967 [34]. This strategic tool helps identify the key external factors influencing an organisation and can be applied across various scenarios to assist professionals and senior managers in making strategic decisions. In this study, political and legal drivers were grouped together to streamline the discussion.

Table 1. Classification of drivers based on PESTLE analysis.

Code	Drivers	Article ID	PESTLE Category	Frequency	Rank
D1	Adequate formulation and application of governmental CDW regulations and policies	[32,35–51]	Political/legal	18	1
D2	Training and education about CDW management aspects (causes, types, LWTs, cost savings, environmental impact)	[19,30–32,36,39,40,42,46,49,52–56]	Social	15	2
D3	Provision of innovative low-waste technologies (LWTs)	[29,36,39,42,51,52,54,57–60]	Technological	11	3
D4	Adequate formulation and application of CDW protocols, programmes, and strategies	[32,39,48,53,59–64]	Political/legal	10	4
D5	Commitment of client and contractor by allocating time and money for CDW management	[36,37,39,40,42,43,51,54,65,66]	Economic	10	4
D6	Attitude and behaviour towards CDW management	[35,36,41,42,49,63,65,67,68]	Social	9	6
D7	Data availability about CDW management	[32,37,57,60,62,65,69–71]	Technological	9	6
D8	Adequate and efficient CDW recycling infrastructure	[19,37,42,53,63,65,72,73]	Technological	8	8
D9	CDW landfilling volume restrictions and increased fees	[19,36,44,50,53,68,74]	Political/legal Economic	7	9
D10	Mature market for reused and recycled CDW	[19,30–32,42,52]	Social Economic	6	10
D11	Mandatory requirement for site waste management plans (SWMPs)	[45,66,70,75–77]	Political/legal	6	10
D12	Cooperation and integration between stakeholders in value chain	[31,56,68,77,78]	Social	5	12
D13	Governmental funding and financial support	[35,38,42,50,56]	Economic Political/legal	5	12
D14	Use of waste prediction tools	[66,71,79–81]	Technological	5	12
D15	Adoption of offsite construction	[43,45,53,76]	Technological	4	15
D16	Adoption of waste traceability system	[29,47,55,60]	Technological	4	15
D17	Standardisation of design	[19,59,66,77]	Technological	4	15
D18	Development of environmental standards for design, construction, and demolition activities	[35,50,59,76]	Political/legal Environmental	4	15
D19	Municipality acting as facilitator	[38,42,50,56]	Political/legal	4	15
D20	Establishing business module for CWM practices	[29,52,53,82]	Economic Technological	4	15

Table 1. *Cont.*

Code	Drivers	Article ID	PESTLE Category	Frequency	Rank
D21	Communication and collaboration platforms for development/implementation of circular economy value chains	[29,56,61]	Technological	3	21
D22	Incentivising project teams by offering benefits and awards	[39,54,66]	Economic	3	21
D23	Skills and experience in construction	[36,77,83]	Social	3	21
D24	R&D in circular economy	[39,41,53]	Technological	3	21
D25	Efficient procurement system	[66,74,77]	Economic Technological	3	21
D26	Pressure from environmental organisations/activists	[38,40]	Environmental	2	26
D27	Availability of waste-sharing platforms	[29,73]	Technological	2	26
D28	Mandatory requirement for selective demolition	[56,84]	Political/legal	2	26
D29	Development of integrated environmental and economic management model	[59]	Environmental Economic	1	29
TOTAL				200	

Based on the frequency of the drivers in each category, it is evident that technological drivers are the most cited, accounting for nearly one-third of all references (30.0%), while environmental drivers are represented by just 3.5%. This indicates a remarkable emphasis on technological factors with minimal attention to environmental factors, even though the primary aim of the circular economy is to reduce environmental impact. While all other driver categories share the common goal of improving the transition towards a circular economy, environmental aspects alone do not appear to be strong enough to incentivise the adoption of CDW management among stakeholders.

Notably, in the paper by Ferriz-Papi et al. [21], the environmental parameters included a limited number of barriers. This is expected and natural because the main purpose of the circular economy concept is to reduce the environmental impact of construction industry activities. Therefore, environmental factors are more likely to act as drivers to achieve these concepts rather than as barriers. Nonetheless, certain challenges, such as the risk of increasing the volume of hazardous waste through recycling and the energy consumption required for recycling treatments, can inhibit the willingness to adopt certain CDW management methods. Regarding the technological parameter, it was also identified as the primary category containing the highest number of barriers and was emphasised the most by the papers reviewed according to Ferriz-Papi et al. [21]. Although technological advancements can largely incentivise stakeholders to adopt CDW management methods, challenges such as the availability of technological infrastructure and the complexity or inadequacy of regulatory frameworks and standardised processes can undermine the value and potential of this category in CDW management. This is similar to other categories, including political, legal, economic, and social aspects. Therefore, it is vital that both barriers and drivers are considered when assessing the current uptake of CDW management in the circular economy in order to support the development of balanced and effective strategies and foster continuous improvement.

3.1.1. Political/Legal Drivers

Government regulations and policies, typically utilising a top-down strategy, are central to this category of drivers. Political decisions play a crucial role in creating frameworks for waste production and management by establishing and implementing targets and policies at a broader level. In this regard, the most significant driver (D1) featured in 18 articles [32,35–51] called for 'adequate' governmental CDW regulations and policies. This need was identified in articles whose research was based primarily in developed nations that have national CDW targets (mainly Italy, the UK, China, and Australia) that cited confusion between what is to be considered mandatory and what the guidelines are, particularly in regional and national contexts—and in countries where gaps in regulation need to be addressed such as Pakistan [43], Saudi Arabia [44], South Africa [36], and Vietnam [41].

The need for 'adequate' regulations and policies (D1) is perpetuated in D4, which heralded the need for 'adequate' CDW protocols, programmes, and strategies that could support organisations to conform to any regulatory requirements [32,39,48,53,59–64], as it was often left to individuals/organisations to develop their own implementation solutions. This was deemed a priority in the review papers covering China [(48,59,61,63)]. Notably, 'solutions' that were identified to be essential to support improved management of CDW included site waste management plans (SWMPs) (D11) [45,66,70,75–77] and establishing a business module for CWM practices (D20) [29,52,53,82]. Seven articles went further and stipulated the need for CDW landfilling volume restrictions and increased fees (D9), primarily in China and the UK where landfill space is at a premium [19,36,44,50,53,68,74]. Finally, regulations governing selective demolition (D28) were advocated to increase the lifespan of buildings whereby retrofit or building reuse is favoured over land tenure.

3.1.2. Economical Drivers

Economic drivers are linked to business profitability, highlighting the benefits of adopting CDW management over traditional construction practices and implementing more economical solutions. The most frequently mentioned driver in this category pertains to the commitment of clients and contractors to allocate time and resources for CDW management (D5). Decision-makers involved in construction projects need to ensure adequate provisions for waste management, including the allocation of sufficient time and resources. Specific operations necessary for waste management include supervision and monitoring [45], the segregation of waste [59,66], transportation [37,42], processing/treatment, and disposal [59,64,65]. Financial support was noted as a relevant driver, through national government (D13) and regional government/municipality schemes (D19). These drivers intensify efforts to reduce CDW costs [36,54] and improve recycling rates [71], thereby fostering greater confidence in recovery initiatives and positioning them at the core of numerous case studies [82]. For instance, Adams et al. [52] stress the importance of financial support for considering end-of-life aspects during the design phase, which would significantly contribute to enhanced reuse and recycling activities, especially given the high cost of landfilling in many countries. Indeed, incentivisation at the project level (D22) was deemed critical [39,54,66] as it was noted that the success of any CDW intervention was largely reliant on individual workers' perception of sustainability, particularly as CDW contributes to cost overruns in projects; the client and contractor were noted for the need to make firm resource commitment for CDW management (D5).

3.1.3. Social Drivers

Social drivers pertain to behavioural and cultural aspects, covering a range of human factors within the construction industry. These include training and education (D2) on the causes, types, technologies, and financial/environmental impacts of CDW, which are considered major drivers in the sector [19,30–32,36,39,40,42,46,49,52–56]. This is affirmed in D23, whereby skills and experience were also listed as pertinent to minimising waste from the design, construction, and demolition stages of a project [36,77,83]. Additionally, public awareness and attitudes play a significant role in shaping cultural perspectives towards waste, such as changing perceptions concerning the inevitability of waste generation and providing adequate education on waste recovery [35,36,51,52,68,85].

In social psychology research, it is suggested that an individual's reaction to related situations and objects is influenced by their attitude, which is a neutral and mental state of willingness organised through experience. This state exerts an influence on their behaviour, either directive or dynamic [86]. Of significance, the attitudes of construction stakeholders were identified as the sixth highest ranked driver (D6) and critical for any CDW management practice to be successfully implemented. The acceptance of secondary materials is influenced by the perceived quality, whether low or high, of recycled products [50,63,65], as well as the preference for virgin materials over recycled ones [41,42,59]. Oke et al. [36] emphasise the necessity for citizens' behaviours to change towards consumption patterns. Additionally, it is acknowledged that a willingness to change acts as a significant driver for the adoption of CDW management methods. According to Tirado et al. [55] and Galvez–Martos et al. [84], improvements in waste management in construction are highly achieved through the utilisation of existing technology, coupled with a commitment to waste reduction.

A marketplace for reused and recycled CDW is needed to mature (D10) as, unfortunately, the construction industry is lagging, and stigma still prevails in the use of recycled materials in buildings, largely due to health and safety, and the lack of reliability/predictability [and costs] involved. Finally, cooperation and integration among stakeholders (D12) are crucial, potentially facilitated by an efficient corporation among governmental bodies, policymakers, and enterprises [38,77,78], communication and collaboration platforms across the entire circular economy value chain (D21; [56]), or through a waste-sharing platform (D27; [29]).

3.1.4. Technological Drivers

Technological drivers relate to the accessibility of infrastructure, tools, and innovative processes for CDW management. A number of modern methods of construction (MMC) were identified as drivers in the review articles; MMC is a term used to describe innovative construction processes, techniques, and technologies that differ from traditional onsite construction methods, typically including offsite and/or digital solutions to design and construct buildings in a more efficient, sustainable, and cost-effective manner. Low-waste technologies (LWTs) were identified as the third highest-ranked factor (D3) with 11 articles [29,36,39,42,51,52,54,57–60]; LWTs are seen to optimise resource consumption to ensure waste minimisation [20]. However, authors are yet to agree if MMC is a component of LWTs or vice versa (see Table 2). A key MMC technology listed as a driver was offsite construction (D15; [43,45,53,76]), particularly in countries such as Ethiopia [76] and Pakistan [43] where uptake is limited.

Table 2. Types of LWTs.

Types of LWTs	References
Soft technologies (i.e., information and communication technologies)	[87–90]
Hard technologies (i.e., innovative construction tools and equipment)	[20,91,92]
Modern methods of construction (MMC, i.e., offsite construction, BIM, 3D printing)	[91,93,94]

A waste traceability system was also put forward (D16), and whilst clearly not a novel idea, there is an inherent need in the construction sector to ensure accountability and responsibility as well as data availability on CDW management (D7); a waste sharing platform (D27); and/or waste prediction tools (D14). These approaches could perhaps be encapsulated in the need for more R&D in the circular economy (D24), which was only featured in articles published from Australia, Spain, and Vietnam [39,41,53], a surprising fact given the global need for improved practice. Adequate and efficient CDW recycling infrastructure (D8), coupled with a marketplace for reused and recycled CDW materials (D10) were identified as other technological drivers (see also social factors). Improved access to waste treatment facilities equipped with modern innovative technologies, especially in remote areas, could significantly enhance waste recovery targets and divert CDW from landfills. Additionally, in situations where waste disposal as an end-of-pipe solution is the only available option, the presence of engineered landfill infrastructure allows for secure final waste disposal, thereby minimising environmental impacts [29,42,44,64].

3.1.5. Environmental Drivers

Whilst the authors have only categorised three primary environmental drivers in this review, D18, D26, and D29, it must be noted that the premises of all drivers stem from the need to address environmental sustainability. CDW generates several significant environmental issues, including reduced landfill capacity due to the growing volume of disposed waste, substantial consumption of resources, contamination and pollution leading to severe health problems, and an increased consumption of energy for the transportation of waste and the production of new materials to replace the discarded ones [11,95,96]. Consequently, the effective management of CDW is becoming increasingly crucial to safeguard the natural ecosystem and public health. Two articles noted the increasing pressure from environmental organisations/activists (D26) as a strong driver [38,40], particularly from a diminishing landfill space, the overmining of raw materials, and general pollution and contamination perspectives. Many construction stakeholders do not prioritise environmental issues in the management of CDW. According to Serpell et al. [85], small- and medium-sized enterprises focus solely on cost considerations, neglecting the social and environmental pillars of sustainability in waste reduction. In contrast, large organisations address all three pillars of sustainability due to the requirement for sustainable projects by clients and the necessity of accountability reporting.

Notably, it was suggested by article [59] that an integrated environmental and economic management model was needed to address CDW globally so as to make any inroads on the circular economy challenge (D29). Whilst this was only cited by one review article [59], it is reasonable to suggest a clear link between this driver and D18 and D17 which call for design environmental standards.

3.2. Drivers Analysis by Geographic Region

It can be observed that some of the identified drivers (Table 1) appear more frequently in articles published in particular continents and countries than in others (Figure 5). Drivers focusing on the improvement of waste management regulations, policies, and environmental performance are pressuring issues in Europe (D1, D9, D11, D13, D19, D26, D28). This can be attributed to the fact that the foundation and basic fundamentals of managing CDW are well developed in most European countries; therefore, the focus would be stronger to ensure an adequate and efficient application of these fundamentals to achieve better performance in CDW management. This is in addition to seeking further advancement in the application of offsite construction (D15) and selective demolition techniques (D28).

On the other hand, personal factors including construction skills, experience, and behaviour towards CDW management are more dominant in Asia (D5, D6, D23), with China alone covering around 20% of the total number of selected articles for this review. Enhancing the construction practices of the supply chain (D17, D25), the development of efficient recycling infrastructures (D8), and improvements in CDW management practices from basic waste classification, sorting, and quantification [37] to further developments of waste data management (D7) are also identified as a major concern in Asia.

Low representativity in the articles reviewed was noticeable from the Americas, for which reason they are analysed together. The most relevant drivers for this continent are D2, D10, D16, D26, and D27. The only published papers found are from the USA, Columbia, and Costa Rica. The need for training and education on CDW (D2) aspects is evident in all previous countries [30–32,40]. However, the papers published in the USA also reflect on other drivers such as establishing business models and sharing platforms for CWM practices (D21, D27). These drivers are considered significant to countries that are advanced in solid waste management like the USA [29].

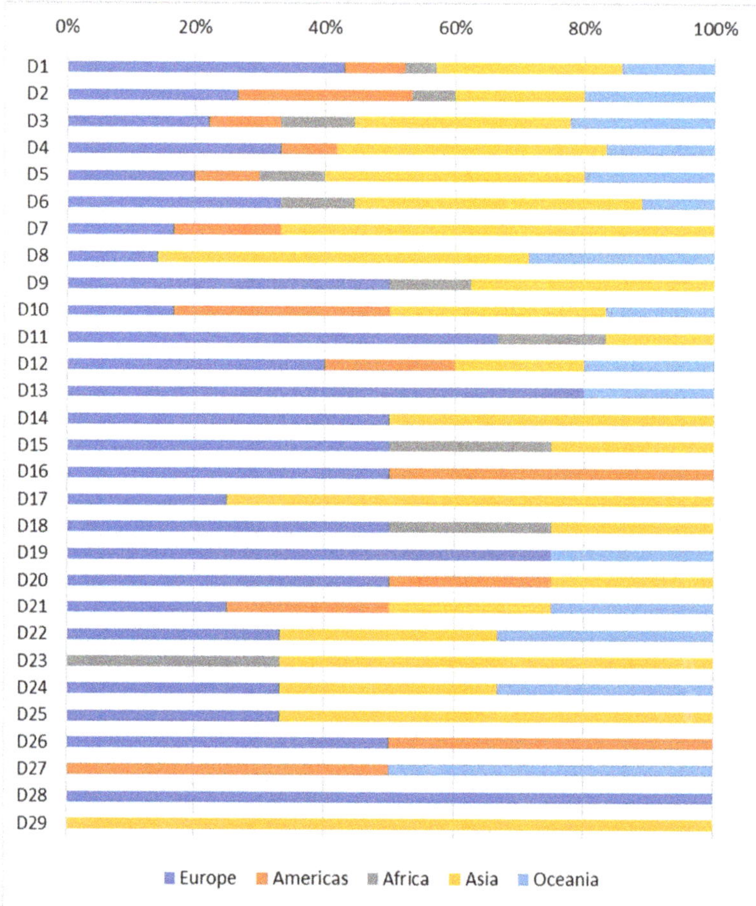

Figure 5. Distribution of drivers per continent.

Regarding Oceania, and considering their small continent size, their representativity in this analysis is substantial. The most relevant drivers that can be identified for this continent are D1–D5, and D8, although other drivers are present as well but with less frequency of occurrence. Making a comparison between Europe and Oceania, it can be observed that they largely follow a similar pattern, with many repeated drivers identified. The notable exception is the need to establish adequate recycling infrastructure (D8) which has a higher relevance than in Oceania; other drivers involving LWTs, stakeholders' commitment in CDW, collaboration platforms, benefits and incentives, and R&D in CDW management (D3, D5, D21, D22, D24) are at the same level of urgency in both continents. Finally, Africa has the shortest representation in this analysis, with only 5%. This continent is present in D1, D2, D3, D5, D6, D9, D11, D15, D18 and D23. The drivers identified are about adequate policies and regulation, innovation, training, stakeholders' commitment, and skills and behaviours.

3.3. Drivers Analysis by Timeframe

Figure 6 demonstrates the nature of the identified drivers in this systematic review during the past decade, from 2013 to 2022. This figure enables the ease of comparison between the article survey years for each of the drivers: the warmer colours (i.e., red, orange, and yellow) represent articles published from 2013, and the colder colours (i.e., green, blue, and black) highlight publications that were published more recently. Thus, it can be observed that in the last 5 years of this analysis timeframe, an increased focus has been placed on the majority of drivers, with a larger prevalence in 2021 and 2022 publications. This matches the growing concern with regard to the need for a circular economy in the construction industry and improved CDW management and recovery during the past few years.

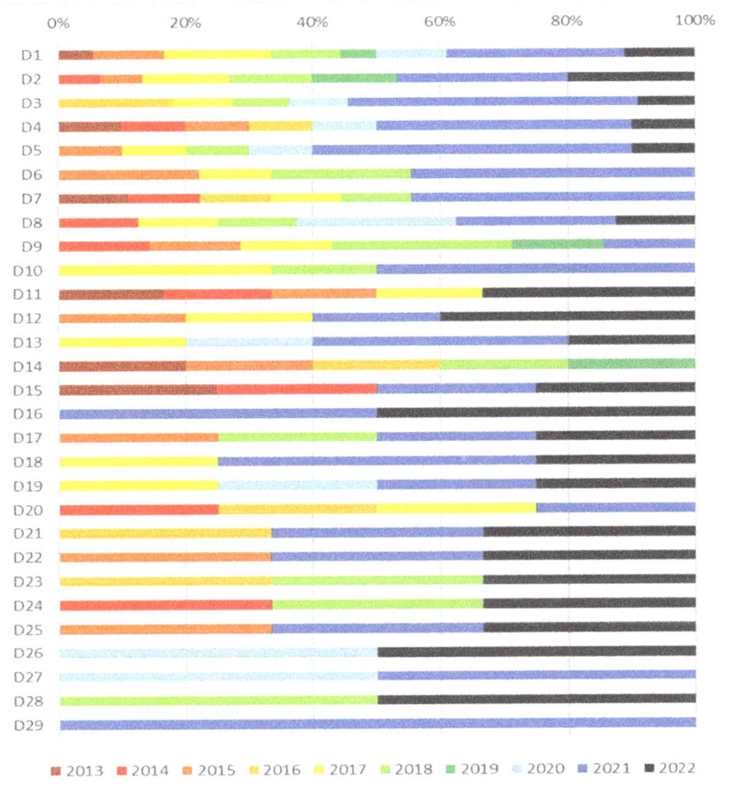

Figure 6. Distribution of drivers per year of publication.

Arguably, there have been different published regulations and policies that have influenced the implementation of a circular economy worldwide in the last decade. This includes, but is not limited to, the Paris Agreement in 2015 [27]; the amendment of the European Waste Framework Directive in 2019 [28]; the new European Circular Economy Action Plan in 2020 [25]; and the zero-waste targets and strategies arising in different countries. These events have largely contributed to the increased attention to a circular economy in the construction sector as in other sectors, by bringing new climate change agreements and goals to the regional and international plane.

Notably, following these policy milestones, three main period intervals can be identified according to the data in Figure 6. The first interval (2013–2015) involved drivers focusing on the establishment of strategies, plans, and protocols as a baseline for attaining a circular economy in construction (D4, D11, D14, D15). The second interval (2016–2019) required advanced solutions such as the creation of new business models, collaboration platforms, and waste prediction tools and putting more emphasis on the development of circular product markets and recycling infrastructure (D8, D10, D14, D20, D21). The final interval (2020–2022), however, involved most of the drivers identified in this paper indicating a globalisation of the circular economy concern. Many waste management aspects are now being considered from a multi-lateral and multi-disciplinary perspective, with evidence of clear interaction and collaboration between countries of different backgrounds and development levels [47,60,61,69,72,84]. More concern is placed on aspects such as the innovation and improvement of existing infrastructure and management procedures, the development of policies and regulations, collaboration and communication, government support, etc.

4. Conclusions

A structured review and meta-analysis of drivers were performed to support circular economy improvements in the construction industry. A total of 55 scientific articles were selected from all countries corresponding to the period of 2013–2022. The analysis of the annual publications trend showed a growing interest in this matter, with 27.3% of articles published in 2021 alone despite any impact the COVID-19 pandemic may have had on the number of articles published during that year.

Most of the articles reviewed were from Europe (31.0%) and Asia (41.0%), with China, the UK, and Australia being the most prolific countries. On the other hand, there were very few articles published on studies in Africa and South America. This suggests many possibilities for that reason such as a lack of priority of CDW in that continent, or even the reduced amount of research in general developed in those countries. Many other factors can have an impact on this, like less urbanised areas, or even the selection criteria for this study (e.g., this research could be written in other languages). Further analysis would be needed to understand these factors and the impact they can have in the development of research around CDW management.

In total, 29 drivers were identified through a frequency analysis of the selected articles and were classified and grouped according to PESTLE analysis (political, economic, social, technological, legal, and environmental), discussing similarities and differences between continents, regions, and countries. The five most repeated drivers were the following:

- D1—adequate formulation and application of government CDW regulations and policies
- D2—training and education about CDW management aspects (causes, types, LWTs, cost savings, environmental impact)
- D3—provision of innovative low-waste technologies (LWTs)
- D4—adequate formulation and application of CDW protocols, programmes, and strategies
- D5—commitment of client and contractor by allocating time and money for CDW management

Different drivers could be recognised as associated with the CDW management practices of some countries and regions and also associated with different levels of development. For example, an improvement of awareness, behaviours, and skills are more considered in Asia together with efficient recycling infrastructures, whereas in Europe there are more considerations about the improvement of political strategies and legal aspects, management procedures, and pressure from environmental organisations. An evolution of the circular economy in construction was distinguished in the period of 2013–2022: initial steps identifying inadequacies, needs, and potential solutions, demanding more guidance and tools for CDW management (2013–2015); the requirement of specialism, the creation of new business models, and the consideration of the full supply chain (2016–2019); and the globalisation of the circular economy concern, involving a multi-lateral and interdisciplinary perspective (2020–2022).

It is evident of the need to move forward towards a more circular and sustainable production system. The construction industry's contribution is paramount to achieving zero-waste targets by 2050. A more regulated, (digitally) connected, collaborative, and skilled sector is demanded, considering the full supply chain in the whole life cycle of construction projects and the waste value chain, which seem to be the steps that will be able to transform the construction industry towards a circular system.

A limitation of this systematic review was the exclusion of papers not written in English. This language restriction might have led to the omission of some relevant research conducted in non-English-speaking regions. Additionally, the review was limited to studies published between 2013 and 2022. While this timeframe captures recent developments, as this past decade has seen increased attention to the circular economy in construction (as discussed in the Section 1), it might have excluded important earlier research; therefore, future research could extend the timeframe of the review. Furthermore, the process of reading, analysing, and abstracting data from the selected papers, despite following a systematic approach, inherently involves some degree of subjective interpretation.

Author Contributions: Conceptualisation, M.A., J.A.F.-P. and A.L.; methodology, M.A., J.A.F.-P. and A.L.; validation, M.A., J.A.F.-P. and A.L.; formal analysis, M.A.; investigation, M.A., J.A.F.-P. and A.L.; data curation, M.A.; writing—original draft preparation, M.A. and A.L.; writing—review and editing, M.A., J.A.F.-P. and A.L.; visualisation, M.A. and A.L.; supervision, M.A. All authors have read and agreed to the published version of the manuscript.

Funding: This research was funded by Horizon Europe (grant agreement number 101058580, 2022) and by Innovate UK as part of the UK Guarantee programme for UK Horizon Europe participation (project number 10038579).

Acknowledgments: We would like to thank Bailey Chapman (University of Salford, UK) and Lucie Grant (Czech Technical University in Prague) for their contribution to the initial steps of the review process that is presented in this article.

Conflicts of Interest: The authors declare no conflicts of interest.

References

1. ISWA. 2015 ISWA Annual Report. 2015. Available online: https://www.iswa.org/annual-reports (accessed on 28 July 2023).
2. DEFRA. UK Statistics on Waste. Available online: https://www.gov.uk/government/statistics/uk-waste-data/uk-statistics-on-waste (accessed on 28 July 2023).
3. Deloitte. Study on Resource Efficient Use of Mixed Wastes Improving Management of Construction and Demolition Waste—Final Report. 2017. Available online: https://ec.europa.eu/environment/pdf/waste/studies/CDW_Final_Report.pdf (accessed on 29 July 2023).
4. Chang, Y.; Li, X.; Masanet, E.; Zhang, L.; Huang, Z.; Ries, R. Unlocking the green opportunity for prefabricated buildings and construction in China. *Resour. Conserv. Recycl.* **2018**, *139*, 259–261. [CrossRef]
5. EPA. Advancing Sustainable Materials Management: 2017 Fact Sheet. 2017. Available online: https://www.epa.gov/sites/default/files/2019-11/documents/2017_facts_and_figures_fact_sheet_final.pdf (accessed on 29 July 2023).
6. Ouda, O.K.M.; Peterson, H.P.; Rehan, M.; Sadef, Y.; Alghazo, J.M.; Nizami, A.S. A Case Study of Sustainable Construction Waste Management in Saudi Arabia. *Waste Biomass Valorization* **2018**, *9*, 2541–2555. [CrossRef]
7. Akenji, L.; Bengtsson, M.; Emily Briggs Chiu, A.; Daconto, G.; Fadeeva, Z. *Sustainable Consumption and Production: A HANDBOOK for Policymakers*; Briggs, E., Ed.; Global Edition; UNEP: Nairobi, Kenya, 2015; 212p.
8. IEA; UNEP. *2019 Global Status Report for Buildings and Construction: Towards a Zero-Emissions, Efficient and Resilient Buildings and Construction Sector*; International Energy Agency: Hamburg, Germany, 2019. Available online: https://www.unep.org/resources/publication/2019-global-status-report-buildings-and-construction-sector (accessed on 29 July 2023).
9. UNEP; IRP. Global Resources Outlook 2019 Natural Resources for the Future We Want. 2019. Available online: https://wedocs.unep.org/handle/20.500.11822/27517 (accessed on 29 July 2023).
10. UNEP. Sand and Sustainability: Finding New Solutions for Environmental Governance of Global Sand Resources. 2019. Available online: https://wedocs.unep.org/handle/20.500.11822/28163 (accessed on 29 July 2023).
11. Ferronato, N.; Torretta, V.; Ragazzi, M.; Rada, E.C. Waste Mismanagement in Developing Countries: A Case Study of Environmental Contamination. *UPB Sci. Bull. Ser. D* **2017**, *79*, 185–196.
12. World Bank. What a Waste 2.0: A Global Snapshot of Solid Waste Management to 2050. 2018. Available online: https://openknowledge.worldbank.org/bitstream/handle/10986/30317/9781464813290.pdf?sequence=12&isAllowed=y (accessed on 29 July 2023).
13. WRAP. Achieving Good Practice Waste Minimisation and Management Guidance for Construction Clients, Design Teams and Contractors. Available online: https://www.eauc.org.uk/file_uploads/wrap2.pdf (accessed on 29 July 2023).
14. Mahpour, A.; Mortaheb, M.M. Financial-Based Incentive Plan to Reduce Construction Waste. *J. Constr. Eng. Manag.* **2018**, *144*, 4018029. [CrossRef]
15. Hao, J.; Yuan, H.; Liu, J.; Chin, C.S.; Lu, W. A model for assessing the economic performance of construction waste reduction. *J. Clean. Prod.* **2019**, *232*, 427–440. [CrossRef]

16. Osmani, M. Construction Waste. In *Waste*; Academic Press: Cambridge, MA, USA, 2011; pp. 207–218.
17. World Economic Forum. What Is the Circular Economy and Why Does It Matter That It Is Shrinking? 14 June 2022. Available online: https://www.weforum.org/agenda/2022/06/what-is-the-circular-economy/ (accessed on 7 July 2023).
18. Alhawamdeh, M.; Lee, A. Construction Waste Minimisation: A Narrative Review. *Int. J. Environ. Sustain.* **2021**, *18*, 1. [CrossRef]
19. Huang, B.; Wang, X.; Kua, H.; Geng, Y.; Bleischwitz, R.; Ren, J. Construction and demolition waste management in China through the 3R principle. *Resour. Conserv. Recycl.* **2018**, *129*, 36–44. [CrossRef]
20. Zhang, X.; Wu, Y.; Shen, L. Application of low waste technologies for design and construction: A case study in Hong Kong. *Renew. Sustain. Energy Rev.* **2012**, *16*, 2973–2979. [CrossRef]
21. Ferriz-Papi, J.A.; Lee, A.; Alhawamdeh, M. Examining the Challenges for Circular Economy Implementation in Construction and Demolition Waste Management: A Comprehensive Review Using Systematic Methods. *Buildings* **2024**, *14*, 1237. [CrossRef]
22. Levy, Y.; Ellis, T.J. Informing Science Journal A Systems Approach to Conduct an Effective Literature Review in Support of Information Systems Research. *Inf. Sci. J.* **2006**, *9*, 181–212.
23. Peters, M.D.; Godfrey, C.M.; Khalil, H.; McInerney, P.; Parker, D.; Soares, C.B. Guidance for conducting systematic scoping reviews. *JBI Evid. Implement.* **2015**, *13*, 141–146. [CrossRef] [PubMed]
24. Singh, V.K.; Singh, P.; Karmakar, M.; Leta, J.; Mayr, P. The journal coverage of Web of Science, Scopus and Dimensions: A comparative analysis. *Scientometrics* **2021**, *126*, 5113–5142. [CrossRef]
25. European Commission. Circular Economy Action Plan. Available online: https://environment.ec.europa.eu/strategy/circular-economy-action-plan_en (accessed on 30 July 2023).
26. United Nations Environment Programme, International Solid Waste Association. *Global Waste Management Outlook 2024—Beyond an age of waste: Turning Rubbish into a Resource*; United Nations Environment Programme: Nairobi, Kenya, 2024. Available online: https://wedocs.unep.org/20.500.11822/44939 (accessed on 13 May 2024).
27. UN Climate Change. The Paris Agreement Related News Related Documents Related Links What is the Paris Agreement? Available online: https://unfccc.int/process-and-meetings/the-paris-agreement (accessed on 30 July 2023).
28. European Commission. European Waste Framework Directive. 2019. Available online: https://eur-lex.europa.eu/legal-content/EN/TXT/HTML/?uri=CELEX:32019D1004&qid=1690732569938 (accessed on 30 July 2023).
29. Guerra, B.C.; Shahi, S.; Mollaei, A.; Skaf, N.; Weber, O.; Leite, F.; Haas, C. Circular economy applications in the construction industry: A global scan of trends and opportunities. *J. Clean. Prod.* **2021**, *324*, 129125. [CrossRef]
30. Guerra, B.C.; Bakchan, A.; Leite, F.; Faust, K.M. BIM-based automated construction waste estimation algorithms: The case of concrete and drywall waste streams. *Waste Manag.* **2019**, *87*, 825–832. [CrossRef] [PubMed]
31. Abarca-Guerrero, L.; Leandro-Hernandez, A.G. Material management practices for construction waste reduction. *WIT Trans. Ecol. Environ.* **2017**, *223*, 551–557.
32. Guerra, B.C.; Leite, F. Circular economy in the construction industry: An overview of United States stakeholders' awareness, major challenges, and enablers. *Resour. Conserv. Recycl.* **2021**, *170*, 105617. [CrossRef]
33. EPA. *Sustainable Management of Construction and Demolition Materials*; US EPA: Washington, DC, USA, 2023. Available online: https://www.epa.gov/smm/sustainable-management-construction-and-demolition-materials (accessed on 7 July 2023).
34. Aguilar, F.J. Scanning the Business Environment. Available online: https://lccn.loc.gov/67011688 (accessed on 13 May 2024).
35. Cristiano, S.; Ghisellini, P.; D'Ambrosio, G.; Xue, J.; Nesticò, A.; Gonella, F.; Ulgiati, S. Construction and demolition waste in the Metropolitan City of Naples, Italy: State of the art, circular design, and sustainable planning opportunities. *J. Clean. Prod.* **2021**, *293*, 125856. [CrossRef]
36. Oke, A.; Aigbavboa, C.; Aghimien, D.; Currie, N. Construction Professionals Perception of Solid Waste Management in the South African Construction Industry. In *Advances in Human Factors, Sustainable Urban Planning and Infrastructure*; Charytonowicz, J., Falcão, C., Eds.; Springer: Berlin/Heidelberg, Germany, 2019; pp. 496–505.
37. Yuan, H. Barriers and countermeasures for managing construction and demolition waste: A case of Shenzhen in China. *J. Clean. Prod.* **2017**, *157*, 84–93. [CrossRef]
38. Ghaffar, S.H.; Burman, M.; Braimah, N. Pathways to circular construction: An integrated management of construction and demolition waste for resource recovery. *J. Clean. Prod.* **2020**, *244*, 118710. [CrossRef]
39. Shooshtarian, S.; Hosseini, M.R.; Kocaturk, T.; Arnel, T.; Garofano, N.T. Circular economy in the Australian AEC industry: Investigation of barriers and enablers. *Build. Res. Inf.* **2023**, *51*, 56–68. [CrossRef]
40. Torres-Guevara, L.E.; Prieto-Sandoval, V.; Mejia-Villa, A. Success Drivers for Implementing Circular Economy: A Case Study from the Building Sector in Colombia. *Sustainability* **2021**, *13*, 1350. [CrossRef]
41. Van Tuan, N.; Kien, T.T.; Huyen, D.T.T.; Nga, T.T.V.; Giang, N.H.; Dung, N.T.; Isobe, Y.; Ishigaki, T.; Kawamoto, K. Current status of construction and demolition waste management in Vietnam: Challenges and opportunities. *Int. J. GEOMATE* **2018**, *15*, 23–29. [CrossRef]
42. Shooshtarian, S.; Caldera, S.; Maqsood, T.; Ryley, T.; Khalfan, M. An investigation into challenges and opportunities in the Australian construction and demolition waste management system. *Eng. Constr. Archit. Manag.* **2022**, *29*, 4313–4330. [CrossRef]
43. Khan, A.R.; Ditta, A.; Mehmood, M.S.; MaoSheng, Z.; Natalia, M. Determinants and implications of environmental practices for waste management and the minimization in the construction industry: A case study of Pakistan. *Environ. Sci. Pollut. Res.* **2021**, *28*, 58221–58231. [CrossRef] [PubMed]

44. Blaisi, N.I. Construction and demolition waste management in Saudi Arabia: Current practice and roadmap for sustainable management. *J. Clean. Prod.* **2019**, *221*, 167–175. [CrossRef]
45. Saez, P.V.; Merino, M.d.R.; González, A.S.-A.; Porras-Amores, C. Best practice measures assessment for construction and demolition waste management in building constructions. *Resour. Conserv. Recycl.* **2013**, *75*, 52–62. [CrossRef]
46. Crawford, R.H.; Mathur, D.; Gerritsen, R. Barriers to Improving the Environmental Performance of Construction Waste Management in Remote Communities. *Procedia Eng.* **2017**, *196*, 830–837. [CrossRef]
47. Giorgi, S.; Lavagna, M.; Wang, K.; Osmani, M.; Liu, G.; Campioli, A. Drivers and barriers towards circular economy in the building sector: Stakeholder interviews and analysis of five European countries policies and practices. *J. Clean. Prod.* **2022**, *336*, 130395. [CrossRef]
48. Wu, H.; Duan, H.; Wang, J.; Wang, T.; Wang, X. Quantification of carbon emission of construction waste by using streamlined LCA: A case study of Shenzhen, China. *J. Mater. Cycles Waste Manag.* **2015**, *17*, 637–645. [CrossRef]
49. Li, J.; Tam, V.W.Y.; Zuo, J.; Zhu, J. Designers' attitude and behaviour towards construction waste minimization by design: A study in Shenzhen, China. *Resour. Conserv. Recycl.* **2015**, *105*, 29–35. [CrossRef]
50. Ajayi, S.O.; Oyedele, L.O. Policy imperatives for diverting construction waste from landfill: Experts' recommendations for UK policy expansion. *J. Clean. Prod.* **2017**, *147*, 57–65. [CrossRef]
51. Pellegrini, L.; Campi, S.; Locatelli, M.; Pattini, G.; Di Giuda, G.M.; Tagliabue, L.C. Digital Transition and Waste Management in Architecture, Engineering, Construction, and Operations Industry. *Front. Energy Res.* **2020**, *8*, 576462. [CrossRef]
52. Adams, K.T.; Osmani, M.; Thorpe, T.; Thornback, J. Circular economy in construction: Current awareness, challenges and enablers. *Proc. Inst. Civil. Eng. Waste Resour. Manag.* **2017**, *170*, 15–24. [CrossRef]
53. Gangolells, M.; Casals, M.; Forcada, N.; Macarulla, M. Analysis of the implementation of effective waste management practices in construction projects and sites. *Resour. Conserv. Recycl.* **2014**, *93*, 99–111. [CrossRef]
54. Alhawamdeh, M.; Lee, A. A behavioral framework for construction waste minimization: The case of Jordan. *Int. J. Environ. Sustain.* **2021**, *17*, 9–32. [CrossRef]
55. Tirado, R.; Aublet, A.; Laurenceau, S.; Habert, G. Challenges and Opportunities for Circular Economy Promotion in the Building Sector. *Sustainability* **2022**, *14*, 1569. [CrossRef]
56. Christensen, T.B.; Johansen, M.R.; Buchard, M.V.; Glarborg, C.N. Closing the material loops for construction and demolition waste: The circular economy on the island Bornholm, Denmark. *Resour. Conserv. Recycl. Adv.* **2022**, *15*, 200104. [CrossRef]
57. Luscuere, L.M. Materials Passports: Optimising value recovery from materials. *Proc. Inst. Civil. Eng. Waste Resour. Manag.* **2017**, *170*, 25–28. [CrossRef]
58. Won, J.; Cheng, J.C.P.; Lee, G. Quantification of construction waste prevented by BIM-based design validation: Case studies in South Korea. *Waste Manag.* **2016**, *49*, 170–180. [CrossRef]
59. Wang, G.; Krzywda, D.; Kondrashev, S.; Vorona-Slivinskaya, L. Recycling and Upcycling in the Practice of Waste Management of Construction Giants. *Sustainability* **2021**, *13*, 640. [CrossRef]
60. zu Castell-Rüdenhausen, M.; Wahlström, M.; Astrup, T.F.; Jensen, C.; Oberender, A.; Johansson, P.; Waerner, E.R. Policies as Drivers for Circular Economy in the Construction Sector in the Nordics. *Sustainability* **2021**, *13*, 9350. [CrossRef]
61. Tam, V.W.Y.; Lu, W. Construction Waste Management Profiles, Practices, and Performance: A Cross-Jurisdictional Analysis in Four Countries. *Sustainability* **2016**, *8*, 190. [CrossRef]
62. Noor, R.N.H.R.M.; Ridzuan, A.R.M.; Endut, I.R.; Noordin, B.; Shehu, Z.; Ghani, A.H.A. The quantification of local construction waste for the current construction waste management practices: A case study in Klang Valley. In Proceedings of the 2013 IEEE Business Engineering and Industrial Applications Colloquium (BEIAC), Langkawi, Malaysia, 7–9 April 2013; pp. 183–188.
63. Bao, Z.; Lu, W. Developing efficient circularity for construction and demolition waste management in fast emerging economies: Lessons learned from Shenzhen, China. *Sci. Total Environ.* **2020**, *724*, 138264. [CrossRef] [PubMed]
64. Condotta, M.; Zatta, E. Reuse of building elements in the architectural practice and the European regulatory context: Inconsistencies and possible improvements. *J. Clean. Prod.* **2021**, *318*, 128413. [CrossRef]
65. Torgautov, B.; Zhanabayev, A.; Tleuken, A.; Turkyilmaz, A.; Mustafa, M.; Karaca, F. Circular Economy: Challenges and Opportunities in the Construction Sector of Kazakhstan. *Buildings* **2021**, *11*, 501. [CrossRef]
66. Ajayi, S.O.; Oyedele, L.O.; Bilal, M.; Akinade, O.O.; Alaka, H.A.; Owolabi, H.A.; Kadiri, K.O. Waste effectiveness of the construction industry: Understanding the impediments and requisites for improvements. *Resour. Conserv. Recycl.* **2015**, *102*, 101–112. [CrossRef]
67. Ramos, M.; Martinho, G. Influence of construction company size on the determining factors for construction and demolition waste management. *Waste Manag.* **2021**, *136*, 295–302. [CrossRef] [PubMed]
68. AAjayi, S.O.; Oyedele, L.O.; Akinade, O.O.; Bilal, M.; Owolabi, H.A.; Alaka, H.A.; Kadiri, K.O. Reducing waste to landfill: A need for cultural change in the UK construction industry. *J. Build. Eng.* **2016**, *5*, 185–193. [CrossRef]
69. Shojaei, A.; Ketabi, R.; Razkenari, M.; Hakim, H.; Wang, J. Enabling a circular economy in the built environment sector through blockchain technology. *J. Clean. Prod.* **2021**, *294*, 126352. [CrossRef]
70. Saez, P.V.; de Guzman Baez, A.; Navarro, J.G.; del Rio Merino, M. Redefining construction and demolition waste management systems: Best practices on civil engineering works. *Balt. J. Road Bridge Eng.* **2014**, *9*, 171. [CrossRef]
71. Lu, W.; Webster, C.; Peng, Y.; Chen, X.; Chen, K. Big data in construction waste management: Prospects and challenges. *Detritus* **2018**, *4*, 129–139. [CrossRef]

72. Oluleye, B.I.; Chan, D.W.M.; Saka, A.B.; Olawumi, T.O. Circular economy research on building construction and demolition waste: A review of current trends and future research directions. *J. Clean. Prod.* **2022**, *357*, 131927. [CrossRef]
73. Low, J.K.; Wallis, S.L.; Hernandez, G.; Cerqueira, I.S.; Steinhorn, G.; Berry, T.A. Encouraging Circular Waste Economies for the New Zealand Construction Industry: Opportunities and Barriers. *Front. Sustain. Cities* **2020**, *2*, 35. [CrossRef]
74. Umar, U.A.; Shafiq, N.; Ahmad, F.A. A case study on the effective implementation of the reuse and recycling of construction & demolition waste management practices in Malaysia. *Ain Shams Eng. J.* **2021**, *12*, 283–291.
75. Ajayi, S.O.; Oyedele, L.O.; Bilal, M.; Akinade, O.O.; Alaka, H.A.; Owolabi, H.A. Critical management practices influencing on-site waste minimization in construction projects. *Waste Manag.* **2017**, *59*, 330–339. [CrossRef] [PubMed]
76. Tafesse, S.; Girma, Y.E.; Dessalegn, E. Analysis of the socio-economic and environmental impacts of construction waste and management practices. *Heliyon* **2022**, *8*, e09169. [CrossRef] [PubMed]
77. Wuni, I.Y.; Shen, G.Q. Developing critical success factors for integrating circular economy into modular construction projects in Hong Kong. *Sustain. Prod. Consum.* **2022**, *29*, 574–587. [CrossRef]
78. Kabirifar, K.; Mojtahedi, M.; Changxin Wang, C.; Tam, V.W.Y. Effective construction and demolition waste management assessment through waste management hierarchy; a case of Australian large construction companies. *J. Clean. Prod.* **2021**, *312*, 127790. [CrossRef]
79. Akinade, O.O.; Oyedele, L.O. Integrating construction supply chains within a circular economy: An ANFIS-based waste analytics system (A-WAS). *J. Clean. Prod.* **2019**, *229*, 863–873. [CrossRef]
80. Cheng, J.C.P.; Ma, L.Y.H. A BIM-based system for demolition and renovation waste estimation and planning. *Waste Manag.* **2013**, *33*, 1539–1551. [CrossRef]
81. Bernardo, M.; Gomes, M.C.; de Brito, J. Demolition waste generation for development of a regional management chain model. *Waste Manag.* **2016**, *49*, 156–169. [CrossRef] [PubMed]
82. Ding, Z.; Yi, G.; Tam, V.W.Y.; Huang, T. A system dynamics-based environmental performance simulation of construction waste reduction management in China. *Waste Manag.* **2016**, *51*, 130–141. [CrossRef] [PubMed]
83. Rahmat, A.S.; Noor, H.R.N.R.M.; Endut, I.R.; Ridzuan, R.A.M. The practice of construction waste management in Pulau Pinang. *AIP Conf. Proc.* **2016**, *1774*, 030018.
84. Gálvez-Martos, J.L.; Styles, D.; Schoenberger, H.; Zeschmar-Lahl, B. Construction and demolition waste best management practice in Europe. *Resour. Conserv. Recycl.* **2018**, *136*, 166–178. [CrossRef]
85. Serpell, A.; Kort, J.; Vera, S. Awareness, actions, drivers and barriers of sustainable construction in Chile. *Technol. Econ. Dev. Econ.* **2013**, *19*, 272–288. [CrossRef]
86. Davies, J.; Foxall, G.R.; Pallister, J. Beyond the Intention–Behaviour Mythology: An Integrated Model of Recycling. *Mark. Theory* **2002**, *2*, 29–113. [CrossRef]
87. Nikakhtar, A.; Hosseini, A.A.; Wong, K.Y.; Zavichi, A. Application of lean construction principles to reduce construction process waste using computer simulation: A case study. *Int. J. Serv. Oper. Manag.* **2015**, *20*, 461–480. [CrossRef]
88. María, M.R.; Nicolás, M.; Amparo, V.M. The Role of Information Technologies to Address Data Handling in Construction Project Management. *J. Comput. Civil Eng.* **2016**, *30*, 04015064.
89. Liu, Z.; Osmani, M.; Demian, P.; Baldwin, A. A BIM-aided construction waste minimisation framework. *Autom. Constr.* **2015**, *59*, 1–23. [CrossRef]
90. Won, J.; Cheng, J.C.P. Identifying potential opportunities of building information modeling for construction and demolition waste management and minimization. *Autom. Constr.* **2017**, *79*, 3–18. [CrossRef]
91. Martin, L.; Perry, F. Chapter 11—Sustainable Construction Technology Adoption. In *Sustainable Construction Technologies*; Tam, V.W.Y., Le, K.N., Eds.; Butterworth-Heinemann: Waltham, MA, USA, 2019; pp. 299–316. Available online: https://www.sciencedirect.com/science/article/pii/B9780128117491000092 (accessed on 13 May 2024).
92. Pan, M.; Linner, T.; Pan, W.; Cheng, H.; Bock, T. A framework of indicators for assessing construction automation and robotics in the sustainability context. *J. Clean. Prod.* **2018**, *182*, 82–95. [CrossRef]
93. Tam, V.W.Y.; Fung, I.W.H.; Sing, M.C.P.; Ogunlana, S.O. Best practice of prefabrication implementation in the Hong Kong public and private sectors. *J. Clean. Prod.* **2015**, *109*, 216–231. [CrossRef]
94. Mesároš, P.; Mandičák, T. Factors affecting the use of modern methods and materials in construction. *IOP Conf. Ser. Mater. Sci. Eng.* **2015**, *71*, 012053. [CrossRef]
95. Wang, J.; Wu, H.; Tam, V.W.Y.; Zuo, J. Considering life-cycle environmental impacts and society's willingness for optimizing construction and demolition waste management fee: An empirical study of China. *J. Clean. Prod.* **2019**, *206*, 1004–1014. [CrossRef]
96. Wang, T.; Wang, J.; Wu, P.; Wang, J.; He, Q.; Wang, X. Estimating the environmental costs and benefits of demolition waste using life cycle assessment and willingness-to-pay: A case study in Shenzhen. *J. Clean. Prod.* **2018**, *172*, 14–26. [CrossRef]

sustainability

MDPI

Review

Legal Barriers in Sustainable Agriculture: Valorization of Agri-Food Waste and Pesticide Use Reduction

Rosalinda Nicastro [1,†], Mattia Papale [1,†], Giovanna Marta Fusco [1], Annalinda Capone [2], Biagio Morrone [2] and Petronia Carillo [1,*]

[1] Department of Environmental, Biological and Pharmaceutical Sciences and Technologies, University of Campania "Luigi Vanvitelli", 81100 Caserta, Italy; rosalinda.nicastro@unicampania.it (R.N.); mattia.papale@studenti.unicampania.it (M.P.); giovannamarta.fusco@unicampania.it (G.M.F.)
[2] Department of Engineering, University of Campania "Luigi Vanvitelli", 81031 Aversa, Italy; annalinda.capone@unicampania.it (A.C.); biagio.morrone@unicampania.it (B.M.)
* Correspondence: petronia.carillo@unicampania.it; Tel.: +39-0823-274562
† These authors contributed equally to this work.

Abstract: The transition to sustainability in agriculture faces significant challenges, especially to balance environmental goals with the practical demands of food production. This paper examines two different case studies that reveal the complexities of agricultural regulation. The first case focuses on the valorization of agri-food residual biomasses, highlighting the potential to transform food waste into valuable bioproducts such as bioenergy and biofertilizers. Despite the clear environmental and economic benefits, the absence of specific European regulations hinders the widespread adoption of these practices. Without clear rules for achieving "end-of-waste" status, the development and marketing of bio-based products remain restricted. The second case study examines the European Union's unsuccessful effort to implement the Sustainable Use of Pesticides Regulation (SUR), which aimed to reduce pesticide use by 50% by 2030. Although the regulation sought to align agricultural practices with the EU's Green Deal, it triggered widespread protests from farmers concerned about the potential economic losses and decreased productivity. These two cases, one showing under-regulation and the other over-regulation, highlight the need for balanced and practical regulatory frameworks that promote sustainability without imposing unrealistic demands on stakeholders. This paper ends with recommendations to harmonize regulations across Europe, ensuring that both innovation in agricultural waste management and practical pesticide reduction strategies are implemented in a way that supports farmers and producers, minimizing economic disruptions and encouraging sustainable agricultural practices.

Keywords: organic waste; bioproducts; end-of-waste status; sustainable agricultural practices; agrochemicals; tractor protests; sustainable use of pesticides regulation (SUR)

1. Introduction

Sustainability is increasingly becoming a global priority, with every productive sector striving to develop environmentally, socially, and economically responsible practices. However, the sustainable transformation of agriculture is particularly complex. Each year, the agri-food sector is a significant contributor to greenhouse gas (GHG) emissions of anthropogenic origin, accounting for approximately 17.9 billion tonnes of CO_2 equivalent (CO_2eq), or nearly one-third of global emissions [1]. Around 7 billion tons are directly related to agricultural production. In this computation, the whole chain was evaluated considering agri-food production, deforestation, land use, food processing, transportation, packaging, and waste disposal, which plays a significant role in this footprint [1]. In particular, food waste is a critical problem, with approximately 1.6 billion tons of agricultural food (15.3% of all food produced) lost at the farm level, with as much as 40% lost during farming and post-farming operations. Notably, 58% of this food waste occurs in middle- and high-income regions [2]. In addition to these factors, the widespread use of agrochemicals, including synthetic pesticides and fertilizers, exacerbates the environmental burden of the agricultural sector. Agrochemicals contribute to soil and water contamination through runoff. At the same time, their production, application, and degradation processes release significant amounts of GHGs, including nitrous oxide (N_2O), which is 298 times more noxious than carbon dioxide [3,4]. This highlights the dual challenge of agriculture: while it is essential for global food security, it is also a significant source of pollution and GHG emissions. Reducing the reliance on agrochemicals is therefore critical to achieving long-term sustainability goals in the sector.

Only adopting the circular economy principles, mainly through regenerative agriculture, can help the sector meet growing global food demands while improving its economic performance and minimizing environmental impact [5]. The circular economy is a key part of the European Union (EU)'s strategy for sustainable growth. It focuses on reducing waste, reusing materials, and improving resource efficiency. Within the European Green Deal, the Circular Economy Action Plan (CEAP) aims to close material loops and reduce reliance on finite resources. This approach not only could support environmental goals but also would drive innovation, creates jobs, and strengthens the EU's economy, helping achieve climate neutrality by 2050. However, a delay in clear legislative action has hindered the shift toward sustainable agricultural practices, particularly regarding regulatory frameworks supporting circular economy models in agriculture.

The EU, through initiatives such as the European Green Deal [6] and the Farm to Fork Strategy [7], has sought to address these challenges by encouraging sustainable agricultural practices. Despite these efforts, the successful implementation of such policies faces two significant challenges: insufficient regulation in some areas and overly stringent regulation in others. On the one hand, the absence of clear, standardized regulatory frameworks for valorizing agricultural waste has hindered the full integration of agri-food residues into a circular economy.

Agricultural waste, which includes residual biomasses from the production and processing of food, represents an untapped resource for bio-based products such as bioenergy, bioplastics, and biofertilizers. Despite the promising potential of these products, regulatory frameworks, such as the EU's Directive 2008/98/EC on Waste [8], have struggled to offer clear guidelines for the valorization of such materials [9]. The "end-of-waste" status, crucial for turning waste into worthy usable products, remains difficult to attain for many agricultural by-products, stifling innovation in this sector.

This regulatory gap contrasts with the EU's broader ambitions for a circular economy, where waste products are meant to be repurposed into new materials, thus reducing landfill use and lowering GHG emissions. Regulation (EU) 2019/1009, for example, focuses on providing a legal pathway for specific recovered products like struvite and biochar [10]. Yet, the agricultural waste sector needs more comprehensive regulation that would allow for a broader range of bio-based materials to enter the market [11]. This lack of clear regulatory direction is a missed opportunity not only for environmental gains but also for economic growth within the agricultural sector, where waste could be transformed into value-added products.

On the other hand, ambitious regulatory efforts aimed at reducing the environmental footprint of agriculture have also faced significant resistance, particularly when regulations are perceived as too stringent or detached from the realities of agricultural practice. A case in point is the EU's proposed SUR, which sought to reduce chemical pesticides by 50% by 2030 [12]. While the SUR was part of a broader strategy to align with the objectives of the Green Deal and the Farm to Fork Strategy, its introduction triggered protests across the European agricultural community.

Farmers, particularly in regions heavily reliant on pesticides for crop protection, argued that such a drastic reduction would severely impact crop yields, economic stability, and overall food security. The regulation's provisions, which included banning pesticides in sensitive areas and adopting low-risk alternatives, were seen as overly ambitious and difficult to implement without sufficient technological or economic support. The widespread protests culminated in tractor demonstrations across Europe, symbolizing the agricultural sector's opposition to what many viewed as regulatory overreach [13]. Despite the environmental intentions behind the SUR, the European Parliament eventually shelved the proposal, acknowledging the need for a more balanced approach that would not disproportionately harm agricultural productivity [14].

These two contrasting cases, one of under-regulation and the other of over-regulation, illustrate the complexities inherent in pursuing sustainable agricultural policies. The lack of coherent regulation for agricultural waste valorization has left a vast potential for untapped innovation. At the same time, the overzealous regulatory push to reduce pesticide use through the SUR led to significant resistance and delayed progress. Both examples underscore the need for a middle path that balances environmental objectives with the economic and practical realities farmers face. Although bioenergy production and pesticide reduction may initially appear unrelated, both are essential components of a sustainable agricultural system. By integrating waste management practices, mainly through the valorization of agricultural residues to produce bioproducts for agriculture or bioenergy, farmers can generate additional revenue while reducing their reliance on synthetic pesticides, ultimately enhancing productivity and income. This dual approach not only mitigates environmental impact but also creates synergies between the agricultural and circular economy sectors, demonstrating the interconnected nature of sustainability challenges. Moreover, they can access increased European Community support, such as eco-schemes, which specifically encourage sustainable practices and offer financial incentives through subsidies to promote sustainability in agriculture. This dual approach would help farmers to adopt eco-friendly practices while providing economic support to ease the transition toward more sustainable farming models. Thus, addressing these issues together fosters the innovation and cross-sector collaboration crucial for promoting effective sustainable agricultural practices.

2. Case Study 1: The Role of the Circular Economy in Agri-Food Waste Valorization

The transition to a circular economy has become a critical pathway for addressing agricultural production's environmental and economic challenges. In the agri-food sector, circular economy principles focus on reducing waste by transforming agricultural by-products into useful resources such as bioenergy, biofertilizers, and bioplastics. This approach not only maximizes resource efficiency but also contributes to sustainability by decreasing reliance on synthetic inputs and minimizing environmental impact. These practices not only contribute to reducing the environmental footprint of farming activities but also offer opportunities for creating new revenue streams from what was once considered waste [15,16].

Despite the evident potential of these solutions, the adoption of circular economy practices in agriculture has been hampered by regulatory and market barriers. The lack of standardized regulatory frameworks tailored to agricultural biomass impedes progress. Strong cooperation between the agricultural and energy sectors is sought for bioenergy production to work. Energy companies need to secure feedstock from farms, giving farmers a steady market for their waste materials. In return, farmers are paid for their leftover biomass, providing extra income, and supporting long-term sustainability. This strict collaboration would encourage innovation in waste management and reduce the need for fossil fuels. Clear policies supporting this cooperation could connect agricultural waste with the energy sector in an easier way, helping both industries work together more efficiently. This case study explores the evolution of circular economy principles in the agricultural sector, examining how they are being applied in the valorization of agri-food waste and the challenges that persist due to regulatory gaps and market limitations.

2.1. The Circular Economy Principles: From an Aspiration to a First Body of Practices

In 1966, the economist and philosopher Boulding [17] was the first to introduce the concept of "scarcity of resources" and underline the urgency for industrialized countries to move from an open or "cowboy economy", that is an open system in which the natural environment is typically perceived as limitless, to a closed economy capable of self-regeneration (circular economy). For the philosopher, this was a prerequisite for maintaining the sustainability of human life on Earth [17]. In fact, according to the Food and Agriculture Organization (FAO), approximately one-third of the food produced worldwide for human consumption is either lost or wasted annually [18]. This implies a substantial depletion of natural resources used throughout the food supply chain and poses a significant risk to global food security [19].

However, until the last two decades, there had been a lack of regulatory instruments specifically used to recover the untapped potential of raw materials and residual biomasses. In fact, only in 2005 the European Commission (EC) presented a Biomass Action Plan that was designed to increase the use of renewable energy from forestry, agriculture, and waste materials [9,20]. For the first time, EU policy recognized that residual biomasses represent a resource rather than a worthless and unmanageable waste, and their smart and virtuous use can help address climate change by reducing GHG. Moreover, their use for producing electricity, heating, and transport fuels could diversify EU energy supply, stimulating jobs and economic growth. Nevertheless, the biomass action plan did not consider the possibility of producing value-added biomaterials.

2.2. The Concept of By-Products and Waste Hierarchy

The Italian national regulatory framework on the circular economy, the Legislative Decree 152/2006, known as the "Environmental Consolidation Act" (Testo Unico Ambientale), set out the legislative framework applicable to all matters concerning environmental protection and regulated for the first time the production of "by-products". At the European level, the concept of "by-products" was set only in 2008 in Directive 2008/98/EC on waste [8] but only for materials like construction and demolition waste, scrap metals, textiles, wastepaper, and glass. The "by-products" were defined as a "substance or object, resulting from a production process, the primary aim of which is not the production of that item". However, the term "by-product" is increasingly used due to the growing need to highlight that food or organic waste can be used as substrates for the development of new value-added products. Article 6 of the 2008 Waste Framework Directive [8], as implemented in the 2011 Waste Directive Regulations [21], outlines that "end-of-waste" status can be achieved when recovered materials meet specific criteria. These include being commonly used for certain purposes, having a market, fulfilling technical standards, and posing no risks to health or the environment. This helps keep waste out of landfills, turning it into useful products that meet regulations and market demand. The 2011 EC (Waste Directive) Regulations [21] also implement the "polluter pays principle" introduced by the Organisation for Economic Co-Operation and Development [22]. This principle is reflected throughout the regulations, emphasizing that the costs of managing waste should be borne by the waste producer or holder (whether private individuals or companies) to ensure that the environment remains in an "acceptable state", as specified in Article 14 of the 2008 Directive [8]. As a logical transposition of the "polluter pays principles", the "extended producer responsibility" (EPR) was also introduced in the directive on waste. It is an additional environmental protection strategy aimed at reducing the total environmental impact of a product and its packaging. In fact, it ensures that producers must take responsibility for their products' entire life cycle and packaging, particularly regarding their possible take-back, recycling, and final disposal. The directive on waste also defines a "waste hierarchy", that is, an order of priority in waste prevention and re-use vs. its recovery or disposal. It is the cornerstone of EU waste policies and legislation, which aims to reduce the production and impact of waste and manage and improve residual biomass resource efficiency. The hierarchy is presented as an inverted pyramid of the main options to be adopted before a material becomes waste for reuse and/or as the extension of products' life, and disposal at the bottom as the last choice for waste management (Figure 1).

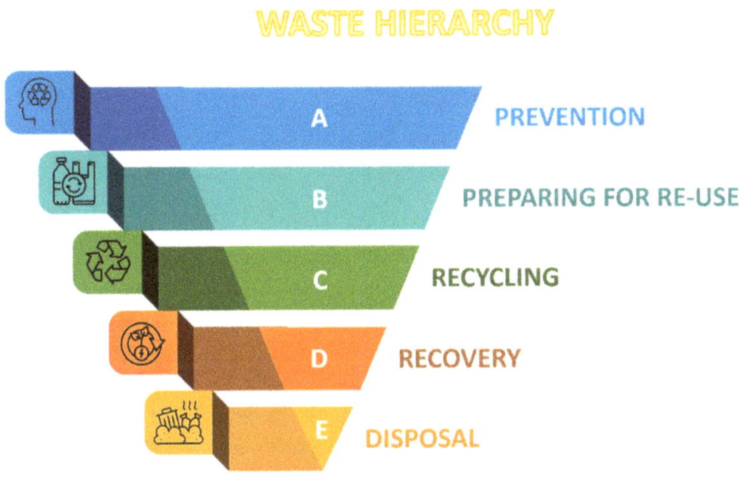

Figure 1. Waste hierarchy inverted pyramid.

2.3. *Europe 2020 Strategy: The First Integrated Approach to Climate and Energy Policy*

In 2011 the EC introduced the Europe 2020 Strategy, focusing on smart, sustainable, and inclusive growth. As part of this initiative, member states committed to a 20% reduction in GHG emissions, a 20% increase in the share of renewable energy in the EU's energy mix, and a 20% improvement in energy efficiency by 2020 [23]. This 20-20-20 framework represents a comprehensive approach to addressing climate change, enhancing the EU's energy security, and boosting its competitiveness. The targets considered the varying starting points and potential for renewable energy growth among member states, with national goals ranging from 10% for Malta to 49% for Sweden, ensuring the EU collectively met the 20% goal. According to this agreement, 17% and 16% of Italy's and Ireland's final energy consumption should have been covered by renewable sources in 2020, respectively, while an expansion target of only 18% was envisaged for Germany by 2020. Bongardt and Torres [24] stated that this strategy was conceived to get Europe out of the global economic and financial crisis that started in 2008. However, even before it was implemented, the EU's economic and financial crisis became a sovereign debt crisis, putting the entire Eurozone at risk. This underscored the necessity for stronger European economic cooperation to tackle the key drivers of the crisis, namely the competitiveness gaps between member states and fiscal imbalances, as well as to prevent monetary spillovers, especially within the eurozone. However, the Europe 2020 Strategy overlooks the potential in the agri-food sector for residual biomasses, which meet all legal requirements, to be transformed and recognized as valuable by-products for agriculture with end-of-waste status, rather than being limited to bioenergy production substrates.

The targets of the Europe 2020 Strategy have already been verified by the European Environment Agency [25]. EU-27 GHG emissions were 31% lower than in 1990, while the EU achieved a 21.3% share of renewables. Due to the COVID-19 pandemic, the EU's primary and final energy consumption was below the target by 5% and 3% margins. In 2020, only 21 member states successfully met their national targets. As a result, the remaining countries, including Bulgaria, Cyprus, Finland, Germany, Ireland, and Malta, would need to utilize flexibility mechanisms, such as purchasing emission quotas from other EU nations, in order to fulfill their legal obligations.

The missed opportunity to attain the Europe 2020 targets could significantly contribute to the strategy's goals of sustainable growth further promoting resource efficiency and reducing environmental impacts. The transformation of these biomasses into bioproducts for agriculture not only supports the circular economy but also fosters innovation in agricultural practices, aligning with the smart growth objectives. Additionally, by providing farmers with new income streams through the valorization of residues, this approach contributes to inclusive growth, bolstering rural economies and job creation. Therefore, the valorization of agricultural residues represents a practical way to further extend the achievements of the Europe 2020 strategy.

2.4. European Strategies for the Bio-Economy and Circular Economy

In 2012, the first European bioeconomy strategy was proposed to address the reuse of residual biological biomasses and their conversion into bio-based products [26]. The strategy aimed to be environmentally friendly, promote sustainability, and offer economic benefits by creating new job opportunities and markets. The bioeconomy strategy has been updated with several documents in the following years [27] mainly to improve policy coherence, to identify and resolve trade-offs, and to enable different countries to design transition pathways according to their specificities and priorities. A report on the progress made in its implementation from 2018 [27] and its action plan, aimed at identifying gaps for possible future EU bioeconomy action and initiatives, was published in 2022 [28].

In 2014, a new policy framework implementing a circular system to re-use valuable residual biomasses and eliminate waste [29] was laid down with the Communication "Towards a circular economy: A zero waste programme for Europe". The main proposed actions set common EU targets to achieve by 2030 compared to 2011: recycling municipal waste of 65%, recycling packaging waste of 75%, reducing landfill disposal of 10%, and material-specific targets for different packaging materials.

The renewal pathway continued in 2015 with the publication of the first CEAP, "Closing the Loop—An EU Action Plan for the Circular Economy" [30], whose main strategy was to produce no waste because today's residual products are tomorrow's raw materials. This could create a closed loop (Figure 2). New raw materials should be developed aiming at biodegradability, recyclability, and/or compostability, allowing them to be reintroduced into the natural environment. This would help stimulate job creation, economic growth, and investment, while supporting the transition to a carbon-neutral, resource-efficient, and competitive economy. The Action Plan also outlines a policy framework incorporating existing policies and legal tools to achieve targeted goals.

In 2015, the United Nations (UN) General Assembly signed the 2030 Agenda for Sustainable Development, an action program for people, the planet, and prosperity [31]. It defined 17 Sustainable Development Goals (SDGs), among which SDG 12 deals with Sustainable Consumption and Production, included in a broad action program that identified 169 targets or goals.

Accordingly, in 2016, the EC launched the European Platform on Food Losses and Waste [32], to share the best practices and develop a common methodology and indicators to measure food waste. The EC emphasized the importance of involving stakeholders, such as public authorities, businesses, trade unions, consumers, and civil society, to facilitate the exchange of best practices. In March 2017, the European Circular Economy Stakeholder Platform was established as a collaborative initiative between the EC and the European Economic and Social Committee (EESC) [32]. This platform contributed to enacting in 2018 four amending Directives, forming the so-called "Circular Economy Package", that "boost economic performance while reducing resource use" in four specific sectors: dir. (EU) 2018/849 on end-of-life vehicles, batteries, accumulators and electronic devices [33]; dir. (EU) 2018/850 on landfill of waste [34]; dir. (EU) 2018/851 on waste in general [35]; dir. (EU) 2018/852 on packaging and packaging waste [36]. In particular, the dir. (EU) 2018/851, amending the first directive on waste [8], contains two important definitions of "bio waste", as something "that means biodegradable garden and park waste, food and kitchen waste from households, offices, restaurants, wholesalers, canteens, caterers and retail premises and comparable waste from food processing plants"; and on "food waste", as something "that means all food as defined in Article 2 of the regulation on "General Food Law" [37]".

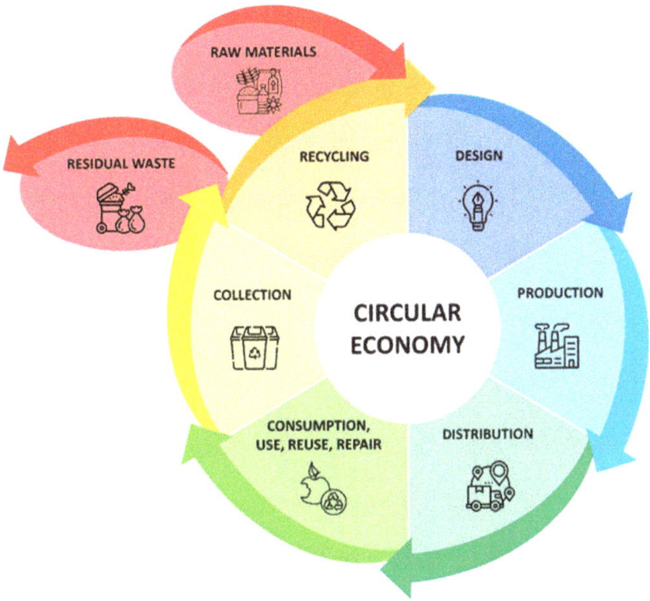

Figure 2. The circular economy's strategy is to produce no waste because today's residual products are tomorrow's raw materials.

Again, the EU introduced the European Green Deal between December 2019 and January 2020 to mitigate market instability and environmental threats. It is a set of policy initiatives to foster the transition towards a climate-neutral economy, whose goal is to reduce GHG emissions by at least 55% by 2030 and achieve full carbon neutrality by 2050 [6].

In March 2020, a new CEAP was introduced as part of the efforts to achieve the objectives set by the European Green Deal [9]. The new plan expands on the previous one, with updated priority areas that now include electronics and ICT, batteries and vehicles, packaging, plastics, textiles, construction, buildings, food, water, and nutrients. A key component for achieving the goals of the European Green Deal is the Farm to Fork Strategy (F2F), launched in May 2020, which focuses on making food systems more fair, healthy, sustainable, and environmentally friendly [7]. Since policy coherence among diverse EU legislation is crucial, to enable the EU to reach these broad goals, the strategy also outlined a range of concrete targets: in particular, by 2030, 25% of total farmland has to be under organic farming, the use of chemicals and certain pesticides in agriculture should be reduced by 50%, and EU sales of antimicrobials for farmed animals and agriculture should be reduced by 50%.

Following the same long-term inspiring principles of the Farm to Fork Strategy (F2F), and in compliance with the European green deal, the new Common Agricultural Policy (CAP) 2023–2027 [38] committed some of its objectives to preserve landscapes and biodiversity, reduce climate change, protect food and health quality, care for the environment, but also to improve the position of farmers in the food chain and ensure fair incomes for them because real sustainable development is based on social, economic, and environmental pillars. Among the ten main objectives of the CAP 2023-27, the "food value chain" seeks to enhance the EU agricultural sector's ability to meet societal demands for high-quality, safe, and nutritious food produced sustainably, while also minimizing food waste. However, no specific legislation still focuses on developing new waste-derived biomaterials for agricultural practices that reduce the use of chemicals, including fertilizers.

Table 1 summarizes the key European and national policies that during previous years have promoted the use of agri-food residues through bio-based product development and bioenergy generation.

Table 1. Laws and directives for the promotion of agri-food residues in bioproducts and bioenergy.

Year	Law/Directive	Description
2006	Legislative Decree 152/2006 (Environmental Code)	Italian legislation established a comprehensive framework for environmental protection, including the regulation of waste and the reuse of by-products in agricultural processes.
2008	Directive 2008/98/EC (Waste Framework Directive)	It established the EU's legal framework for waste management, including criteria for when waste ceases to be waste, mainly through recovery and recycling processes.
2011	EU 2011 Waste Directive	It reinforced the principles of waste minimization, recycling, and recovery and introduced the "polluter pays" principle, emphasizing waste producers' responsibility.
2012	European Bioeconomy Strategy	It encouraged the sustainable use of biological resources from land and sea, including agricultural residues, for the production of food, feed, bio-based products, and bioenergy.
2014	EC Communication "Towards a Circular Economy"	It proposed strategies to shift towards a circular economy, including reusing agricultural waste for producing bio-based materials and chemicals.
2015	EU Circular Economy Action Plan "Closing the Loop"	It targeted the recycling and reuse of agricultural waste to create biodegradable materials and bio-based products, highlighting the potential of agri-waste valorization.
2018	Regulation (EU) 2018/848	It introduced new rules for organic production and labeling, allowing for the use of bio-based products derived from agricultural waste in organic farming systems.
2019	Regulation (EU) 2019/1009	It set EU-wide standards for marketing fertilizing products, including bio-based materials such as composts, digestates, and recovered phosphorus, encouraging the circular use of agricultural residues.
2020	New Circular Economy Action Plan	It expanded on the 2015 plan by targeting agricultural waste management through innovation in bioproducts and bio-based materials, fostering the development of biofertilizers and composts from residues.
2023	Common Agricultural Policy (CAP) 2023–2027	It introduced eco-schemes and rural development measures to support sustainable farming, including the use of agricultural residues for producing bioproducts and fertilizers and reducing reliance on synthetic inputs.

2.5. *Circular Economy Developments beyond Europe*

While this case study focuses on the European Union, which is widely seen as a global leader in sustainability and circular economy practices, it is important to acknowledge that other countries have also made notable progress. For instance, Japan has long been recognized for its waste management and recycling leadership. Its Circular Economy Vision 2020 builds on this foundation by promoting circularity across various sectors. Japan's Plastic Resource Circulation Act (2022) [39] explicitly targets reducing single-use plastics and encourages using sustainable, alternative materials to minimize environmental impact. Similarly, South Korea has invested heavily in circular economy initiatives through its Resource Circulation Plan (2018–2027) [40]. This plan emphasizes increasing recycling rates and advancing resource recovery technologies, mainly focusing on extracting rare metals from electronic waste. Japan and South Korea view circularity as essential to their sustainability agendas, including their ambitious goals of achieving carbon neutrality by 2050. These countries are setting solid examples of how circular economy principles can be integrated into national sustainability efforts.

Australia's Circular Economy (Waste Reduction and Recycling) Act 2021 [41] is a key step in the country's move towards a more sustainable economy. The law targets industries like mining, agriculture, and construction, focusing on reducing waste, reusing materials, and improving resource recovery. Victoria, the Australian state, has been leading the way with its Recycling Victoria policy [42], which aims to recycle 80% of waste by 2030 and phase out certain single-use plastics. These efforts are all part of Australia's larger plan to rely less on new raw materials and encourage more sustainable practices across different sectors. New Zealand's commitment to a circular economy is reflected in its Emissions Reduction Plan [43], which aims for a complete transition by 2050. The plan focuses on recovering resources, promoting circular design, and increasing the use of bio-based products, especially in construction. New Zealand has introduced programs that make producers responsible for managing the environmental impact of their products from the time they are made until they are disposed of. The country is also working to reduce waste from construction and demolition by requiring the use of recycled materials and encouraging more sustainable building methods.

Latin America has made significant progresses in circular economy efforts, mainly through the work of the Latin America and Caribbean Circular Economy Coalition, established in 2021 [44]. The coalition promotes regional cooperation and dialog to advance circular practices in diverse sectors such as agriculture, waste management, and manufacturing. Colombia has introduced national policies to enhance resource efficiency, particularly in urban waste management and industrial practices. Chile is focusing on eco-design and extended producer responsibility, encouraging industries to reduce waste during the design phase of products. Brazil, with its high level of urbanization and industrial activity, is fostering circular agriculture initiatives, especially in cities like São Paulo, where organic waste recycling and regenerative farming techniques are gaining traction [45].

Although there is not a federal circular economy policy in the U.S., several states are leading the way. California is a key player with its extended producer responsibility (EPR) laws [46], which hold manufacturers responsible for the entire lifecycle of their products, including disposal. Programs like the California Mattress Recycling Council and PaintCare require companies to manage the recycling of their products at the end of their use. New York City has set an ambitious zero-waste goal to eliminate waste sent to landfills by 2030 through mandatory composting, recycling programs, and public awareness efforts [47]. On the federal level, circular practices are being explored in areas such as renewable energy, waste management, and sustainable packaging. The U.S. Environmental Protection Agency (EPA) also promotes circular economy principles through its Sustainable Materials Management (SMM) program [48], which encourages the productive use of resources throughout their entire lifecycle. As part of this effort, the EPA launched the National Strategy for Reducing Food Loss and Waste and Recycling Organics to address food waste through prevention, reuse, and recycling initiatives [49]. This strategy supports the transition to a circular economy by diverting organic waste from landfills, reducing greenhouse gas emissions, and promoting the recovery of valuable resources from organic materials.

3. Case Study 2: The Challenges of Pesticide Regulation in Sustainable Agriculture

While the principles of a circular economy emphasize waste reduction and resource efficiency, a critical aspect of sustainable agriculture revolves around managing and reducing harmful chemical inputs, particularly pesticides. The widespread reliance on pesticides has long been central to modern agricultural practices, crucial in maintaining crop yields and mitigating pest damage. In addition, these agrochemicals, much like pharmaceuticals, depend heavily on the correct dosage, appropriate usage, and adherence to legal guidelines to be effective and safe. When applied according to regulations, pesticides can offer significant advantages, such as safeguarding crop yields and preventing losses caused by pests and diseases. Under these controlled conditions, their benefits (e.g., enhanced productivity and food security) often outweigh potential downsides [50].

However, the misuse or excessive application of pesticides has raised serious environmental and health concerns, particularly about soil and water pollution, the loss of biodiversity, and human health risks. This has led to increased advocacy for stricter regulations, with initiatives like the EU's Farm to Fork Strategy pushing for a reduction in pesticide use. Nevertheless, it is essential to differentiate between the regulated, responsible use of pesticides and the damage caused by substances banned due to their hazardous effects. Moreover, the export of pesticides banned in the EU to developing countries underscores the urgent need for global regulatory alignment to safeguard human health and the environment [51].

3.1. *Historical Excursus on the EU's Regulation for Pesticide Reduction*

The EU has long been a leader in global initiatives to regulate pesticide use and mitigate their adverse environmental and health impacts. Starting in the 1970s, the EU recognized the growing concerns around pesticide use, particularly the harmful effects of certain chemicals. One of the earliest significant actions was the introduction of Directive 79/117/EEC in 1979, which banned a group of hazardous pesticides, including organochlorine compounds such as DDT, due to their long-lasting environmental presence and significant risks to human health [52,53].

The CAP, established in 1962 to boost European agricultural productivity, initially prioritized increasing yields [54]. However, the CAP evolved to incorporate sustainability into its core objectives over time. Beginning in the 1990s, CAP reforms increasingly integrated environmental considerations, particularly through agri-environmental schemes that incentivized farmers to adopt practices aimed at reducing chemical inputs, including pesticides. This shift represented a broader alignment of agricultural subsidies with the principles of sustainable farming, reflecting the EU's commitment to promoting environmentally responsible agriculture.

During the 1990s, the EU took further steps to strengthen its regulatory framework for pesticides with the introduction of Directive 91/414/EEC. This directive established a comprehensive process for the evaluation and authorization of plant protection products, marking a significant step toward harmonizing pesticide regulations across member states. The directive mandated rigorous scientific risk assessments, focusing particularly on the impact of active substances on both human health and the environment. This harmonization was crucial in creating a unified approach to pesticide regulation within the EU [55].

The adoption of the 2000 Water Framework Directive expanded the EU's environmental objectives by addressing the contamination of water bodies by agricultural pollutants, including pesticides. This directive required member states to implement protective standards for water resources, specifically targeting runoff and residues resulting from intensive pesticide use [56]. In 2005, the EU introduced Regulation 396/2005, which standardized the setting of Maximum Residue Levels (MRLs) for pesticides in food and feed. This regulation was a key development in ensuring food safety and protecting consumers across all EU countries by harmonizing the limits for pesticide residues [57].

The EU further intensified its efforts to reduce pesticide dependence with the 2006 Thematic Strategy on the Sustainable Use of Pesticides. This initiative focused on promoting Integrated Pest Management (IPM), a comprehensive approach to pest control that minimizes chemical use by encouraging alternative methods such as biological control and crop rotation [58]. This strategy laid the groundwork for the 2009 Directive 2009/128/EC, commonly known as the Sustainable Use Directive (SUD). The SUD required member states to develop National Action Plans (NAPs), setting clear targets for reducing pesticide use and promoting IPM [59]. In 2009, Regulation (EC) No 1107/2009 replaced Directive 91/414/EEC, introducing stricter rules for the approval and monitoring of pesticides. This regulation placed a strong emphasis on protecting vulnerable groups, such as children and pregnant women, from the potential adverse effects of chemical substances. It also aimed to ensure that only pesticides deemed safe for human health, non-target organisms, and the environment would be authorized for use within the EU [60].

A significant milestone in EU pesticide regulation occurred in 2011 when the EU ratified the Rotterdam Convention, an international treaty aimed at regulating the trade of hazardous chemicals, including pesticides. The convention requires a Prior Informed Consent (PIC) procedure, ensuring that importing countries are fully informed about the risks associated with pesticide use [61]. This ratification underscored the EU's strong commitment to maintaining its global leadership in pesticide safety and regulation.

In recent years, the EU has continued to advance its sustainability agenda, as with the introduction of the European Green Deal in 2020 [6]. This ambitious plan, which aims to make the EU climate-neutral by 2050, identified agriculture as a key sector for reform. As part of this initiative, the Farm to Fork Strategy was launched, setting explicit targets to reduce the use of chemical pesticides by 50% by 2030. The strategy also promotes organic farming and sustainable food production systems, emphasizing the need for a transition to food systems that are both environmentally sustainable and safe for human health. By pushing for reductions in the use of hazardous pesticides and encouraging nature-based solutions, the Farm to Fork Strategy aims to significantly reduce the EU's reliance on chemical inputs [7].

The CAP continues to evolve, with the 2023–2027 reform placing an even stronger emphasis on environmental sustainability and climate action. This latest reform further integrates sustainability goals by conditioning subsidies on eco-schemes, where farmers are rewarded for adopting practices that reduce chemical inputs, including pesticides [38].

The EC's proposal for the SUR has been one of the most ambitious regulatory efforts towards binding targets for pesticide reduction. Introduced in 2022, the SUR aimed to reduce chemical pesticides by 50% by 2030, aligning with the broader objectives of the European Green Deal and the Farm to Fork Strategy [12]. The SUR is intended to replace the existing Directive 2009/128/EC [59] and introduce stricter regulations to promote sustainable pesticide use across EU member states. The regulation also sought to encourage the adoption of low-risk and non-chemical alternatives for pest management, such as IPM techniques. However, it encountered strong opposition from the agricultural sector, particularly in regions where pesticide use is essential for sustaining crop productivity. The complexity of balancing environmental protection and agricultural productivity posed a significant challenge, leading to fears of food insecurity and declining competitiveness for European agricultural products in the global market. Farmers, especially in areas with high pest pressure, have raised concerns that such drastic reductions would severely compromise crop yields and economic sustainability, especially without viable, cost-effective alternatives. Given that pesticides play a crucial role in ensuring food security, a rapid reduction in their use without parallel technological innovations or sustainable substitutes could threaten both agricultural output and the livelihoods of farmers [62]. The concerns extended beyond production issues. One of the most controversial aspects of the SUR was the proposed ban on pesticide use in "sensitive areas" such as Natura 2000 sites [63], urban green spaces, and other protected areas. Farmers in these regions felt that such restrictions, while environmentally motivated, failed to consider the practical needs of agricultural production.

3.2. *The Challenges of Implementing SUR: Economic and Practical Barriers*

After the introduction of the SUR proposal on 22 June 2022, the EC requested its review, asking the European Parliament's Environment, Public Health, and Food Safety Committee (ENVI) to conduct an evaluation within six months. This delayed the vote on the regulation to the following year. The Council's request was driven by concerns that the data supporting the impact analysis were based on conditions before the outbreak of the war in Ukraine in 2022, which significantly altered food security concerns in the EU and globally [64].

The revised document, released in July 2023, became a focal point of discussion during the European Council meeting on 25 July 2023, where critiques focused not only on the vague definition of "sensitive areas" but also on the proposed ban on pesticide use in these zones [65]. Despite updates, the document still failed to clearly define these areas, which include non-agricultural spaces such as public parks, private gardens, sports fields, and Natura 2000 sites, while excluding airports and industrial zones. Farmers in regions affected by these restrictions, particularly around Natura 2000 sites and urban green spaces, expressed concern that, while the measures were environmentally motivated, they did not sufficiently consider the practical needs of agricultural production. Several stakeholders recommended that the regulations offer member states more flexibility in implementing intervention strategies and achieving pesticide reduction targets, allowing them to use historical data.

Another concern was the potential negative economic impact due to the administrative costs linked to CAP funding access. The SUR was set to be funded by the new 2023–2027 CAP, both through the eco-schemes under the First Pillar and the rural development under the Second Pillar. The CAP funds for these two measures are part of the 2021–2027 multiannual financial framework, whose allocations include €44.7 billion for eco-schemes and €95.5 billion for rural development, including €8.1 billion from the Next Generation EU recovery fund. One of the key focuses of this funding is to support investments in promoting environmental and climate-friendly farming practices, including precision farming technologies, which are crucial for meeting the EU's pesticide reduction targets and promoting sustainable agricultural practices. However, for Small and Medium-sized Enterprises (SMEs) in agriculture, the administrative costs of applying for and accessing CAP funds may disproportionately affect their operations. The EC estimated that these costs could rise by around €180 per year, largely due to mandatory training on sustainable alternatives to pesticides. This increase might affect the competitiveness of EU agricultural products in global markets, although the Commission had insufficient data to fully quantify the broader economic impact of these costs [65].

Moreover, with the new CAP, the Good Agricultural and Environmental Conditions (GAEC), also known as conditionalities, have been implemented. Farmers must adhere to these conditions to receive the basic payment. Notably, GAEC 7 and GAEC 8 have generated the most controversy. GAEC 7, similar to the SUR, aims to achieve a 50% reduction in pesticide use by 2030, as outlined in the Farm to Fork Strategy. It makes crop rotation mandatory, requiring farmers to plant a different crop genus on the same land each year, utilizing IPM techniques. Additionally, GAEC 8, which came into effect in January 2024, requires that arable farms leave at least 4% of their land fallow, contributing to the biodiversity goals of the Green Deal's Biodiversity Strategy. This measure aims primarily to protect biodiversity, which is severely threatened by the widespread use of agrochemicals [66]. Indeed, GAEC 7, GAEC 8, and the proposed SUR introduce binding regulations to facilitate the transition to more sustainable agriculture. Still, many believe these "green transition laws" are economically unsustainable, impractical, and incompatible with enhancing the competitiveness of agricultural products [67]. More broadly, the perceived loss of competitiveness due to the SUR, the CAP's GAECs, and the Green Deal has been a significant concern for EU farmers.

On 24 October 2023, the European Parliament's Committee on Environment, Public Health and Food Safety (ENVI) approved the revised SUR document, with 47 votes for, 37 against, and 2 abstentions [68]. Despite these efforts, many farmers worried about the economic impact, particularly regarding competition with non-EU countries that had less strict pesticide regulations. Therefore, the committee committed to examining differences in pesticide use between EU and non-EU imports by the end of 2025, aiming to ensure fairness in global markets [69]. In November 2023, the European Parliament rejected the Commission's proposal during a plenary vote, with 299 votes opposing, 207 supporting, and 121 abstaining. This decision effectively halted further amendments to the SUR, raising concerns about the feasibility of reaching the EU's pesticide reduction objectives by 2030.

While many farmers welcomed the outcome, a considerable portion of EU citizens remain in favor of fully eliminating synthetic chemical pesticides [69].

3.3. Further Challenges to the SUR from Political Issues and Climate Crisis

The approval of the SUR was complicated by broader socioeconomic problems already affecting Europe, such as disputes over economic policies, including Brexit, and supply chain disruptions caused by the COVID-19 pandemic. The situation worsened with the war in Ukraine, which strongly hit energy markets, pushing energy prices up by 86%. This sharp increase put further financial pressure on farmers, who were already dealing with higher costs for seeds, pesticides, and fertilizers. In fact, the cost of fertilizers, especially those imported from Russia, escalated dramatically due to the latter's role as a major global exporter of nitrogen and phosphate fertilizers [70], as shown in Figure 3.

Moreover, since the outbreak of the conflict in February 2022, the global food markets, particularly for essential crops such as wheat and sunflower oil, have faced significant disruptions. In fact, Ukraine ranks as the world's ninth-largest grain producer and fifth-largest exporter [71]. It also accounts for nearly half of the global sunflower oil production, with the EU depending on Ukraine for 38% of its imports [72]. The Russian invasion has significantly disrupted these exports, causing instability in international markets and raising concerns over food security and price volatility across Europe. In an attempt to alleviate the crisis, the EU decided to waive tariffs on Ukrainian agricultural products, offering a temporary lifeline to Ukraine's farming industry. However, this sparked controversy among European farmers, especially those from Eastern European countries, who argued that Ukrainian imports were not subject to the same strict regulations, particularly regarding pesticide use, resulting in unfair competition. Many farmers protested that the arrival of cheaper agricultural products from Ukraine, which did not meet the EU's strict pesticide regulations, put their livelihoods and local markets at risk. Consequently, in May 2023, the EC introduced temporary restrictions on grain imports from Ukraine to address the concerns in affected countries [73].

Although these restrictions were removed in September 2023, tensions continued to rise as some member states took unilateral action, introducing national laws to limit low-cost imports from Ukraine. This underscored the ongoing struggle within the EU to balance support for Ukraine with the interests of local farmers, who remained concerned about market distortions caused by the war [74].

In early 2024, the EU reinstated tariff exemptions for Ukrainian agricultural products to further complicate the situation, sparking renewed protests among farmers, particularly in Poland and France. Farmers in these countries argued that the lower-cost imports of poultry, eggs, and cereals from Ukraine severely undercut their ability to compete, as Ukrainian goods were produced at lower costs, largely due to differences in production standards and the absence of EU regulations on pesticides and fertilizers [75].

The European context in 2023, already under pressure due to the sharp increase in agricultural prices caused by the conflict in Ukraine, faced further challenges from a series of extreme weather events. These climate-related incidents inflicted significant short-term damage on agriculture, including the destruction of crops and damage to farming infrastructure. In addition to direct physical damage, these events intensified supply chain disruptions, further exacerbating the already fragile agricultural economy in Europe. These climate-related challenges included the largest wildfire ever recorded in Europe, one of the wettest years in recent memory, intense marine heatwaves that harmed coastal environments, and severe floods affecting Italy, Greece, and Slovenia [76]. The year 2023 was the second hottest year on record, with average temperatures 1 °C above the norm for 11 months and 7% more rainfall than usual, contributing to the above-mentioned severe flooding. The combined effects of these conditions also led to a reduction in crop diversity and a significant delay in planting seasons across multiple regions.

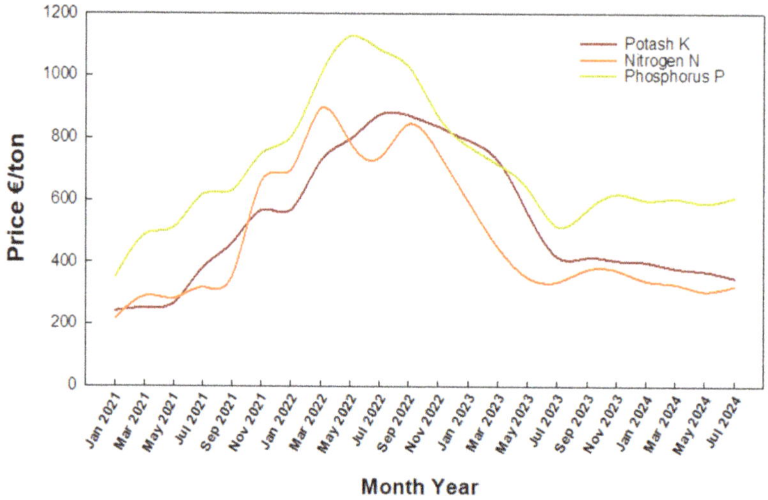

Figure 3. The variation in the price of fertilizers aggregated by nutrient in Europe from January 2021 to July 2024. Image created using data from Agri-food market data [77].

In Italy, 378 extreme weather events were documented, a 22% increase compared to the previous year, with agricultural damages reaching €6 billion [78]. These escalating incidents point to a future where extreme weather will be a regular occurrence, severely impacting agricultural productivity if adaptation strategies are not implemented. Climate change is projected to raise production costs for EU farmers by 1% to 7% by 2050, while food price inflation could increase by up to 2% by 2035. Implementing strategies to lower GHG emissions could help reduce these impacts [79].

Furthermore, drought severely impacted water availability, especially in Southern Europe during 2023. When combined with heatwaves, drought significantly threatened agriculture, reducing crop yields and threatening farmers' livelihoods. The persistence of these drought episodes also led to increased competition for water resources between agriculture and other sectors, further straining agricultural operations. Projections suggest that without adequate measures, economic losses from drought-related agricultural damage could rise from €4.8 billion in 2015 to €28.6 billion annually by 2100 [80].

3.4. The Tractor Protests Upend the EU Agenda: The Withdrawal of the SUR

Starting in December 2023, a wave of farmer protests spread across much of the EU. The protests began in Germany, where thousands of farmers blocked roads in Berlin with their tractors. The primary cause was the German government's decision to discontinue subsidies for diesel fuel in 2024. This issue was particularly pressing for farmers due to escalating costs following the war in Ukraine [81]. German farmers contended that the decision would have further jeopardized their already fragile financial standing and make their products less competitive, as it would increase production costs significantly [81].

By January 2024, the protests expanded to other EU nations, beginning in France and quickly spreading to Italy, the Netherlands, Poland, Romania, Hungary, and beyond. While each country faced its own domestic challenges, such as Germany's diesel subsidy cuts, several shared concerns united the protests across member states. These included soaring production costs, competition from Ukrainian imports, and, most prominently, the EU's environmental policies. Among these policies were the proposed 50% reduction in pesticide use under the Sustainable Use Regulation (SUR) and the mandate to leave 4% of farmland uncultivated [67,82].

In January 2024, farmers in France took to the streets, blocking highways with tractors and threatening to shut down Paris. Their protests centered around slow bureaucracy causing delays in CAP subsidy payments, competition from Ukrainian imports, rising fuel costs, and EU green regulations, which were considered excessive and financially burdensome [83]. In Eastern European countries like Poland, Romania, and Hungary, farmers protested at border crossings with Ukraine, denouncing the "unfair competition" from Ukrainian products produced at lower costs and under less stringent pesticide regulations [13]. On 27 February 2024, thousands of Polish farmers organized a large protest in Warsaw, marching through the city's streets to voice their opposition to the EU's Green Deal and the increasing imports of low-cost food from Ukraine. The farmers criticized the Green Deal for imposing environmental regulations, such as reductions in pesticide use, which they argued significantly increased their operational costs. They also expressed frustration with the influx of Ukrainian agricultural products, claiming that cheaper imports were driving down domestic prices, making it harder for them to remain competitive [84]. In Italy, protests were centered on the EU's Green Deal and CAP conditionalities, particularly the obligation to leave 4% of land fallow, rotate crops, and reduce the use of agrochemicals. Italian farmers also protested rising diesel prices and low returns within the food supply chain. One unique aspect of the Italian protests was the defense of "Made in Italy" products, which they believed were threatened by introducing synthetic meat and insect-based foods into the European market [83]. The protests continued for weeks across the continent. By the end of February 2024, around a thousand tractors had surrounded Brussels, where protesters reached the EC just as EU agriculture ministers were meeting to discuss the ongoing crisis [67].

As a result of the protests, farmers managed to secure certain concessions, both at the national level and from the EC. The Commission was forced to meet some of their demands to quell discontent, loosening some of the regulations designed to facilitate the green transition. For example, on 31 January 2024, farmers won a major concession when the EC granted a derogation from GAEC 8, which mandates that 4% of farmland must be left fallow to receive CAP payments [13]. Additionally, the EU imposed limits on Ukrainian imports to address concerns over competitiveness and accusations of unequal competition [85].

Despite these early concessions, protests continued for several more weeks. Among the main issues was the proposed SUR, which had become a focal point of protests in countries such as Belgium, France, Italy, and the Netherlands. After the European Parliament rejected the proposal in November 2023 and decided not to return the text to the Commission for further revision, the ongoing tractor protests played a crucial role in the final collapse of the SUR. On 6 February 2024, after weeks of intense protests, EC President Ursula von der Leyen announced the withdrawal of the SUR, describing it as a "symbol of polarization". The Commission ultimately decided to shelve the regulation by March 2024, acknowledging the need for a more inclusive dialog and less divisive solutions [86]. However, the official withdrawal of the SUR proposal occurred on 6 May 2024 [87], leaving the Sustainable Use of Pesticides Directive (Directive 2009/128) [59] in place. Table 2 summarizes the key EU legislation and directives to reduce pesticide use in agriculture while promoting sustainable practices and food safety.

Table 2. The relevant legislation for the reduction in pesticide use.

Year	Law/Directive	Description
1979	Directive 79/117/EEC	It is the first EU directive to ban certain dangerous pesticides, including DDT, establishing the framework to reduce harmful pesticide use in agriculture.
1991	Directive 91/414/EEC	It introduced an authorization process for placing plant protection products on the market, harmonizing pesticide regulation across the EU.
2000	Directive 2000/60/EC (Water Framework Directive)	It set water quality standards, including controlling pesticide contamination from agricultural sources to protect aquatic ecosystems and human health.
2005	Regulation (EC) 396/2005	It established MRLs for pesticides in or on food and feed, reducing exposure to harmful substances through agricultural products.
2009	Directive 2009/128/EC (Sustainable Use of Pesticides Directive)	It aimed to achieve a more sustainable use of pesticides by promoting IPM and alternative approaches, significantly reducing the risks and impacts of the use of pesticides.
2015	Regulation (EU) 2015/2030	It strengthened the control over the approval and use of pesticides, focusing on high-risk substances and promoting non-chemical alternatives.
2020	Farm to Fork Strategy	Part of the European Green Deal, this strategy set ambitious targets for reducing the overall use and risk of chemical pesticides by 50% by 2030, supporting a transition to more sustainable food systems.
2022	Proposed Sustainable Use of Pesticides Regulation (SUR)	A proposed regulation to enforce stricter measures for reducing pesticide use, including mandatory targets for cutting chemical pesticide use by 50% by 2030.
2023	CAP 2023–2027	It introduced eco-schemes and financial incentives for farmers who adopt sustainable pest control practices, aiming to reduce the use of synthetic pesticides through IPM.
2024	Withdrawal of SUR Proposal	Following public opposition, the EC withdrew the SUR proposal, indicating ongoing debates around pesticide regulation.

4. Balancing Pesticide Reduction, Agri-Food Waste Reuse, Financial Mechanisms, and Stakeholder Perspectives

Notwithstanding the withdrawal of the SUR, on 8 July 2024, the EC reported significant progress towards the Farm to Fork pesticide reduction targets, claiming a 46% reduction in the use and risk of chemical pesticides compared to 2015–2017 and a 25% decrease in hazardous pesticide use from 2018 to 2022 [88]. While these data suggest the strategy is on track, environmental groups like PAN Europe raised concerns about the accuracy of the reported reductions, noting discrepancies with Eurostat sales data [89]. For example, pesticide sales in France have tripled for certain harmful substances like PFAS pesticides over the last 13 years, and the use of highly toxic pesticides in the Netherlands has increased by 66% since 2010. These inconsistencies point to deeper financial and regulatory issues. The current methodology (e.g., the HRI1 indicator) has been criticized for underrepresenting highly toxic pesticides used in small amounts while overestimating the impact of lower-risk substances [89]. This inaccurate reporting not only fails to reflect the real economic and environmental risks of pesticide use, but also prevents the implementation of essential support for farmers, who will face serious impacts on productivity and income as they reduce pesticide use.

However, on 26 June 2024, the EC's Joint Research Centre published an interesting report based on the "Proceedings of the Workshop on Alternative Business Models for Pesticide Reduction". This report explored innovative business models and financial strategies to reduce pesticide use while maintaining agricultural productivity and food security. The report, written by Rennick et al. [90], highlights the potential of alternative approaches, such as outcome-based services and insurance policies, to support farmers in adopting more sustainable practices. These models offer financial incentives that help reduce pesticide use while minimizing the financial risks to farmers. For example, outcome-based models compensate farmers if their productivity drops, encouraging them to reduce pesticide use without fear of significant financial loss.

In particular, the report's data [90] showed that reducing pesticide use can lead to yield losses of 10–15% in crops such as wheat, barley, potatoes, and tomatoes, where pest and weed control are critical to maintaining high productivity. For instance, lower pesticide use in cereals like wheat and barley can increase weed pressure and pest damage, which impacts yields. Similarly, pest control plays a significant role in keeping yields stable in vegetable crops such as tomatoes and potatoes, while reducing pesticide use may cause a substantial decrease in productivity. However, these yield losses are often (or should be) compensated by government programs and higher market prices for pesticide-free or organic products, meaning that the overall impact on farmers' income is typically limited to a 2–5% reduction. French grapevine farmers, for example, managed to cut their fungicide use by 35%, with some reducing it by over 55%, while staying profitable due to insurance schemes that support sustainable practices [91]. In Switzerland, moving towards pesticide-free farming for crops like potatoes also resulted in lower costs for labor and machinery, helping to cope with the economic impact.

Indeed, stakeholders, including farmers, policymakers, and environmental groups, agree that the lack of financial support for farmers transitioning to sustainable practices is a significant challenge. Moving to organic farming or IPM often increases labor and equipment costs and potential productivity drops. Farmers need better financial incentives to make these changes sustainable. Policymakers must also ensure that regulatory plans, like the Farm to Fork Strategy, include strong economic measures to safeguard farmers' livelihoods.

Furthermore, cross-sector collaboration is essential for achieving sustainable agricultural practices and enhancing financial opportunities for farmers. Bioenergy feedstock sales clearly show how collaboration between the agricultural and energy sectors can benefit farmers financially. According to the EC's report [79], selling agricultural residues for bioenergy feedstock can increase farmers' income by 5–10%. This would provide an additional revenue stream, helping offset the costs of adopting more sustainable practices. For instance, farmers can sell crop residues, such as straw or corn stover, to bioenergy producers, creating new market opportunities while reducing reliance on chemical inputs. These combined incentives, ranging from compensation programs to insurance models, demonstrate that pesticide reduction can meet both environmental and financial goals when supported by proper financial mechanisms.

Cross-sector collaboration with energy producers, policymakers, and farmers will be crucial to ensure that bioenergy markets are accessible and profitable. Regulatory frameworks should support the development of these markets, while targeted grants and subsidies can help farmers transition to sustainable practices without risking financial losses. The report emphasizes the need for further research on the scalability of these business models. It suggests that future policies should align economic and environmental objectives to ensure the long-term sustainability of agriculture.

Finally, PAN Europe [78] has stressed the importance of more transparent and accurate tools, such as the NODU system used in France, which measures pesticide use relative to the area treated, providing a clearer view of real-world pesticide usage. Integrating financial support with accurate tracking methods is essential to achieving sustainability goals without compromising farmers' economic stability.

5. Conclusions

The case studies of the circular economy in agriculture and the Sustainable Use Regulation (SUR) highlight two critical challenges the EU faces in balancing environmental goals with practical implementation. In the first case concerning the circular economy, while the Regulation (EU) 2019/1009 addresses certain waste products like struvite and biochar, there remains a significant gap in regulatory frameworks tailored specifically to agricultural circularity. The lack of broader instruments to guide the circular use of agricultural by-products limits the potential for systemic sustainability in the sector. That said, in the case of the SUR, the EU's ambitious goal of reducing pesticide use by 50% by 2030 became a point of contention due to the perceived lack of consultation with farmers, who felt the policy was economically and operationally unfeasible. The following protests reflected the dangers of implementing drastic regulatory changes without sufficient support or viable alternatives for stakeholders. As a result, the withdrawal of the SUR in February 2024 underscored the need for more gradual transitions, paired with technological innovations and economic incentives to ease the burden on farmers.

Indeed, the EU Waste Directive, the CEAP, and the CAP reforms are crucial to promoting sustainability in agriculture. However, there is potential to better coordinate these policies to address both waste management and pesticide reduction, which are crucial for sustainable farming. The EU Waste Directive provides the framework for turning agricultural waste into valuable products, but it could go further by linking this process with pesticide reduction efforts. Expanding the "end-of-waste" criteria to include products that reduce the need for chemical inputs would encourage farmers to adopt practices combining waste management with more sustainable farming methods. At the same time, the CEAP focuses on reusing waste but does not fully address how this can work together with reducing pesticide use. This plan could help reduce reliance on chemical pesticides while offering farmers new revenue streams by encouraging the development of bio-based fertilizers and biopesticides from organic waste. This would create a more vital link between waste management and sustainable agriculture. Finally, the CAP reforms (2023–2027) already reward environmentally friendly practices, but there is room to improve. CAP subsidies could be more directly tied to innovations that support waste promotion and pesticide reduction, providing farmers with the financial support to adopt these practices while maintaining economic stability. These important frameworks must be better connected to address waste management and pesticide reduction. Aligning these policies would create a more practical approach to sustainable agriculture that benefits both the environment and the livelihoods of farmers.

Therefore, these two cases demonstrate the importance of creating flexible and inclusive policies that consider the real-world economic pressures on farmers and provide clear pathways for integrating sustainable practices. Achieving long-term environmental goals requires collaboration with the agricultural community, sufficient financial support, and incremental policy adjustments that ensure sustainability does not come at the expense of agricultural viability.

Author Contributions: Conceptualization, P.C.; resources, R.N., M.P., G.M.F. and A.C.; data curation, R.N., M.P. and P.C.; writing—original draft preparation, R.N., M.P., P.C. and B.M.; writing—review and editing, R.N., B.M. and P.C. All authors have read and agreed to the published version of the manuscript.

Funding: This work was supported by the projects GREENHORT (PRIN 2022—grant 2022WHTNZT) and BBHORT (PRIN 2022 PNRR—grant P2022P52XK) funded by the Italian Ministry of University and Research (MUR), and the CIBUS project (CUP B27G2200029002) funded by Regione Campania.

Institutional Review Board Statement: Not applicable.

Informed Consent Statement: Not applicable.

Data Availability Statement: Not applicable.

Conflicts of Interest: The authors declare that there are no conflicts of interest to report regarding the present study.

References

1. Ritchie, H. How Much of Global Greenhouse Gas Emissions Come from Food? 2021. Available online: https://ourworldindata.org/greenhouse-gas-emissions-food (accessed on 23 December 2023).
2. WWF. Driven to Waste: The Global Impact of Food Loss and Waste on Farms. 2021. Available online: https://wwfeu.awsassets.panda.org/downloads/driven_to_waste___the_global_impact_of_food_loss_and_waste_on_farms.pdf (accessed on 20 December 2023).
3. Hassan, M.U.; Aamer, M.; Mahmood, A.; Awan, M.I.; Barbanti, L.; Seleiman, M.F.; Bakhsh, G.; Alkharabsheh, H.M.; Babur, E.; Shao, J.; et al. Management Strategies to Mitigate N_2O Emissions in Agriculture. *Life* **2022**, *12*, 439. [CrossRef] [PubMed]
4. Shabir, I.; Dash, K.K.; Dar, A.H.; Pandey, V.K.; Fayaz, U.; Srivastava, S.; Nisha, N. Carbon footprints evaluation for sustainable food processing system development: A comprehensive review. *Future Foods* **2023**, *7*, 100215. [CrossRef]
5. Pomoni, D.I.; Koukou, M.K.; Vrachopoulos, M.G.; Vasiliadis, L. Circular economy: A multilevel approach for natural resources and wastes under an agri-food perspective. *Water Energy Nexus* **2024**, *7*, 103–123. [CrossRef]
6. COM. The European Green Deal. Striving to Be the First Climate-Neutral Continent. 2019. Available online: https://commission.europa.eu/strategy-and-policy/priorities-2019-2024/european-green-deal_en (accessed on 18 December 2023).
7. EU. Communication from the Commission to the European Parliament, the Council, the European Economic and Social Committee and the Committee of the Regions. A Farm to Fork Strategy for a Fair, Healthy and Environmentally-Friendly Food System COM/2020/381 Final. 2020. Available online: https://eur-lex.europa.eu/legal-content/EN/TXT/?uri=CELEX:52020DC0381 (accessed on 5 September 2024).
8. EC. Directive 2008/98/EC of the European Parliament and of the Council of 19 November 2008 on Waste and Repealing Certain Directives. 2008. Available online: https://eur-lex.europa.eu/legal-content/EN/TXT/?uri=celex:32008L0098 (accessed on 18 December 2023).
9. COM. A New Circular Economy Action Plan For a Cleaner and More Competitive Europe. 2020. Available online: https://eur-lex.europa.eu/legal-content/EN/TXT/?qid=1583933814386&uri=COM:2020:98:FIN (accessed on 20 December 2023).
10. EC. Regulation (EU) 2019/1009 of the European Parliament and of the Council of 5 June 2019 Laying Down Rules on the Making Available on the Market of EU Fertilising Products and Amending Regulations (EC) No 1069/2009 and (EC) No 1107/2009 and Repealing Regulation (EC) No 2003/2003. 2019. Available online: https://eur-lex.europa.eu/legal-content/EN/TXT/?uri=celex:32019R1009 (accessed on 20 December 2023).
11. Haque, F.; Fan, C.; Lee, Y.-Y. From waste to value: Addressing the relevance of waste recovery to agricultural sector in line with circular economy. *J. Clean. Prod.* **2023**, *415*, 137873. [CrossRef]
12. EC. Sustainable Use of Pesticides Regulation (SUR). 2022. Available online: https://food.ec.europa.eu/plants/pesticides/sustainable-use-pesticides_en#:~:text=The%20European%20Commission%20adopted%20on,to%20Fork%20and%20Biodiversity%20strategies (accessed on 4 September 2024).
13. Tanno, S.; Liakos, C. Farmers' Protests Have Erupted across Europe. Here's Why. 2024. Available online: https://edition.cnn.com/2024/02/03/europe/europe-farmers-protests-explainer-intl/index.html (accessed on 4 September 2024).
14. Liboreiro, J.; Fortuna, G. Von der Leyen Withdraws Contentious Pesticide Law amid Right-Wing Backlash and Farmer Protests. 2024. Available online: https://www.euronews.com/my-europe/2024/02/06/von-der-leyen-announces-withdrawal-of-contentious-pesticide-law-the-first-defeat-of-the-gr (accessed on 4 September 2024).
15. Morrone, B. Residual Biomass Resources: An Invaluable Reservoir of Energy and Matter. In *Innovations in Sustainable Energy and Cleaner Environment*; Gupta, A.K., De, A., Aggarwal, S.K., Kushari, A., Runchal, A., Eds.; Springer: Singapore, 2020; pp. 149–170.
16. Carillo, P.; Morrone, B. Converting Residual Biomasses into Energy through Anaerobic Digestion. In *Municipal Solid Waste: Management Strategies, Challenges and Future Directions*; Waste and Waste Management Series; Tzortzakis, N., Ed.; Nova Science Publishers: Hauppauge, NY, USA, 2017; ISBN 978-1-53611-865-0.
17. Boulding, K. The Economics of the Coming Spaceship Earth. In *Environmental Quality in a Growing Economy, Resources for the Future*; Jarrett, H., Ed.; Johns Hopkins University Press: Baltimore, MD, USA, 1966; pp. 3–14.
18. FAO. *The State of Food and Agriculture—Moving Forward on Food Loss and Waste Reduction*; Food and Agriculture Organization of the United Nations: Rome, Italy, 2019; Available online: http://www.fao.org/3/ca6030en/ca6030en.pdf (accessed on 8 September 2024).
19. Nicastro, R.; Carillo, P. Food Loss and Waste Prevention Strategies from Farm to Fork. *Sustainability* **2021**, *13*, 5443. [CrossRef]
20. COM. Biomass Action Plan, European Commission. 2005. Available online: https://eur-lex.europa.eu/LexUriServ/LexUriServ.do?uri=COM:2005:0628:FIN:EN:PDF (accessed on 18 December 2023).
21. EC. Waste Directive Regulations S.I. No. 126/2011. 2011. Available online: https://faolex.fao.org/docs/pdf/ire103799.pdf (accessed on 8 September 2024).
22. OECD. The Polluter-Pays Principle—OECD Analyses and Recommendations. 1992. Available online: https://one.oecd.org/document/OCDE/GD(92)81/En/pdf#:~:text=The%20Polluter-Pays%20Principle,%20as,is%20in%20an%20acceptable%20state%22 (accessed on 20 December 2023).
23. COM. EUROPE 2020 A Strategy for Smart, Sustainable and Inclusive Growth. 2011. Available online: https://eur-lex.europa.eu/legal-content/EN/TXT/?uri=celex:52010DC2020 (accessed on 19 December 2023).
24. Bongardt, A.; Torres, F. The competitiveness rationale, sustainable growth and the need for enhanced economic coordination. In *Intereconomics: A Review of European Economic Policy*; Sciendo: Warsaw, Poland, 2010; Volume 45, pp. 136–141, ISSN 1613-964X.

25. EEA. Trends and Projections in Europe 2021. 2021. Available online: https://www.eea.europa.eu/publications/trends-and-projections-in-europe-2021 (accessed on 20 December 2023).
26. COM. Innovating for Sustainable Growth: A Bioeconomy for Europe. 2012. Available online: https://eur-lex.europa.eu/legal-content/EN/TXT/?uri=celex:52012DC0060 (accessed on 19 December 2023).
27. EC. A Sustainable Bioeconomy for Europe—Strengthening the Connection between Economy, Society and the Environment—Updated Bioeconomy Strategy. 2018. Available online: https://data.europa.eu/doi/10.2777/792130 (accessed on 19 December 2023).
28. COM. EU Bioeconomy Strategy Progress Report—European Bioeconomy Policy: Stocktaking and Future Developments. 2022. Available online: https://knowledge4policy.ec.europa.eu/publication/report-com2022283-eu-bioeconomy-strategy-progress-report-european-bioeconomy-policy_en (accessed on 19 December 2023).
29. COM. Towards a Circular Economy: A Zero Waste Programme for Europe. 2014. Available online: https://eur-lex.europa.eu/legal-content/EN/TXT/?uri=celex:52014DC0398 (accessed on 19 December 2023).
30. COM. Closing the Loop—An EU Action Plan for the Circular Economy. 2015. Available online: https://eur-lex.europa.eu/legal-content/EN/TXT/?uri=CELEX:52015DC0614 (accessed on 19 December 2023).
31. UN. Transforming Our World: The 2030 Agenda for Sustainable Development. 2015. Available online: https://sdgs.un.org/2030agenda (accessed on 20 December 2023).
32. EESC. European Circular Economy Stakeholder Platform—A Joint Initiative by the European Commission and the European Economic and Social Committee. 2019. Available online: https://data.europa.eu/doi/10.2864/448047 (accessed on 19 December 2023).
33. EC. Directive (EU) 2018/849 of the European Parliament and of the Council of 30 May 2018 Amending Directives 2000/53/EC on End-of-Life Vehicles, 2006/66/EC on Batteries and Accumulators and Waste Batteries and Accumulators, and 2012/19/EU on Waste Electrical and Electronic Equipment. 2018. Available online: https://eur-lex.europa.eu/legal-content/EN/TXT/?uri=CELEX:32018L0849 (accessed on 19 December 2023).
34. EC. Directive (EU) 2018/850 of the European Parliament and of the Council of 30 May 2018 Amending Directive 1999/31/EC on the Landfill of Waste. 2018. Available online: https://eur-lex.europa.eu/legal-content/EN/TXT/?uri=celex:32018L0850 (accessed on 19 December 2023).
35. EC. Directive (EU) 2018/851 of the European Parliament and of the Council of 30 May 2018 Amending Directive 2008/98/EC on Waste. 2018. Available online: https://eur-lex.europa.eu/legal-content/EN/TXT/?uri=celex:32018L0851 (accessed on 19 December 2023).
36. EC. Directive (EU) 2018/852 of the European Parliament and of the Council of 30 May 2018 Amending Directive 94/62/EC on Packaging and Packaging Waste. 2018. Available online: https://eur-lex.europa.eu/eli/dir/2018/852/oj (accessed on 19 December 2023).
37. EC. Regulation (EC) No 178/2002 of the European Parliament and of the Council of 28 January 2002 Laying Down the General Principles and Requirements of Food Law, Establishing the European Food Safety Authority and Laying Down Procedures in Matters of Food Safety. 2002. Available online: https://eur-lex.europa.eu/legal-content/EN/ALL/?uri=celex:32002R0178 (accessed on 19 December 2023).
38. EC. The Common Agricultural Policy: 2023–27. 2022. Available online: https://agriculture.ec.europa.eu/common-agricultural-policy/cap-overview/cap-2023-27_en (accessed on 20 December 2023).
39. Ministry of the Environment, Japan. The Plastic Resource Circulation Act. 2022. Available online: https://www.meti.go.jp/english/press/2022/01114_001.html (accessed on 2 October 2024).
40. Lee, K.; Cha, J. Towards Improved Circular Economy and Resource Security in South Korea. *Sustainability* **2021**, *13*, 17. [CrossRef]
41. Victoria State Government. Circular Economy (Waste Reduction and Recycling) Act 2021. 2021. Available online: https://www.legislation.vic.gov.au/in-force/acts/circular-economy-waste-reduction-and-recycling-act-2021 (accessed on 2 October 2024).
42. Victorian Government. Recycling Victoria: A New Economy. 2020. Available online: https://www.vic.gov.au/sites/default/files/2020-02/Recycling%20Victoria%20-%20Factsheet.pdf (accessed on 2 October 2024).
43. Ministry for the Environment. Emissions Reduction Plan. 2022. Available online: https://environment.govt.nz/publications/aotearoa-new-zealands-first-emissions-reduction-plan (accessed on 2 October 2024).
44. Latin America and Caribbean Circular Economy Coalition. Circular Economy in Latin America and the Caribbean: A Shared Vision. 2021. Available online: https://coalicioneconomiacircular.org/wp-content/uploads/2022/02/ENGLISH-Circular-economy-in-Latin-America-and-the-Caribbean.pdf (accessed on 2 October 2024).
45. Van Hoof, B.; Núñez Reyes, G.; De Miguel, C.J. Scaling Up Circular Economy Initiatives in Latin America and the Caribbean. Production, Productivity, and Management. ECLAC, December 2023. 2023. Available online: https://repositorio.cepal.org/server/api/core/bitstreams/cd15f230-3122-4b5d-9d8f-b7a72eb5ac2d/content (accessed on 2 October 2024).
46. CalRecycle. California Plastic Pollution Prevention and Packaging Producer Responsibility Act (SB 54). 2022. Available online: https://calrecycle.ca.gov/packaging/packaging-epr/ (accessed on 2 October 2024).
47. City of New York. Zero Waste Plan. 2023. Available online: https://www.nyc.gov/assets/dsny/downloads/resources/reports/zero-waste-plan/zero-waste-plan.pdf (accessed on 2 October 2024).
48. United States Environmental Protection Agency (EPA). Sustainable Materials Management. 2024. Available online: https://www.epa.gov/smm (accessed on 2 October 2024).

49. United States Environmental Protection Agency (EPA). National Strategy for Reducing Food Loss and Waste and Recycling Organics. 2024. Available online: https://www.epa.gov/circulareconomy/national-strategy-reducing-food-loss-and-waste-and-recycling-organics (accessed on 2 October 2024).
50. EFSA. Pesticides. 2024. Available online: https://www.efsa.europa.eu/en/topics/topic/pesticides (accessed on 5 September 2024).
51. European Parliament—European Union; Sarkar, S.; Dias, J.; Gil, B.; Keeley, J.; Möhring, N.; Jansen, K. *The Use of Pesticides in Developing Countries and Their Impact on Health and the Right to Food*; European Parliament: Strasbourg, France, 2021.
52. European Commission. Directorate General for Health and Food Safety. In *Development of Future Scenarios for Sustainable Pesticide Use and Achievement of Pesticide-Use and Risk Reduction Targets Announced in the Farm to Fork and Biodiversity Strategies by 2030: Final Report*; Feijao, C., d'Angelo, C., Flanagan, I., Abellan, B., Gloinson, E.R., Smith, E., Traon, D., Gehrt, D., Teare, H., Dunkerley, F., Eds.; European Union: Maastricht, The Netherlands, 2022; Available online: https://data.europa.eu/doi/10.2875/565045 (accessed on 5 September 2024).
53. EU. Council Directive 79/117/EEC of 21 December 1978 Prohibiting the Placing on the Market and Use of Plant Protection Products Containing Certain Active Substances. 1978. Available online: https://eur-lex.europa.eu/legal-content/EN/ALL/?uri=celex:31979L0117 (accessed on 5 September 2024).
54. EC. Timeline—History of the CAP. 2021. Available online: https://www.consilium.europa.eu/en/policies/cap-introduction/timeline-history-of-cap/ (accessed on 5 September 2024).
55. EU. Council Directive 91/414/EEC of 15 July 1991 Concerning the Placing of Plant Protection Products on the Market. 1991. Available online: https://eur-lex.europa.eu/legal-content/EN/ALL/?uri=celex:31991L0414 (accessed on 5 September 2024).
56. EU. Directive 2000/60/EC of the European Parliament and of the Council of 23 October 2000 Establishing a Framework for Community Action in the Field of Water Policy. 2000. Available online: https://eur-lex.europa.eu/eli/dir/2000/60/oj (accessed on 5 September 2024).
57. EU. Regulation (EC) No 396/2005 of the European Parliament and of the Council of 23 February 2005 on Maximum Residue Levels of Pesticides in or on Food and Feed of Plant and Animal Origin and Amending Council Directive 91/414/EEC Text with EEA Relevance. 2005. Available online: https://eur-lex.europa.eu/legal-content/EN/ALL/?uri=CELEX:32005R0396 (accessed on 5 September 2024).
58. EU. Communication from the Commission to the Council, the European Parliament, the European Economic and Social Committee and the Committee of the Regions—A Thematic Strategy on the Sustainable Use of Pesticides {COM(2006) 373 final} {SEC(2006) 894} {SEC(2006) 895} {SEC(2006) 914}. 2006. Available online: https://eur-lex.europa.eu/legal-content/EN/TXT/?uri=celex:52006DC0372 (accessed on 5 September 2024).
59. EU. Directive 2009/128/EC of the European Parliament and of the Council of 21 October 2009 Establishing a Framework for Community Action to Achieve the Sustainable Use of Pesticides. 2009. Available online: https://eur-lex.europa.eu/eli/dir/2009/128/2009-11-25 (accessed on 5 September 2024).
60. EC. Regulation (EC) No 1107/2009 of the European Parliament and of the Council of 21 October 2009 Concerning the Placing of Plant Protection Products on the Market and Repealing Council Directives 79/117/EEC and 91/414/EEC. 2009. Available online: https://eur-lex.europa.eu/eli/reg/2009/1107/oj/eng (accessed on 5 September 2024).
61. EU. Council Decision 2011/162/EU of 14 March 2011 Establishing the Position to Be Taken by the European Union within the Fifth Meeting of the Conference of the Parties of the Rotterdam Convention as Regards the Amendments to Annex III to the Rotterdam Convention on the Prior Informed Consent Procedure for Certain Hazardous Chemicals and Pesticides in International Trade. 2011. Available online: https://eur-lex.europa.eu/legal-content/EN/TXT/?uri=CELEX:32011D0162 (accessed on 5 September 2024).
62. Schneider, K.; Barreiro-Hurle, J.; Rodriguez-Cerezo, E. Pesticide reduction amidst food and feed security concerns in Europe. *Nat. Food* **2023**, *4*, 746–750. [CrossRef] [PubMed]
63. EC. Council Directive 92/43/EEC of 21 May 1992 on the Conservation of Natural Habitats and of Wild Fauna and Flora. 1992. Available online: https://eur-lex.europa.eu/legal-content/EN/TXT/?uri=celex:31992L0043 (accessed on 4 September 2024).
64. EU. Council Decision (EU) 2022/2572 of 19 December 2022 Requesting the Commission to Submit a Study Complementing the Impact Assessment of the Proposal for a Regulation of the European Parliament and of the Council on the Sustainable Use of Plant Protection Products and Amending Regulation (EU) 2021/2115 of the European Parliament and of the Council, and to Propose Follow-Up Actions, If Appropriate in View of the Outcomes of the Study. 2022. Available online: https://eur-lex.europa.eu/eli/dec/2022/2572/oj (accessed on 5 September 2024).
65. EU Council. Commission Response to Council Decision (EU) 2022/2572 of 19 December 2022 Requesting That the Commission Submit a Study Complementing the Impact Assessment of the Proposal for a Regulation of the European Parliament and of the Council on the Sustainable Use of Plant Protection Products and Amending Regulation (EU) 2021/2115. 2023. Available online: https://data.consilium.europa.eu/doc/document/ST-11644-2023-INIT/en/pdf (accessed on 5 September 2024).
66. Mezzacapo, E. Mind the Gap: Assessing Member States' Implementation of Farm to Farm-to-Fork Targets within the 2023–2027 Common Agricultural Policy. *Eur. J. Risk Regul.* **2024**, *15*, 265–279. [CrossRef]
67. The Guardian. They're Drowning Us in Regulations': How Europe's Furious Farmers Took on Brussels and Won. 2024. Available online: https://www.theguardian.com/environment/2024/feb/10/theyre-drowning-us-in-regulations-how-europes-furious-farmers-took-on-brussels-and-won (accessed on 5 September 2024).

68. PAN Europe. Committee Vote on Pesticides Regulation SUR. 2023. Available online: https://www.pan-europe.info/press-releases/2023/10/envi-committee-vote-pesticides-regulation-sur (accessed on 5 September 2024).
69. EP. LEGISLATIVE TRAIN 08.2024. 2024. Available online: https://www.europarl.europa.eu/legislative-train/carriage/sustainable-use-of-pesticides-%E2%80%93-revision-of-the-eu-rules/report?sid=8301 (accessed on 5 September 2024).
70. World Economic Forum. This Is How War in Europe Is Disrupting Fertilizer Supplies and Threatening Global Food Security. 2023. Available online: https://www.weforum.org/agenda/2023/03/ukraine-fertilizer-food-security/ (accessed on 5 September 2024).
71. Devadoss, S.; Ridley, W. Impacts of the Russian invasion of Ukraine on the global wheat market. *World Dev.* **2024**, *173*, 106396. [CrossRef]
72. Chepeliev, M.; Maliszewska, M.; Pereira, M.F.S.e. The War in Ukraine, Food Security and the Role for Europe. *EuroChoices* **2023**, *22*, 4–13. [CrossRef]
73. News Europen Parliament. Deal on Trade Support for Ukraine with More Protection for EU Farmers. 2024. Available online: https://www.europarl.europa.eu/news/en/press-room/20240405IPR20228/deal-on-trade-support-for-ukraine-with-more-protection-for-eu-farmers (accessed on 2 October 2024).
74. EU. COMMISSION STAFF WORKING DOCUMENT Ukraine 2023 Report Accompanying the document Communication from the Commission to the European Parliament, the Council, the European Economic and Social Committee and the Committee of the Regions 2023 Communication on EU Enlargement Policy. 2023. Available online: https://eur-lex.europa.eu/legal-content/EN/TXT/?uri=CELEX:52023SC0699 (accessed on 5 September 2024).
75. EUNews. EU Agrees on Extension of Stop on Ukrainian Import Duties. But with Safeguards for European Farmers. 2024. Available online: https://www.eunews.it/en/2024/03/20/eu-agrees-on-extension-of-stop-on-ukrainian-import-duties-but-with-safeguards-for-european-farmers/ (accessed on 5 September 2024).
76. COM. Proposal for a Decision of the European Parliament and of the Council on the Mobilisation of the European Union Solidarity Fund to Provide Assistance to Italy, Slovenia, Austria, Greece and France Relating to Six Natural Disasters Occurred in 2023. 2023. Available online: https://data.consilium.europa.eu/doc/document/ST-12870-2024-INIT/en/pdf (accessed on 5 September 2024).
77. EC. Prices. Market Data on National and European Agriculture Provided by the European Commission's Agricultural and Rural Development Directorate. 2024. Available online: https://agridata.ec.europa.eu/extensions/DataPortal/prices.html (accessed on 8 September 2024).
78. Copernicus. 2023 Is the Hottest Year on Record, with Global Temperatures Close to the 1.5 °C Limit. 2024. Available online: https://climate.copernicus.eu/copernicus-2023-hottest-year-record (accessed on 8 September 2024).
79. Kotz, M.; Levermann, A.; Wenz, L. The economic commitment of climate change. *Nature* **2024**, *628*, 551–557. [CrossRef] [PubMed]
80. Naumann, G.; Cammalleri, C.; Mentaschi, L.; Feyen, L. Increased economic drought impacts in Europe with anthropogenic warming. *Nat. Clim. Chang.* **2021**, *11*, 485–491. [CrossRef]
81. Alkousaa, R.; Stein, S. German Farmers Protest with Tractors against Austerity Measures. 2023. Available online: https://www.reuters.com/world/europe/german-farmers-protest-with-tractors-against-austerity-measures-2023-12-18/ (accessed on 5 September 2024).
82. Boutelet, C.; Bran, M.; Girard, L.; Morel, S.; Stroobants, J.-P. Farmers' Anger Is Mounting across Europe. 2024. Available online: https://www.lemonde.fr/en/economy/article/2024/01/24/farmers-anger-is-mounting-across-europe_6460470_19.html (accessed on 5 September 2024).
83. Martelli, S. La Protesta dei Trattori Assedia l'Europa, Ecco cosa Chiedono gli Agricoltori. 2024. Available online: https://www.ilsole24ore.com/art/la-protesta-trattori-assedia-l-europa-ecco-cosa-chiedono-agricoltori-AFs5EXXC?refresh_ce=1 (accessed on 5 September 2024).
84. Le Monde. Thousands of Polish Farmers March against EU Climate Measures and Ukrainian Imports. 2024. Available online: https://www.lemonde.fr/en/economy/article/2024/02/27/thousands-of-polish-farmers-march-against-eu-climate-measures-and-ukrainian-imports_6564449_19.html (accessed on 5 September 2024).
85. Popławski, K. Unequal Competition. The Consequences of the Liberalisation of EU–Ukraine Road Carriage for Central Europe. 2024. Available online: https://www.osw.waw.pl/en/publikacje/osw-commentary/2024-04-04/unequal-competition-consequences-liberalisation-eu-ukraine (accessed on 5 September 2024).
86. Le Monde. EU's von der Leyen Backtracks on Cutting Pesticide Use. 2024. Available online: https://www.lemonde.fr/en/european-union/article/2024/02/06/eu-s-von-der-leyen-bows-to-protesting-farmers-on-pesticide-use_6497763_156.html (accessed on 5 September 2024).
87. EU. Withdrawal of Commission Proposals. 2024. Available online: https://eur-lex.europa.eu/eli/C/2024/3117/oj (accessed on 5 September 2024).
88. EC. EU Trends in the Use and Risk of Chemical Pesticides and in the Use of More Hazardous Pesticides. 2024. Available online: https://food.ec.europa.eu/plants/pesticides/sustainable-use-pesticides/farm-fork-targets-progress/eu-trends_en (accessed on 1 October 2024).
89. PAN Europe. PAN Europe. EU Commission Spreads Unscientific Information about Pesticide Reduction. 2024. Available online: https://www.pan-europe.info/blog/eu-commission-spreads-unscientific-information-about-pesticide-reduction (accessed on 1 October 2024).

90. Rennick, G.; Berling, A.; Chappell, T.M.; Vicent, A.; Wolf, M.; De Toffol, M.; Hloben, P.; Mack, G.; Le Mouël, C.; Sebastian, F.; et al. Proceedings of the Workshop on Alternative Business Models for Pesticide Reduction. Publications Office of the European Union. [JRC136924]. 2024. Available online: https://datam.jrc.ec.europa.eu/datam/file/e1d732d6-d5c9-4c4e-9fc5-5352d2f62e58/JRC136924%20small%20size.pdf (accessed on 1 October 2024).
91. Enjolras, G.; Aubert, M. How does crop insurance influence pesticide use? Evidence from French farms. *Rev. Agric. Food Environ. Stud.* **2020**, *101*, 461–485. [CrossRef]

sustainability

MDPI

Article

Exploring Barriers to the Effective Implementation of Integrated Waste Management Plans in Developing Economies: Lessons Learned from South African Municipalities

Phathutshedzo Mukwevho *, Prince Radzuma and Claudine Roos

Unit for Environmental Science and Management, North-West University, Private Bag X6001, Potchefstroom 2520, South Africa; radzumap@gmail.com (P.R.); claudine.roos@nwu.ac.za (C.R.)
* Correspondence: phathu.mukwevho@nwu.ac.za

Abstract: Solid waste management poses a significant challenge globally, especially in developing economies, where increasing waste generation and ineffective management measures are common. Municipalities, responsible for managing waste, often struggle to provide efficient services. To address this, local governments implement policy tools like the Integrated Waste Management Plans (IWMPs), which aim to improve waste management. Despite the existence of such policies, waste pollution continues to rise in countries like South Africa. The purpose of this study is to explore the barriers to the effective implementation of IWMPs, focusing on the Makhado Local Municipality (MLM) in South Africa. Using a qualitative approach, interviews, document reviews, and observations were conducted with stakeholders, targeting 23 participants, of which 15 contributed. Our findings highlight several key barriers, including lack of monitoring, political interference, corruption, insufficient budgeting, poor infrastructure maintenance, lack of skilled personnel, and inadequate community education. These challenges hinder the successful implementation of IWMPs in the MLM. Our study further identified several opportunities at the MLM and general strengths and threats using a SWOT (strengths, weaknesses, opportunities, threats) analysis.

Keywords: barriers; integrated waste management plan (IWMP); solid waste management; implementation; policy instrument; developing economies

1. Introduction

Increasing municipal solid waste (MSW) generation is a major concern worldwide, with least-developed African economies such as South Africa, Egypt, and Nigeria particularly affected due to ineffective waste management practices [1,2]. These practices include littering, illegal dumping, poor waste management, outdated measures, escalating management costs, decreasing landfill space, and limited land availability for new landfills [1,2]. It is anticipated that Africa is headed for a major social and economic transformation in this century, with its population expected to reach 2.4 billion by 2050 [3]. Urbanisation and changing consumer purchasing habits will lead to exponential growth in waste generation [3].

Africa generated 125 million tonnes of MSW in 2012, with 81 million tonnes coming from sub-Saharan Africa [4]. This is expected to increase to 244 million tonnes per year by 2025 [5]. In 2011, South Africa generated approximately 108 million tonnes of waste, which increased to 121 million tonnes in 2017 [1,6,7]. Like other developing economies, South Africa faces significant challenges due to poor waste management practices, primarily driven by issues at the local government level [8,9]. Contributing factors include poor organisational structures, lack of capacity and appropriate skills, inadequate budgeting, ineffective legislation and enforcement, low public awareness, conflicts, and corruption [5].

The Constitution of the Republic of South Africa assigns responsibility for waste management services to local municipalities. Local governments are the backbone of waste management services and are expected to provide efficient and quality waste collection and disposal services. Local authorities are also mandated by the Constitution to protect the environment and manage solid waste [10,11].

Legally, South Africa has evolved from a fragmented legislative framework, which made it challenging for the public and private sectors to comply. This evolution began with the recognition of the right of all citizens to an environment that is not harmful to their health and well-being. Key pieces of legislation include the Environmental Conservation Act (Act 73 of 1989) (repealed), which focused on permitting, controlling, and managing waste disposal sites, and the National Environmental Management Act (Act 107 of 1998), which introduced several guiding principles to environmental legislation, such as the lifecycle approach to waste management, the precautionary principle, and the polluter pays principle [12].

Furthermore, according to Godfrey and Oelofse (2017) [12], waste legislation was initiated with the publication of the White Paper on Integrated Pollution and Waste Management (WPIP&WM) in 2000, preceded by the adoption of the first National Waste Management Strategy (NWMS) in 1999. Eight years after the WPIP&WM, the Waste Act 59 of 2008 was promulgated, providing for the management of every aspect of waste in South Africa. The principle behind Waste Act 59 of 2008 is based on the waste management hierarchy. Following the Act, several regulations, norms, and standards were introduced [7,12].

South Africa has transitioned from waste management practices focused mainly on waste disposal to the adoption of the internationally recognised waste management hierarchy. This hierarchy emphasises waste prevention and minimisation, followed by reducing, reusing, recycling, and finally, disposing as the least preferred option [13].

Governments have been developing policy instruments to mitigate solid waste impacts, with management practices varying greatly across regions, countries, and even within countries. Modern waste management approaches, such as the waste management hierarchy, advocate for reducing waste generation, reusing, recycling, recovering, composting, and safe disposal through landfills [14]. However, these practices are often not applied, especially in developing countries, where a large proportion of solid waste is disposed of or burned [14].

One policy instrument to regulate MSW and its impacts is the Integrated Waste Management Plan (IWMP), which has been developed and implemented by many developing and developed countries. The concept of solid waste management planning has historical roots [15]. Countries such as South Africa, the United States, China, and Brazil have incorporated IWMPs into their policies [16–18]. IWMPs are essential for designing and enhancing waste management practices within organisations [19]. In South Africa, IWMPs are intended as decision-support tools to inform higher-level decision-making instruments, particularly in local government, such as Integrated Development Plans (IDPs), concerning waste-related matters.

In South Africa, IWMP development and implementation is a legal requirement under Section 12 of the National Environmental Management (Waste Act 59 of 2008). Additionally, the Department of Environmental Affairs (DEA) developed guidelines for IWMPs in 2012, which provide guidance on implementation instruments and plans [20]. According to these guidelines, "the primary objective of an IWMP is to integrate and optimise waste management planning to maximise efficiency, reduce environmental impacts and financial costs, and improve the quality of life for all South Africans".

Municipalities are required by the Municipal Systems Act and the Waste Act to integrate IWMPs into their IDPs. South Africa's Municipal Integrated Development Plan (IDP) is a strategic document that outlines a municipality's development priorities and strategies over a five-year period, ensuring coherence between national, provincial, and local government plans.

Despite numerous governmental interventions, waste generation continues to pose a significant threat to human health, environmental quality, and the ecotourism economy [21]. Several studies have found that many completed municipal IWMPs lack successful implementation [22–24].

This study, therefore, aims to explore the barriers to implementing the IWMP system in developing economies, using South Africa's Makhado Local Municipality (MLM) as a case study. According to Hossu et al. (2021) [25], measuring the achievement of outcomes in many environmental plans is challenging, and little is known about whether such plans achieve their goals as intended. This study argues that with little or no quantitative data available to assess the effective implementation of plans like the IWMP, reliance on stakeholder perceptions is crucial to gaining insights. This paper contributes to international knowledge and literature by exploring stakeholder perceptions to identify the challenges of implementing IWMPs in developing economies, using South Africa as a case study. The findings may provide a basis for improving IWMP systems in other regions.

South African municipalities provide a good case study as the country is considered to have one of the most complex and mature environmental management and legal frameworks globally [26] (pp. 1–55). Despite this, waste management issues persist in the country, making it crucial to understand the perceptions of key stakeholders, such as implementing partners, to assess the effectiveness of the IWMP system in developing economies.

2. Review of Existing Similar Theories

Solid waste management has long been on the public agenda and has globally moved to the forefront of policymakers' priorities. This shift is due to decades of increasing waste generation in many countries, which has led to greater reliance on landfilling and, to some extent, the burning of waste as a management strategy. These outdated waste management measures have persisted for too long and are no longer acceptable or viable worldwide. Additionally, inadequate waste management practices have exacerbated socioecological issues, such as the rise in human diseases and climate change [16].

The world faces unprecedented production and resource demand due to continuous population growth, coupled with urbanisation, industrialisation, and other factors, which have significantly influenced the increase in the volume of solid waste generated globally [24]. Global waste generation in large cities has surpassed 2 billion tons annually and is expected to reach 3.4 billion tons by 2050. Africa, including South Africa, contributes significantly to this figure [7]. Due to the same factors mentioned by Mautla (2022) [24], global waste generation is expected to outpace population growth by more than double by 2050 [27,28]. Similarly, the rise in waste generation in Africa is likely to overshadow the global decrease [4].

In response to global waste generation and its associated challenges, such as poor waste management and outdated measures, many policymakers are adopting new strategies and sustainable practices. The most recently adopted strategies include zero-waste, circular economy (CE), sustainable materials management, and extended producer responsibility (EPR) schemes. Although countries like the United States, China, and South Africa have embraced these strategies, challenges remain with integrating them into solid waste policy and planning [15].

Many countries have adopted integrated waste management planning (IWMP) into their management systems. The concept of solid waste management planning has historical roots, and internationally, it provides mechanisms for improving waste management and related behaviours [15]. IWMP is essential for designing and enhancing waste management systems within organisations. As a signatory to several international protocols and conventions, South Africa is committed to promoting an integrated waste management (IWM) approach [24]. Similar to South Africa, policymakers and planners in other countries have invested millions of dollars and years of research into establishing and improving IWMP systems and their implementation [19].

In South Africa, the development and implementation of a municipal IWMP within the IWM system is a legal requirement under Section 12 of the National Environmental Management (Waste Act 59 of 2008). Monitoring the effective implementation of IWMPs is crucial for identifying opportunities and challenges for continuous improvement [29]. IWMP is one of the key instruments in the Waste Act, enabling local governments to better manage their waste. Previous studies have highlighted the importance of developing IWMPs, including detailed implementation plans. Furthermore, these studies emphasise the need for further exploration into the extent of IWMP implementation within municipalities [24,30]. Viljoen et al. (2021) [31] recommended further research focusing on the challenges faced by municipalities in rural or remote areas and how these challenges can be overcome.

3. Methods

3.1. Data Collection

Document reviews, semi-structured open-ended interviews, and observations were conducted between August and October 2023. For the document reviews, the 2018 MLM IWMP Implementation report, MLM IDP, MLM waste by-law, and MLM IWMP MEC endorsement were reviewed.

Participants for the interviews were recruited via telecommunications, during which the researcher explained the background, importance of this study, and the role of their participation. The interviews were conducted one-on-one and face-to-face at agreed dates, times, and locations between the researcher and the participants. Voice recordings of the interviews were made with permission and using a cell phone.

South Africa is a constitutional democracy with a three-tier system of government, where the national, provincial, and local levels of government have legislative and executive authority in their respective spheres, which are defined in the Constitution as distinctive, interdependent, and interrelated [32]. The national and provincial governments assume a more strategic and focused role, providing support and resources to municipalities. The National Department of Forestry, Fisheries, and the Environment (DFFE) serves as the custodian of the environment at the national level and develops waste management policies. Provincial Departments of Environment, such as the Limpopo Department of Economic Development, Environment and Tourism, are mandated with managing waste within their provincial jurisdiction and developing waste management policies in accordance with their provincial context, while not contradicting national policies [10]. According to Mlambo and Maserumule (2024) [32], the role of local municipalities is to excel in providing public services, including waste management. District municipalities are responsible for macro-level functions such as planning and promoting integrated development planning, as well as land, economic, and environmental development [32,33].

The South African government actively promotes small and medium-sized enterprises (SMMEs) in the waste sector, with the view that SMMEs will contribute to the green economy through their participation in waste management activities [34]. The MLM IWMP particularly identified recyclers as implementing partners. Other stakeholders involved in waste management in South Africa include representatives from national, provincial, and local governments, all of whom were interviewed as part of the data collection process.

Purposive sampling, rather than representative sampling, was applied to select research participants. The general and broader landscape of solid waste management in South Africa includes key role players, mainly the government at the national, provincial and local spheres (both district and local government). It further includes small and micro–medium Enterprises (SMMEs)/industries, academia, research institutions, civil society organisations (activists and NGO/NPC, among others) and the informal sector, mainly constituted by waste pickers and reclaimers. Government participants from all spheres involved in waste management were interviewed. SMMEs/Industry representatives were also interviewed as part of the list of recycling entities from the MLM database. These participants had a direct participation in either the development and/or implementation of the MLM IWMP. A limitation identified in this study is the limited involvement of academia, research institutions, and civil society organisations in the IWMP development and implementation process. These stakeholders were not interviewed for the MLM as they could not be identified at the time of this study.

The participants interviewed included recyclers and government officials responsible for waste management at the local, district, provincial, and national levels. In purposive sampling, the researcher intentionally selects participants based on specific criteria that align with the research objectives. This method focuses on gathering rich, detailed data from a targeted group of experts or individuals with unique perspectives rather than ensuring that the sample reflects the general population. The MLM database of recyclers (implementing partners) was requested and provided by the municipality. Purposive sampling was used to select recyclers for interviews, focusing on those who had been operational since or before 2018, the year the IWMP was developed and approved, to ensure meaningful contributions. The MLM database contained 18 recyclers, of whom 10 met this selection criteria. Nine were interviewed. One out of ten declined to participate. Six government officials were also interviewed, representing four institutions: the MLM, Vhembe District Municipality, Limpopo Economic Development, Environment and Tourism, and the National Department of Forestry, Fisheries, and the Environment. In total, 15 out of 23 potential participants were interviewed, equating to 65% of the implementing partners, which can be considered a sufficiently representative sample size from the purposefully selected individuals. The researcher targeted active participants who had played a role in the development or implementation of the IWMP. These participants were identified in the MLM IWMP as implementing partners and were considered relevant to this study.

The interview questions focused on the extent to which the goals, objectives, and actions in the MLM IWMP have been achieved, as well as the barriers and opportunities for implementing the MLM IWMP (see Supplementary Materials).

Observations were conducted as an additional method of verification. These observations took place in Limpopo, specifically at the Ha-Rabali Village, Elim, and Makhado, in April, June, and October 2023. During the observations, pictures of the existing waste management infrastructure were taken to complement the interview data.

3.2. Data Analysis

After the data were collected and transcribed, data were recorded in an Excel worksheet for a word-frequency count. Words mentioned most frequently reflected the greatest concerns or matters of importance. Names and other identifiable information were removed from the organised transcripts, and interviewees were manually assigned name codes.

The content analysis technique was used to analyse the response data collected from participating stakeholders. This technique is employed to determine and record the attitudes, views, perceptions, and interests of the stakeholders, either as individuals or in groups [35]. Content analysis examines what the data reveal, followed by organising the outcome information into themes, categories, or codes based on the key questions posed by the researcher [36]. Emergent themes were identified, categorised within those codes from the data, and presented in tables and graphs. The researcher reviewed the interpretations of the data according to the level of focus on each theme. First, the researcher thoroughly examined all responses to gain a general understanding of the content of each one. This was followed by developing themes for each identified meaning, after which a working definition for each theme was provided. This step ensured a more accurate interpretation of each specific meaning relative to other categories [37].

Themes emerged using a manual frequency coding of the barriers to the MLM IWMP implementation. Themes were counted based on the number of mentions by participants, with similar mentions grouped to form themes. Assessing the cases against the themes across the different activities provides a better instrumental understanding of the success of the selected case study across the three determination criteria [38].

As for the document review assessment, a conformity criterion adopted from other studies was used [38–40]. This criterion assessed the degree to which the MLM has successfully implemented its IWMP goals:

(a) Conformity determination: actions were performed or achieved to completion.
(b) Partial conformance: The actions' achievement process was underway but not yet completed.
(c) Non-conformance: actions were not performed or were not in process. This included activities that were not aligned with the goal.

4. Results

4.1. Extent of the Implementation of the MLM IWMP

The 2018 MLM IWMP contains eight goals, along with associated targets and actions to achieve these goals. Table 1 shows the results of evaluating the extent to which these goals, targets, and actions have been implemented using the conformity criteria. The table indicates that out of the eight goals and 55 actions stipulated in the IWMP, 13 actions conformed, meaning they were fully implemented. Sixteen actions partially conformed, and 26 did not conform at all. These results were obtained through document reviews as well as interviews. This means that approximately 76% (42 out of 55) of the actions have not been fully implemented since the development of the MLM IWMP in 2018. Many of these targets and actions remain incomplete.

Table 1. Findings of the MLM IWMP's overall implementation of the goals.

MLM IWMP Implementation Plan Goals	No. of Actions	Conformance	Partial Conformance	Non-Conformance
Goal 1: Promote recycling and recovery of waste	11	3	3	5
Goal 2: Ensure the effective and efficient delivery of waste services	7	4	2	1
Goal 3: Ensure that legislative tools are developed to deliver on the Waste Act and other applicable legislation	5	2	3	0
Goal 4: Sound budgeting and financing of waste management services	5	0	1	4
Goal 5: Ensure the safe and proper disposal of waste	3	0	1	2
Goal 6: Education and awareness	7	1	4	2
Goal 7: Compliance and enforcement	8	3	2	3
Goal 8: Reporting on monitoring the implementation plan for Goal 8: Compliance and enforcement	9	0	0	9
Total averages	55	13	16	26

4.2. Barriers (Weaknesses) to Implementing the MLM IWMP

Table 2 shows the results obtained from interviews regarding the barriers (weaknesses) and opportunities for implementing the MLM IWMP. From the responses, eight key themes have been identified, which are discussed further below.

Table 3 shows a summary of the results obtained from interviews regarding the opportunities for implementing the MLM IWMP.

Table 2. Themes identified from stakeholder response on the barriers (weaknesses) towards the implementation of the MLM IWMP.

Themes on Barriers (Weakness)	Number of Mentions
Lack of compliance and enforcement	6
Poor planning, transport breakdowns, and poor infrastructure maintenance	6
Unreliable service and collection frequency challenges	5
Lack of skilled human capacity and unskilled technical position appointments	5
Political interference and corruption	4
Insufficient budget for waste infrastructure and resources	4
Inadequate community education and information awareness	3
Absence or lack of monitoring on the implementation of the IWMP by the MLM	2

Table 3. Themes identified from stakeholder response on the opportunities towards the implementation of the MLM IWMP.

Opportunities	Number of Mentions
Waste diversion from landfill and landfill airspace saving	10
SMME development, training and support	9
Create clean environment and illegal dumping conversion into parks	8
Job creation	6
Establishment of waste processing plant by the MLM	3
Revenue collection	3
Enhance tourism	1

5. Discussion

5.1. Absence or Lack of Monitoring on the Implementation of the IWMP by the MLM

A study conducted by Molaba (2019) [22], which critically analysed the IWMPs of local municipalities within the Fezile Dabi District Municipality, found that the implementation of the IWMPs was below 20%, as major targets were either not achieved or not met on time. The challenges identified included inadequate knowledge among waste management officials, lack of evidence for monitoring or reporting IWMP implementation progress, and insufficient financial resources [22]. The results of our study align with Molaba's (2019) [22] findings, revealing that the MLM does not monitor its IWMP. This is evident from the nine actions related to Goal 8 (Table 1), which failed to conform, meaning they were not carried out at all. Goal 8 of the MLM IWMP pertains to reporting and monitoring the implementation of the plan. At the time of this study, no annual monitoring report, as required under Section 13(3) of the Waste Act, was found. Moreover, the lack of monitoring emerged as a theme from respondents' feedback, with one stating that *"a five-year annual draft report is in the final stage and will be completed before the IWMP review process starts in October 2023"*.

Respondents also pointed out the absence of monitoring regarding the effectiveness of MLM's waste management programs, such as education and awareness initiatives. These findings are consistent with a study by Mautla (2022) [24], which highlighted inadequate reporting and monitoring of IWMPs at the municipal level. Similarly, Molaba (2019) [22] found that the failure of the provincial department to monitor implementation contributes to poor performance by local governments in implementing the IWMP.

5.2. Political Interference and Corruption

Our results further suggest that the MLM is facing significant challenges in managing the interface between politicians and officials. This theme was mentioned four times by respondents. Once again, this aligns with other studies highlighting political interference in the administration of basic service delivery as a serious problem in many municipalities, including the MLM [22]. Due to a lack of knowledge, councillors' support in implementing IWMPs is often very poor in many municipalities [22]. According to Naidoo (2009) [41], politics significantly influence municipalities, and such influence cannot be underestimated. Solid waste management, particularly in terms of budget allocation, is not a priority in South Africa compared to other services [41].

Although there was reluctance among interviewees to detail how political interference affects IWMP implementation and service delivery, some mentioned that political preferences sometimes take precedence over service delivery requirements. This study revealed that political interference is often linked to corruption. Some respondents indicated that *"when it comes to the appointment of entrepreneurs who were expected to recover waste directly from landfill sites, regardless of their successful application, they still have no access to recover waste from the landfill due to political interference"*.

However, this study could not directly establish how or to what extent political interference impacts IWMP implementation and service delivery. Political interference is not a new issue in developing countries. Resnick (2014) [42] highlighted the troubling trend of recentralisation in developing economies such as Kampala, Uganda, and Senegal, where political and fiscal powers were stripped from city authorities aligned with opposition parties, negatively impacting efforts to address issues like urban flooding, waste collection, and infrastructure [42]. Similarly, Makanyeza et al. (2013) [43] identified councillor interference in administration and a lack of political and administrative leadership as causes of poor service delivery in Kenya and South Africa [43].

5.3. *Insufficient Waste Management Budget*

Poor financial management, characterised by inadequate funding, misappropriation of funds, and poor waste service recovery, deters effective Solid Waste Management (SWM) by interfering with institutional behaviour, equipment, infrastructure, and labour management [44]. Poor waste tariff allocation and financial constraints in the SWM sectors of many municipalities in South Africa underscore the seriousness of this challenge [44]. These challenges have made operational expenses, such as capital expenditure, employee remuneration, maintenance, and fuel, unrealistic to meet. Inadequate financial accounting of allocated funds, compounded by corruption, leads to poor management of the waste cycle due to a lack of investment based on service demands.

Insufficient funding, inadequate budget allocation, and budgetary constraints appear to be common reasons impacting the implementation of IWMPs [24]. Molaba (2019) [22] notes that while the IWMP may be in place, the availability of funds or resources remains a significant challenge for municipalities. The results in this study are no different, indicating that the MLM presents insufficient funding as a major problem. Furthermore, some respondents indicated that *"this is concerning because the municipality collects a waste revenue of about 9 million ZAR to 10 million ZAR annually, and it is claimed to be the best-performing municipality in revenue collection in the Vhembe District Municipality"*.

One important reason waste management poses a challenge for authorities in developing countries is the burden on municipal budgets due to the high costs associated with its management. Municipalities have struggled to manage waste effectively because substantial expenditures are required to provide the service [45]. The absence of financial support, the unwillingness of users to pay for the service, and the inadequate use of economic instruments have hampered the delivery of appropriate waste management services in South African municipalities, including the MLM [45].

Our results further suggest that the increasing costs of maintaining waste management-related equipment, infrastructure, fleet, and fuel have worsened and compromised MLM's budget availability for waste services. Fakoya (2019) [8] noted that municipalities in developing countries typically spend between 20% and 50% of their total expenditure on solid waste management service delivery. Although the MLM charges private waste disposal fees based on volume, our study found that the MLM does not implement cost-reflective tariffs; rather, it charges less than the costs associated with managing the landfill site. Like other rural local municipalities, the MLM faces community resistance to annual increases in cost-reflective tariffs.

5.4. *Poor Planning, Transport Breakdown and Poor Infrastructure Maintenance*

The ability of the MLM to provide effective waste management services has been worsened by the deterioration of solid waste infrastructure. This theme was mentioned by respondents approximately six times, as shown in Table 2. According to some respondents, the recent procurement of a new fleet was expected to alleviate equipment and infrastructure pressures. However, there is still high demand, particularly in rural areas where some communities are not regularly serviced. Waste from these communal areas requires frequent collection, as skip bins fill up quickly. Other challenges affecting the implementation of the MLM IWMP include poor maintenance, inadequate planning, and a lack of available waste infrastructure, such as waste transfer stations and transport or equipment to reduce travel distances, given that the MLM landfill site is located far from many residents. One transfer station located in the Dzanani region was observed to be operational. Bartone and Bernstein (1993) [46] suggest that where collection and transportation distances exceed 15 to 20 km or travel times exceed 30 min, transfer stations should be considered. The existing waste buyback centres observed are owned and maintained by the private sector.

The vehicles and equipment used in waste collection processes are also considered inefficient due to these long distances. Some vehicles are not designed for frequent stop-and-go operations and are not located near disposal sites. These conditions reduce their effectiveness and often lead to a shorter operational lifespan. This was confirmed during interviews with the MLM officials, who mentioned difficulties accessing some villages in remote areas due to distances and gravel roads.

Similar challenges are also found elsewhere in Africa. Imam et al. (2008) [47] found that in Abuja, Nigeria, a shortage of waste collection vehicles is attributed to funding gaps and inadequate maintenance. Furthermore, challenges such as poor servicing of MSW collection vehicles, inadequate state infrastructure, and ongoing urban migration are common issues faced by sub-Saharan countries [48].

According to the MLM officials, although there is an existing plan and schedule for waste collection, it remains challenging to plan collection routes due to limited vehicles. Often, available vehicles are in the depot for maintenance or repair. These barriers are not unique to the MLM; Viljoen et al. (2021) [31] have documented that many municipalities face obstacles in waste collection due to inadequate equipment, infrastructure, and treatment facilities. These challenges contribute to issues such as littering and illegal dumping, particularly in remote and rural South African towns [31].

Guerrero (2013) [45] found that few cities have transfer station infrastructures where waste is temporarily stored, sorted, and recyclables recovered before disposal as a last resort. These cities include Ambon, Jogjakarta, Beijing, Bangkok, Dar es Salaam, Emfuleni, Langeberg, Pretoria, and Gazipur. However, the timing and methods of waste collection largely depend on the availability of waste transportation facilities and the quality of the roads [45].

5.5. Lack of Skilled Human Capacity and Unskilled Technical Position Appointment

According to Mautla [24], insufficient human resource capacity is a problem in implementing the IWMP in many municipalities in South Africa. Our results concur with this, showing that the shortage of skilled and qualified staff in the MLM is often the reason for inefficient, ineffective, and unsustainable waste management services. The greatest challenge appears to be inadequate knowledge among waste management officials in implementing the IWMP, especially supervisors who are the first line in enforcing the IWMPs [22]. This study also found that the MLM has only two municipal employees directly involved in the day-to-day operations of waste management. Although qualified and experienced, these employees are often supported by unskilled personnel.

Faccio et al. [49] suggested that improving infrastructure and roads, providing additional equipment, and enhancing human capacity positively impact the delivery of waste services. However, municipalities must allocate additional finances for waste collection, transfer, and transport, which are crucial but expensive [49].

5.6. Unreliable Service and Collection Frequency Challenges

According to the MLM IDP, MLM's total population was 416,728 in 2022. The municipality comprises four main formal towns: Louis Trichardt, Vleifontein, Waterval, and Dzanani, along with over 200 villages. The number of households was 116,371, of which 99,104 lacked refuse removal services, equivalent to 85.1% of its total households in 2022 [50]. The MLM generates most of its waste from households and businesses. According to MLM's reports to South Africa's Waste Information System, the MLM collects approximately 190 tons per week from its main townships, about 90 tons weekly from Waterval township, approximately 65 tons from Dzanani township, and around 25 tons from Vleifontein township. The total monthly waste collection amounts to approximately 1425 tons, translating to 17,098 tons per year. This waste is collected using both compactor and skip loader trucks.

Dlamini et al. (2019) [27] argued that the ability of local authorities to provide effective waste management services in South Africa has been hindered by the deterioration of solid waste infrastructure. Waste collection in MLM's Dzanani and Waterval townships mainly occurs through kerbside collection, while other areas utilise communal skips or drop-off centres. Waste is collected once a week in townships, and a similar frequency is intended for other rural areas where waste is collected via drop-off centres. However, this is often not consistently implemented. This inefficiency is evident in skips that remain full and uncollected, resulting in waste accumulating around the skips for several days, as observed. The MLM officials state that they rely on communities or councillors in areas with communal skips to report when skips are full for collection. However, failure to report by communities and councillors leads to waste not being collected, causing nuisance and environmental pollution.

Moreover, according to Fakoya (2014) [8], many rural municipalities make the mistake of purchasing vehicles for waste collection that are not designed for long stop-and-go distances. Their study found that the MLM has only one transfer station located in Dzanani township, where recyclables are recovered, and non-recyclables are transferred to the waste disposal facility, thereby reducing the distance and loads for their compactor and skip loader trucks.

5.7. *Inadequate Community Education and Information Awareness*

According to Molaba (2019) [22], some Integrated Waste Management Plans (IWMPs) in municipalities, including those in Fezile Dabi, lack proper public participation processes. He also notes that consultation often only involves the office of the local municipality's mayor rather than engaging all relevant stakeholders. In the Fezile Dabi municipalities, specifically, Molaba (2019) [22] also found that public comments were not adequately incorporated into IWMPs during their development stages. In the City of Johannesburg metropolitan municipality, Malope (2020) [51] found that there are challenges related to gaining community participation in attending awareness campaigns.

In the context of the MLM, this study highlights several existing awareness programs related to the IWMP, such as Adopt-a-School, Nakisani Vhupo, the Greenest Municipality competition, executive mayor quarterly imbizos, and monthly environmental education and awareness initiatives. These programs aim to educate and engage the community in waste management practices. However, the effectiveness of these initiatives remains a critical question. Respondents noted additional opportunities for the MLM to enhance public awareness through platforms like social media and newsletters, although budget constraints often limit these efforts. Molaba (2019) [22] underscores that insufficient knowledge dissemination remains a significant barrier to effective IWMP implementation in developing economies, including the MLM.

Kubanza and Simatele (2020) [52] argue that education and awareness are crucial for fostering environmental consciousness and conservation efforts. Despite these initiatives, the MLM officials expressed concerns about the community's attitudes and behaviour towards waste management, indicating a need for improvement in public engagement and educational outcomes. While the MLM has implemented various awareness programs related to its IWMP, challenges such as limited public participation, budget constraints for effective communication channels, and inadequate behaviour change among communities still hinder comprehensive waste management initiatives. Addressing these challenges could enhance the overall effectiveness of MLM's waste management strategies and community engagement efforts.

5.8. Lack of Enforcement and Compliance

Radzilani (2019) [53] underscores that compliance and enforcement are major hurdles in many municipalities, attributing the lack of enforcement to resource constraints. This study aligns with Radzilani's (2019) [53] findings, emphasising that misunderstandings regarding roles, responsibilities, and powers within the MLM limit the effective compliance, enforcement, and monitoring of waste management regulations. It is specifically noted that waste management officials have requested training as peace officers but have not been granted this authority, which underscores capacity gaps in enforcement.

Comparative examples from Nigeria, highlighted by Ike et al. (2018) [54], illustrate similar challenges in solid waste management, including poor collection methods, inadequate infrastructure, and noncompliance with government policies and regulations. This noncompliance is often rooted in a lack of comprehensive waste management strategies and insufficient enforcement mechanisms, mirroring the challenges faced in the MLM.

Moreover, Ostrovskaya and Leentvaar (2011) [55] argue that developing countries like Tanzania, Rwanda, and Kenya struggle with the practical interpretation and application of environmental legislation due to its evolving nature and limited resources for enforcement. This resonates with the difficulties outlined in the MLM, where the enforcement of waste by-laws and environmental regulations remains inadequate.

5.9. Gaps in the IWMP Implementation Plan and IDP

Ideally, the status quo report provides a baseline assessment of the waste-related gaps and challenges, which are subsequently used to inform the municipality's IWMP goals, targets, and actions [50]. Using the document review method, this study finds a misalignment between some of the IWMP's goals, targets, and/or actions. A typical example is the target of establishing e-waste recycling, while the action aims to discourage people from disposing of garden waste in landfill sites. It is evident that this target may be difficult to achieve, considering that the implemented action does not contribute to achieving the target. According to Mautla (2022) [24], the successful implementation of the IWMP depends on the quality of the implementation plan, where the goals, targets, and actions in the MLM IWMP may be challenging or unrealistic to implement without proper and extensive consultation.

Furthermore, the IWMP is executed through an implementation plan, which needs to be incorporated into the IDP as a sector plan in accordance with the Waste Act, Section 4(a)(ii) [56]. Upon reviewing the IWMP and its Implementation Plan and IDP, it was established that only four IWMP-linked projects are incorporated into the IDP: the distribution of refuse bags to townships and villages, extending waste collection to villages, refurbishing six drop-off points, and establishing a new landfill site while rehabilitating the old Vondeling landfill site. It is important to note that the rehabilitation of the old Vondeling landfill site is not included in the IWMP but is part of the IDP. Many projects identified in the IWMP from 2018 to 2023 are not integrated into the IDP. This was further identified by Mautla (2022) [24] when evaluating the Drakenstein Local Municipality's monitoring and reviewing of its IWMPs.

5.10. Brief SWOT Analysis for Implementing the MLM IWMP

The SWOT analysis evaluates the internal strengths, weaknesses and external opportunities and threats in an organisation's environment with the aim of identifying resources, capabilities, core competencies, and competitive advantages inherent to the organisation [57]. Leigh, 2009 [58] summarised SWOT meanings as follows: strength is the internal competence and resource availability; weakness is the internal inhibitor of the competence, resources, or qualities necessary for success; opportunity is an external enhancer of performance that can be pursued to gain benefit; and threat is as an external inhibitor of performance that has the potential to reduce accomplishments [58].

Although this study's main aim was to identify the barriers, a SWOT analysis was also conducted to evaluate the IWMP system in general. The themes identified from stakeholder response on the barriers towards the implementation of the MLM IWMP in Table 2 constitute the weaknesses and threats within the SWOT analysis approach. Table 3 further provides the opportunities identified from stakeholder responses. To complete the SWOT analysis of this study, the strengths and threats towards the implementation of municipal integrated waste management planning were identified from the literature review. The strengths identified towards the implementation of the integrated waste management plans include [59–61]:

i. The planning system is properly introduced, including aims, tasks and costs of its realisation for funding purposes.
ii. Communities mobilise each other and create cooperation for separate collections.
iii. Introduction of environmental fees and fines, which are separate from the tax system but further enhance the waste management services cost reflective charge system.
iv. Progress in separate collections is realised, although not all inhabitants may use it.
v. Provision of financial support for new waste facilities and education and awareness programmes from management.
vi. IWMPs identify problematic waste streams and a series of measures to be implemented.
vii. Strong and guided awareness of local government promoting responsible waste management behaviour.
viii. Promulgation of regulation to further strengthen the existing ones.

Threats identified towards the implementation of the integrated waste management plans include [59,62]:

i. Lack of citizen participation and disinterested citizens in the development and implementation of the IWMPs.
ii. Limited availability and inconsistencies of the relevant and reliable data in the development of IWMPs for efficient implementation.
iii. Failure to fulfil the IWMPs set objectives, which may include reducing the amount of waste diversion from landfill facilities.
iv. Lack of waste separation at source facilities.
v. Set of unrealistic IWMP objectives and consequently objectives not fulfilled thereof
vi. High cost of waste transfer in the urban area and high cost of disposal in the rural areas
vii. Business as usual and not in my backyard syndrome
viii. Strikes and vandalism of waste management facilities.

6. Conclusions

The IWMP is designed to be a decision-support tool that informs and influences higher-level decision-making policy instruments on waste-related matters. However, despite the existence of policy instruments such as IWMPs, waste-related issues like generation and management remain challenging in many countries.

We argue that the lack of quantitative data to assess the effective implementation of plans like the IWMP necessitates reliance on key stakeholders' perceptions to understand the barriers hindering the effective implementation of the IWMP system in developing economies. Using South Africa's MLM as a case study, our findings and analysis suggest the following concluding remarks:

i. Of the eight goals and 55 actions outlined in MLM's IWMP, 76% (42 out of 55) have not been fully implemented since the plan's development in 2018. Most of these targets and actions have exceeded their implementation deadlines as specified in the report.
ii. Similar to other municipalities in South Africa and in developing countries, the following key barriers or challenges are hindering the effective implementation of the IWMP:
    a. Lack of monitoring of IWMP implementation by the MLM.
    b. Political interference and corruption.
    c. Insufficient budget for waste infrastructure and resources.
    d. Poor planning, breakdowns in transport, and inadequate maintenance of infrastructure.

e. Lack of skilled human resources and appointments in unskilled technical positions.
f. Unreliable service and collection frequency issues in urban areas, along with a lack of waste collection in rural areas.
g. Insufficient community education and awareness.

Due to these barriers, this study concludes that the IWMP system is not currently effective in achieving its intended objectives in a developing economy such as South Africa. Until these barriers are removed or addressed, they will continue to pose a risk to the effectiveness of the IWMP system as a key waste management policy instrument. In addition, some opportunities have been identified in this study, including waste diversion from landfills and landfill airspace saving; SMME development, training and support; creating a clean environment and illegal dumping conversion into parks; job creation; establishment of a waste processing plant by the MLM; revenue collection; and enhancement of tourism. This paper, therefore, offers valuable insights for other nations facing similar waste-related challenges as South Africa by identifying the obstacles to implementing IWMPs. This can serve as a foundation for improving the effectiveness of IWMP systems elsewhere.

Supplementary Materials: The following supporting information can be downloaded at https://www.mdpi.com/article/10.3390/su16229643/s1, Questionnaire interview in relation to the implementation of Makhado IWMP.

Author Contributions: P.R.: Conceptualization, Methodology, Data curation, Formal analysis, Investigation, Writing—original draft, Writing—reviewing and editing. P.M.: Conceptualization, Supervision, Writing—reviewing and editing. C.R.: Writing—reviewing and editing. All authors have read and agreed to the published version of the manuscript.

Funding: This research received no external funding and the APC was funded by the Unit for Environmental Sciences and Management, North-west University.

Institutional Review Board Statement: This study formed part of a master's study at North-west University and has been approved by the University's Faculty of Natural and Agricultural Sciences Research Ethics Committee and classified as a minimal risk category with ethics number NWU-01234-23-A9.

Data Availability Statement: Data will be made available on request.

Conflicts of Interest: The authors declare that they have no known competing financial interests or personal relationships that could have appeared to influence the work reported in this paper.

References

1. Muzenda, E. A discussion on waste generation and management trends in South Africa. *Int. J. Chem. Environ. Biol. Sci.* **2014**, *2*, 2320–4087.
2. Sango, T.; Kissoon, S.; Basson, L.; Munganga, G. Development of the waste economy in the Western Cape: A decision support tool for integrated municipal waste management. In *Proceedings of the 20th WasteCon Conference;* Institute of Waste Management of Southern Africa: Cape Town, South Africa, 2014.
3. Hall, C.; Dawson, T.P.; Macdiarmid, J.I.; Matthews, R.B.; Smith, P. The impact of population growth and climate change on food security in Africa: Looking ahead to 2050. *Int. J. Agric. Sustain.* **2017**, *15*, 124–135. [CrossRef]
4. Debrah, J.; Godfred, T.; Dinis, M. Barriers and Challenges to Waste Management Hindering the Circular Economy in Sub-Saharan Africa. *Urban Sci.* **2022**, *6*, 1–16. [CrossRef]
5. Godfrey, L.; Ahmed, M.T.; Gebremedhin, K.G.; Katima, H.Y.; Oelofse, S.; Osibanjo, H.; Richter, H.; Yonli, H.Y. Solid waste management in Africa: Governance failure or development opportunity? *Reg. Dev. Afr.* **2019**, 1–14.
6. Godfrey, L.; Dambuza, T. Integrated waste management plans—A useful management tool for local government or a bureaucratic burden? In Proceedings of the WasteCon 2006 Biennial International Waste Congress, Cape Town, South Africa, 5–8 September 2006; pp. 5–8.
7. Nyika, J.; Megersa, O.; Dinka, O.; Shivani, M. *Waste management in South Africa In Sustainable Waste Management Challenges in Developing Countries;* IGI Global: Hershey, PA, USA, 2020; pp. 327–351.
8. Fakoya, M.B. Institutional challenges to municipal waste management service delivery in South Africa. *J. Hum. Ecol.* **2014**, *45*, 119–125. [CrossRef]
9. Tsheleza, V.; Nakin, M.D.; Ndhleve, S.; Kabiti, H.M.; Musampa, C.M. Vulnerability of growing cities to solid waste-related environmental hazards: The case of Mthatha, South Africa. *Jàmbá: J. Disaster Risk Stud.* **2019**, *11*, 1–10. [CrossRef] [PubMed]

10. Kubanza, N. The role of community participation in solid waste management in Sub-Saharan Africa: A study of Orlando East, Johannesburg, South Africa. *S. Afr. Geogr. J.* **2021**, *103*, 2151–2418.
11. Kubanza, N.S.; Matsika, R.; Magha, C.G. Exploring the role of local authorities and community participation in solid waste management in sub-Saharan Africa: A study of Alexandra, Johannesburg, South Africa. *Local Environ.* **2022**, *27*, 197–214. [CrossRef]
12. Godfrey, L.; Oelofse, S. Historical review of waste management and recycling in South Africa. *Resources* **2017**, *6*, 57. [CrossRef]
13. Ogola, J.S.; Chimuka, L.; Tshivhase, S. Management of Municipal Solid Wastes: A Case Study in Limpopo Province, South Africa. In *Integrated Waste Management—Volume I*; IntechOpen: London, UK, 2011. Available online: https://www.intechopen.com/chapters/17432 (accessed on 9 September 2024).
14. Ziraba, A.K.; Haregu, T.N.; Mberu, B. A review and framework for understanding the potential impact of poor solid waste management on health in developing countries. *Arch. Public Health* **2016**, *74*, 55. [CrossRef]
15. Anshassi, M.; Laux, S.; Townsend, G.T. Approaches to integrate sustainable materials management into waste management planning and policy. *Resour. Conserv. Recycl.* **2019**, *148*, 55–66. [CrossRef]
16. Clarke, M.; Read, A.; Phillips, P. Integrated waste management planning and decision-making in New York City. *Resour. Conserv. Recycl.* **1998**, *2*, 125–141. [CrossRef]
17. Tam, V. On the effectiveness in implementing a waste-management-plan method in construction. *Waste Manag.* **2007**, *28*, 1072–1080. [CrossRef]
18. De Souza, V.; Bloemhof, J.; Borsato, M. Assessing the eco-effectiveness of a solid waste management plan using agent-based modelling. *Waste Manag.* **2021**, *125*, 235–248. [CrossRef] [PubMed]
19. Hostovsky, C. Evaluation in integrated waste management: Understanding the crisis and improving practice through planning theory. *Can. J. Urban Res.* **2005**, *14*, 81–101.
20. Department of Environmental Affairs (DEA). Guideline for the Development of Integrated Waste Management Plans, Pretoria: Department of Environmental Affairs 2012. Available online: http://iwmp.environment.gov.za/book/export/html/16 (accessed on 13 October 2024).
21. Hoornweg, D.; Bhada-Tata, P. *What a Waste: A Global Review of Solid Waste Management*; Word Bank Group: Washington, DC, USA, 2012.
22. Molaba, L. A Critical Analysis of the Integrated Waste Management Plans of Local Municipalities. Master's Thesis, Northwest University, Potchefstroom, South Africa, 2019.
23. Polasi, T.; Matinise, S.; Oelofse, S. *South African Municipal Waste Management Systems: Challenges and Solutions, South Africa*; Department of Environment, Forestry and Fisheries (DEFF), the Council for Scientific and Industrial Research (CSIR), and The United Nations Environment Programme (UNEP) International Environmental Technology Centre (IETC): Nairobi, Kenya, 2020.
24. Mautla, M. Evaluating the Monitoring and Review of the Integrated Waste Management Plans (IWMPs) of the Drakenstein Local Municipality. Master's Thesis, Northwest University, Potchefstroom, South Africa, 2022.
25. Hossu, C.A.; Iojă, I.C.; Mitincu, C.G.; Artmann, M.; Hersperger, A.M. Assessing planning implementation using the conformance and performance approach: The case of Romania's local environmental action plans. *J. Environ. Manag.* **2021**, *299*, 113680. [CrossRef]
26. Nel, J.G.; Alberts, R. Environmental Management and Environmental Law in South Africa: An introduction. In *Environmental Management in South Africa*, 3rd ed.; Strydom, H., King, N., Retief, F., Eds.; Juta Publishing: Cape Town, South Africa, 2018; pp. 1–55.
27. Dlamini, S.; Simatele, M.S.; Kubanza, N.S. Municipal solid waste management in South Africa: From waste to energy recovery through waste-to-energy technologies in Johannesburg. *Int. J. Justice Sustain.* **2019**, *24*, 249–257. [CrossRef]
28. El-Fadel, M.; Maalouf, A. Challenges of waste management in a developing context: Lessons from Lebanon. In *Sustainable Waste Management Challenges in Developing Countries*; IGI Global: Hershey, PA, USA, 2020; pp. 166–185.
29. Mbambuku, T. An Evaluation of the Gaps and Barriers in Implementing the National Waste Management Policy and its Implementation in Formal and Informal Urban Areas in Ekurhuleni Municipality, South Africa. Master's Thesis, University of South Africa, Pretoria, South Africa, 2012, unpublished.
30. Muswema, A.P. Wrestling with IWMP implementation: Case study challenges and lessons learnt. In Proceedings of the WasteCon 2012, East London International Convention Centre, East London, South Africa, 9–12 October 2012; pp. 362–370.
31. Viljoen, J.; Schenck, C.; Volschenk, L.; Blaauw, P. Household waste management practices and challenges in a rural remote town in the Hantam Municipality in the Northern Cape, South Africa. *Sustainability* **2021**, *13*, 5903. [CrossRef]
32. Mlambo, D.N.; Maserumule, M.H. Constitutional and legislative frameworks for the local sphere of government in South Africa: Analytical and interpretive perspective. *Insight Afr.* **2024**, *16*, 211–229. [CrossRef]
33. Koma, S.B. The state of local government in South Africa: Issues, trends and options. *J. Public Adm.* **2010**, *45*, 111–120.
34. Higgs, C.J.; Hill, T. The role that small and medium-sized enterprises play in sustainable development and the green economy in the waste sector, South Africa. *Bus. Strategy Dev.* **2019**, *2*, 25–31. [CrossRef]
35. Drisko, J.W.; Maschi, T. *Content Analysis. Pocket Guide to Social Work Research Methods*; Oxford University Press: Oxford, UK, 2016.
36. Bowen, G.A. Document Analysis as a Qualitative Research Method. *Qual. Res. J.* **2009**, *9*, 27–40. [CrossRef]
37. Fox, J.; Gutierrez, D.; Haas, J.; Braganza, D.; Berger, C. A phenomenological investigation of centering prayer using conventional content analysis. *Pastor. Psychol.* **2015**, *64*, 803–825. [CrossRef]
38. Alberts, R.C.; Retief, F.P.; Cilliers, D.P.; Roos, C.; Hauptfleisch, M. Environmental impact assessment (EIA) effectiveness in protected areas. *Impact Assess. Proj. Apprais.* **2021**, *39*, 290–303. [CrossRef]

39. Moolman, H.J.; Alberts, R.C.; Retief, F.P.; Mukwevho, P.; Roos, C.; Malherbe, W. The Effectiveness of Water Use Authorisation Systems as a Policy Implementation Instrument. *Water Research Commission*. 2023. Available online: https://www.researchgate.net/profile/Jurie-Moolman/publication/381552284_The_effectiveness_of_water_use_authorisation_systems_as_a_policy_implementation_instrument/links/6673f73fd21e220d89c10529/The-effectiveness-of-water-use-authorisation-systems-as-a-policy-implementation-instrument.pdf (accessed on 13 October 2024).
40. Mukwevho, P. Evaluating the Effectiveness of Air Quality Management Plans as a Governance Instrument in South Africa. Ph.D. Thesis, North-West University, Potchefstroom, South Africa, 2023.
41. Naidoo, K. An Analysis of Municipal Solid Waste Management in South Africa Using the UMsunduzi Municipality as a Case Study. Master's Thesis, University of Natal, Pietermaritzburg, South Africa, 2009.
42. Resnick, D. Urban governance and service delivery in African cities: The role of politics and policies. *Dev. Policy Rev.* **2014**, *32*, 3–17. [CrossRef]
43. Makanyeza, C.; Kwandayi, H.P.; Ikobe, B.N. Strategies to improve service delivery in local authorities. *Int. J. Inf. Technol. Bus. Manag.* **2013**, *15*, 1–11.
44. Van Niekerk, S.; Weghmann, V. Municipal Solid Waste Management Services in Africa. In *PSIRU Working Paper*; PSI: Ferney-Voltaire, France, 2019. Available online: https://www.world-psi.org/sites/default/files/documents/research/waste_management_in_africa_2018_final_dc_without_highlights_2019.pdf (accessed on 9 September 2024).
45. Guerrero, L.; Maas, G.; Hogland, W. Solid waste management challenges for cities in developing countries. *Waste Manag.* **2013**, *33*, 220–232. [CrossRef]
46. Bartone, C.R.; Bernstein, J.D. Improving municipal solid waste management in third world countries. *Resour. Conserv. Recycl.* **1993**, *8*, 43–54. [CrossRef]
47. Imam, A.; Mohammed, B.; Wilson, D.C.; Cheeseman, C.R. Solid waste management in Abuja, Nigeria. *Waste Manag.* **2008**, *28*, 468–472. [CrossRef]
48. Parrot, L.; Sotamenou, J.; Dia, B.K. Municipal solid waste management in Africa: Strategies and livelihoods in Yaoundé, Cameroon. *Waste Manag.* **2009**, *29*, 986–995. [CrossRef]
49. Faccio, M.; Persona, A.; Zanin, G. Waste collection multi-objective model with real-time traceability data. *Waste Manag.* **2011**, *31*, 2391–2405. [CrossRef] [PubMed]
50. Makhado Local Municipality (MLM). Integrated Development Plan 2022. Available online: https://www.makhado.gov.za/sstaff/pages/sites/makhado/documents/idp/DRAFT%20INTERGRATED%20DEVELOPMENT%20PLAN%20%202022%20-2027.pdf (accessed on 13 October 2024).
51. Malope, L. Assessing the City of Johannesburg's Waste Management Plan in Reducing Solid Waste. Master's Thesis, University of Johannesburg: Johannesburg, South Afirca, 2020.
52. Kubanza, N.S.; Simatele, M.D. Sustainable solid waste management in developing countries: A study of institutional strengthening for solid waste management in Johannesburg, South Africa. *J. Environ. Plan. Manag.* **2020**, *63*, 175–188. [CrossRef]
53. Radzilani, T. The Quality of Integrated Waste Management Plans for Metropolitan Municipalities in South Africa. Master's Thesis, North West University, Potchefstroom, South Africa, 2019.
54. Ike, C.C.; Ezeibe, C.C.; Anijiofor, S.C.; Daud, N.N. Solid waste management in Nigeria: Problems, prospects, and policies. *J. Solid Waste Technol. Manag.* **2018**, *44*, 163–172. [CrossRef]
55. Ostrovskaya, E.; Leentvaar, J. Enhancing Compliance with Environmental Laws in Developing Countries: Can Better Enforcement Strategies Help? In Proceedings of the Ninth International Conference on Environmental Compliance and Enforcement, Whistler, BC, Canada, 20–24 June 2011. Available online: https://www.researchgate.net/profile/Elena-Ostrovskaya-2/publication/259702201_Enhancing_Compliance_with_Environmental_Laws_in_Developing_Countries_Can_Better_Enforcement_Strategies_Help/links/0046353a12e3cdcd5c000000/Enhancing-Compliance-with-Environmen (accessed on 9 September 2024).
56. Department of Environmental Affairs (DEA). National Environmental Management: Waste Act, 29 of 2008, Pretoria: Department of Environmental Affairs. 2008. Available online: https://www.gov.za/documents/national-environmental-management-waste-act (accessed on 13 October 2024).
57. Sammut-Bonnici, T.; Galea, D. SWOT analysis. In *Wiley Encyclopedia of Management*; John Wiley & Sons, Ltd.: Hoboken, NJ, USA, 2015; pp. 1–8. Available online: https://kobbytamakloe.wordpress.com/wp-content/uploads/2020/07/swot-analysis-kjm.pdf (accessed on 9 September 2024).
58. Leigh, D. SWOT analysis. In *Handbook of Improving Performance in the Workplace: Volumes 1–3*; John Wiley & Sons, Ltd.: Hoboken, NJ, USA, 2009; pp. 115–140.
59. Wismer, S.; Lopez de Alba Gomez, A. Evaluating the Mexican Federal District's integrated solid waste management programme. *Waste Manag. Res.* **2010**, *29*, 480–490. [CrossRef] [PubMed]
60. Kulczycka, J.; Generowicz, A.; Kowalski, Z. Strength and weakness of municipal and packaging waste system in Poland. In *Integrated Waste Management—Volume I*; Intech Open: London, UK, 2011. Available online: https://books.google.co.za/books?hl=en&lr=&id=qeGdDwAAQBAJ&oi=fnd&pg=PA79&dq=Strength+and+weakness+of+municipal+and+packaging+waste+system+in+Poland.+In+Integrated+Waste+Management%E2%80%94Volume+&ots=cy7fpp01ZY&sig=3cVLO83IHMa6BjItIICpJzrPRnE&redir_esc= (accessed on 9 September 2024).

61. Yuan, H. A SWOT analysis of successful construction waste management. *J. Clean. Prod.* **2013**, *39*, 1–8. [CrossRef]
62. Zotos, G.; Karagiannidis, A.; Zampetoglou, S.; Malamakis, A.; Antonopoulos, I.S.; Kontogianni, S.; Tchobanoglous, G. Developing a holistic strategy for integrated waste management within municipal planning: Challenges, policies, solutions and perspectives for Hellenic municipalities in the zero-waste, low-cost direction. *Waste Manag.* **2009**, *29*, 1686–1692. [CrossRef]

sustainability

MDPI

Article

Successful Project—Limited Transfer: Learnings from the Local Circularity Experiment WiedergeBORN

Andreas Obersteg

Institute of Urban Planning and Regional Development, HafenCity Universität Hamburg, 20457 Hamburg, Germany; andreas.obersteg@hcu-hamburg.de

Abstract: The paper presents an evaluation of the strategic experimental project "WiedergeBORN", conducted by Stadtreinigung SRH, Hamburg's public waste management company, with stakeholders and citizens in the Osdorfer Born large housing estate. Based on an analysis of the case study, which included document analysis, observation and interviews, the article delineates the genesis of the project, its main stakeholders, and its objectives, measures, and outcomes. Success and failure factors of the project and the possible transfer of lessons learned are then categorized and discussed using the four pillars of accountability as an analytical framework. The place-based approach, which considered local conditions, the close cooperation between key stakeholders, and the early involvement of local actors and citizens, supported the successful development and implementation of measures to improve waste management and cleanliness. Furthermore, the integration of environmental, social, and educational aspects in the measures and the cooperation with actors from these fields were demonstrated to lead to positive outcomes. The transfer of the project's results and approach remained limited due to a lack of involvement of strategic actors in the project; this hindered an integration of the participating stakeholders into strategies and into policies at the district or city levels.

Keywords: circular economy; experiment; strategic planning; accountability; participation

1. Introduction

In the discussion of how to address the challenge of climate change, the importance of circular economy (CE) and improved waste management as contributions to climate mitigation has been emphasized. Earlier studies focused on the necessity to include measures to improve waste management in climate action [1]. More recent reports delineate the extended potential of circular economy approaches to contribute to climate mitigation in Germany [2] and the EU [3]. Others argued that CE contributions are key to reaching sustainability goals [4].

While the necessity to include CE in climate action on national and international levels is widely recognized in the scholarly discourse, research has demonstrated a lack of accountable action at the city level and a failure to integrate CE into strategic urban planning [5] (see Section 2).

At the same time, various studies have examined the role of urban experiments as a strategic approach to governing climate change in cities [6–8]. For example, Wirth et al. [9] and Bulkeley et al. [10] analyzed examples of urban living laboratories and how lessons learned from these experiments can be transferred into strategic urban planning. Some of the studies on the role of experiments for strategic planning included experiments in the fields of CE and waste management, but their number is limited. Furthermore, the specific connection of CE and waste management to strategic urban planning has only become the subject of the scientific discussion more recently.

To contribute to this discussion, this article examines a case study at the local level, the strategic experimental project WiedergeBORN, which was conducted by Stadtreinigung

SRH, Hamburg's public waste management company, in Osdorfer Born in Hamburg, a large housing estate, and involved various stakeholders and citizens (see Section 4). The goal of this experimental project was to foster CE by improving waste management locally. The WiedergeBORN project is elaborated and examined in this paper to outline what can be learned from this case about the integration of CE into strategic urban planning. The case study is analyzed using a methodological mix of document analysis, participatory observation, and interviews (Section 3). The aim of the presented research is to analyze and understand the success and failure factors of the strategic project, and to learn from the project for potential transfer and replication. The accountability framework [11] is used to categorize the results of the analysis, to discuss possibilities of replication and scalability beyond the experimental project, and to indicate their potential for integration on the strategic planning level (Section 6).

To answer the research question, the article is subdivided in the following sub-questions:

- What are the lessons learned and what are the success and failure factors of the project (Section 5)?
- How are the lessons learned perceived by the key stakeholders with regard to potential continuation and transfer of the results of the project (Section 6)?
- On a conceptional and methodological level: What are the potentials of the categorization and assessment of the results through the accountability framework (see Section 2) to enable the transfer from the experimental level into strategic planning (Section 7)?

2. Accountability of the Circular Economy in Cities

CE is a relatively new paradigm that is rooted in the broader concept of sustainability [12,13]. More specifically, CE draws upon the concept of Urban Metabolism (UM) [14], which describes cities as open systems exchanging resource flows with their environment [15]. While UM is mainly a descriptive paradigm, CE offers a conceptual framework for managing flows [16]. The Ellen MacArthur Foundation [17], among others, translated CE into policy recommendations and has been widely adopted by the European Union in the EU action plans for the CE 2015 and 2020 [12,18].

Despite its roots in UM, CE-related policies initially mostly ignored the accounting of effects of CE on the (urban) space and vice versa [19]. Noting this gap, scholars have argued that the spatial dimension of CE needs to be stressed when handling flows in urban areas since every action undertaken inevitably implies changes to the land(use) [20,21]. Girardet [20] connects the concept of CE with urban development approaches in his regenerative city concept. Additionally, William's [19] circular cities concept highlights the deficits of CE in the incorporation of spatial and social questions. On this basis, Williams later suggests evolving CE into a circular development approach that should expressly include a spatial perspective for urban planning strategies and, thus, enable a sustainable development towards circularity. Further authors propose that urban planning as a discipline should incorporate CE perspectives into the discourse on spatial development at the city level [5,22,23], while others claim that urban regions are the most suitable scale to act for the concretization and spatialization of CE [24].

The scientific and policy discourses on the necessity of a stronger consideration of spatial aspects in CE policies and actions have inspired cities to develop their own strategies and policies that combine CE and spatial development (e.g., Amsterdam). Consequently, spatial aspects have also been integrated into EU policies on CE, the foremost of which is the 'New CE action plan 2020', and this has led to the creation of the European 'Circular cities and regions initiative CCRI' [18]. The latter has stimulated numerous local and regional activities to develop and test CE (e.g., in the form of living labs), and initiated further EU-funded research. With regard to these recent activities, Williams underlines the need for more research on spatial implications of CE and the role of spatial planning in steering governance processes to support circular development [5,23].

Recent work in the field discusses the process of implementation of CE strategically in cities and links it to the question of how to assess these processes [21,25]. The accountability framework is one method that has been used to assess strategic planning related to climate action. It centers the relationship between the principal (the citizens) and agents (state, public institutions) in its analysis of strategic planning and actions. Furthermore, it enables a polycentric governance perspective on strategic planning processes, meaning that, in the relation between citizens and state, several institutions can represent the public side and, also, interactions among them can be considered [26].

In the application of the accountability framework, Zengerling proposes four pillars to analyze accountability [11]. In the first pillar, "responsibility", the relationship between the principal (the citizens) and agents (public institutions) is examined with regard to the strategic planning project. Clear responsibilities between the stakeholders and the citizens are important for the successful implementation of a strategic project [11,27]. As the case study WiedergeBORN was an experimental strategic project involving several public, private, and civic stakeholders, not only the relation between the principal and agents, but also the responsibility relations between different stakeholders as agents are considered. The second pillar, "transparency", focuses on questions related to communication and information flows between principals and agents, similar to the other pillars between the different stakeholders (agents) in the strategic project. In the third pillar, "assessment", the focus is on data in relation to the strategic project and its objectives. Lastly the fourth pillar, "participation", examines the involvement of citizens in the strategic project and of stakeholders in the decision-making process [11,27]. An assessment of a strategic CE project using this framework allows for a nuanced depiction of the processes of the project implementation, beyond a merely technical or regulatory approach, and provides insights into the potential for transfer of project learnings and results.

3. Methods Applied and Accountable Strategic Planning as Conceptual Frameworks

The case study, the experimental strategic project WiedergeBORN, was analyzed using a methodological mix: participatory observations during meetings, workshops, and public events (see Table 1), and field studies, including spatial observation and analysis of the area (implemented built measures, impact on cleanliness, see Section 4) before, during, and after the project [28,29]. A document analysis was performed, drawing on data reports, unpublished minutes of workshops, plans, stakeholders' policy concepts and newspaper articles. After the end of the project, three semi-structured qualitative interviews with the key stakeholders involved in the experimental project, Stadtreinigung Hamburg SRH and SAGA/Pro Quartier PQ (see Section 4.3), were conducted to reflect on the project's activities, outcomes, lessons learned, and possible transfer and replication of results (see Supplementary Materials File S1: Guideline for interviews with key stakeholders). The interviews were recorded and transcribed. The transcribed interviews were subsequently analyzed using qualitative content analysis [30,31]. The interviews had two aims: to complete and clarify information on the measures and their immediate outcomes (see Section 5), and to evaluate the factors contributing to their success or failure. The interviews, documents, and minutes were then examined following the four pillars of the accountability framework, described in Section 2. The results of the analysis are presented in Section 6.

Table 1. Effective and potential waste disposal for different waste fractions and different neighborhoods.

Waste in kg per Year and per Capita	Osdorfer Born	Ottensen	Altona-Nord	Blankenese	Lurup
Sum of waste collected	367.9	352	243.7	306.9	277.1
Disposed effectively in recyclables bin	20	44.8	8.4	47.1	15.2
Disposed effectively in paper bin	66	68	10.5	49.8	57.6
Disposed effectively in residual waste bin	269.4	216.7	182.4	104.7	176.5

Table 1. *Cont.*

Waste in kg per Year and per Capita	Osdorfer Born	Ottensen	Altona-Nord	Blankenese	Lurup
Disposed effectively in organic waste bin	12.5	22.5	42.4	105.3	27.8
Ideally disposed in recyclables bin	42.3	46.1	25.3	28.5	36.8
Ideally disposed in paper bin	79.8	81.3	23	58	71.5
Ideally disposed in residual waste bin	111.4	116.4	59.6	42.2	56.2
Ideally disposed in organic waste bin	105.1	70.8	109.7	159.9	86.6
Ideally disposed in collection containers (glass, textile, WEEE)	29.3	37.5	26	26	12.8

4. The Case Study "WiedergeBORN"

4.1. The Experimental Project of WiedergeBORN

WiedergeBORN led to direct, measurable improvements in waste management and circularity in the local project focus area. These results hold special significance with regard to the discussion of the increasing importance of CE contributions to the achievement of sustainability goals and, especially, to the reduction of greenhouse gas emissions. This is of specific relevance in Hamburg, as the city has, to date, no circularity strategy, and its city-wide and district-level climate action plans largely lack measurable activities in the field of circular economy.

The results of the regular waste separation surveys of SRH had shown that the sensitivity for waste separation is relatively low in large housing areas like Osdorfer Born compared to other types of areas: dense historic area (Ottensen), dense post-war area (Altona-Nord), lower density multiple-family housing (Lurup), and single-family housing (Blankenese) [32]. Table 1 shows the amount of waste per year and per capita effectively disposed of in the bins for the different waste fractions, and the potential amount of waste per fraction that could be ideally disposed of in the correct bin. Osdorfer Born has the highest amount of waste in total, showing potential for waste avoidance. The gap between effectively and ideally disposed organic waste and recyclables is the highest compared to the other areas. Therefore, the key stakeholders decided to conduct the experimental project in the large housing estate Osdorfer Born [32].

The key stakeholders chose Osdorfer Born for the experimental project because of its characteristics as a large residential area with a population that is culturally mixed, economically rather poor, and with a comparatively low level of education.

Table 2 shows a summary of the main spatial and socio-economic information for different neighborhoods in the district Hamburg-Altona that were compiled in the frame of the REPAiR project and were the base to choose Osdorfer Born as the area for the experimental project. Data sources are statistical profiles for Hamburg's neighborhoods and districts [33], and, for Osdorfer Born, the evaluation report for social development programs [34].

The WiedergeBORN project was conceived within the research and sustainability unit of Stadtreinigung SRH in two Horizon2020 projects: FORCE and REPAiR. SRH adopted ideas for circularity solutions and place-based concepts that were developed in REPAiR, a more research-oriented project, and applied these in the more practically-oriented project, FORCE, which had a dedicated budget for the implementation of concrete measures on the ground. Table 3 shows the timeline of the project. A major motivation for the research and sustainability unit was their recognition of SRH practitioners' needs for tangible project actions and results. Based on an earlier waste separation analysis [32] and a socio-spatial study conducted in REPAiR, as well as the place-based solutions that had been developed in the REPAiR project for different areas, the research and sustainability unit selected the large housing estate Osdorfer Born as the implementation site. SRH chose Osdorfer Born, first, because the waste separation analysis showed a low compliance with waste separation policy in the community, and, second, because of their ambition to work in an area that is considered more challenging. Osdorfer Born is among Hamburg's lowest income communities. Thus, this project was an attempt to demonstrate measurable improvement in an area with a clear waste problem, while also improving service and support to a community with a high level of need. At an early stage, outreach was performed with further stakeholders, such as SAGA, Hamburg's public housing company, and Pro Quartier (PQ), a subsidiary company of SAGA offering services like neighborhood management. They also agreed to participate in the project and in the choice of the project area. The objectives of the project were to improve circularity in the area with a focus on waste management and cleanliness through the development and implementation of place-based solutions with citizens and the involved stakeholders. The project name "WiedergeBORN" refers to Osdorfer Born and is a play on the German word "wiedergeboren", meaning "reborn".

Table 2. Main spatial and socio-economic information for different neighborhoods in the district Hamburg-Altona.

Indicators	Osdorfer Born	Ottensen	Altona-Nord	Blankenese	Rissen	Altona District
Population [n]	10,263	35,370	22,137	13,407	15,192	273,263
Population density [inh/km^2]	11,591	12,654	9981	1733	909	3469
Population over 65 years old [%]	18.0	13.4	10.1	27.2	30.6	18.0
Foreign population [%]	26.4	13.1	18.8	8.3	7.3	16.1
Population with migration background [%]	63.8	26.0	36.0	17.1	16.9	32.0
Annual average income per taxpayer [EUR]	17,480	40,830	29,901	117,139	65,855	48,620
Unemployment rate (between 15 and 65 years old) [%]	11.8	4.4	7.2	1.9	3.5	5.9

Table 3. Timeline of the project WiedergeBORN.

Date	Activity/Event
2017–2019	Co-creative process in REPAiR living lab; first analysis of place-specific waste and circularity problematic and potential in different areas in Hamburg
19–22 June 2018	Student workshops at HCU and TU Delft as part of the REPAiR project to develop solutions to improve circularity in Osdorfer Born and other different areas in Hamburg
January–May 2019	Idea for WiedergeBORN as a project focusing on the specific area of Osdorfer Born was developed by SRH; using input from REPAiR (socio-spatial, waste analysis, eco-innovative solutions) and FORCE (input from other cities: Lisbon and Copenhagen)
5–9 August 2019	International summer school of the FORCE project with a focus on the development of WEEE (Waste of Electrical and Electronic Equipment) solutions for Osdorfer Born
June 2019	SRH establishes an internal team, local project manager starts work
28 August 2019	WiedergeBORN Kick-off Workshop (WS1) with key stakeholders; the core working group is established
September–December 2019	Survey of inhabitants conducted by SRH
September 2019–February 2020	First implementation of project measures and activities
17 February 2020	WiedergeBORN mid-term Workshop (WS2), first evaluation of activities
February 2020–February 2021	Continuation of measures and activities (limited due to COVID-19)
8–10 September 2020	WiedergeBORN Festival
8 February 2021	WiedergeBORN closing Workshop (WS3) with project internal evaluation of the results
Spring 2021	Postprocessing of the project: guideline for SAGA, SRH
Spring 2021	Attempt to achieve EU funding for follow-up project failed
From 2021	Usage of selected experiences and learnings from the project in other areas and project by the different partners (SRH, PQ)

4.2. Osdorfer Born the Case Study Area of the WiedergeBORN Project

Osdorfer Born is located in the northern part of the District of Hamburg-Altona, one of the seven districts in Hamburg that represent the municipal political and administrative level. In Hamburg, the districts take on many of the tasks that municipalities would elsewhere, as Hamburg, as a city-state, is one of the sixteen German federal states. According to the latest statistics available in the area, in 2016, Osdorfer Born had 4750 housing units for a total of 10,263 inhabitants [33]. Newer statistics are only publicly available for the larger administrative urban area Osdorf or for the larger social development area (Fördergebiet Sozialer Zusammenhalt) Osdorfer Born/Lurup [34].

Osdorfer Born was built in the late 1960s as part of the housing development program in reaction to the immense need for flats after the Second World War. It was the first large housing estate project of this kind built in Hamburg. It was meant to be a modern and livable place, offering social, educational, and cultural infrastructure and access to a big park area. A connection to the metro system, initially included in the plans, was not built due to the economic crisis in the 1970s, a fact that contributed to the poor accessibility of the neighborhood that lingers to today. Since the end of the 1970s, the share of households with higher incomes has decreased and the area has attracted more households with lower incomes due to its affordability. Osdorfer Born has become, like other comparable neighborhoods, a first place to find accommodation for many persons moving to Hamburg.

Until the 1990s, due to a lack of investment in renovation and modernization of the housing stock and public spaces, a process of physical downgrading occurred [35]. In reaction to this development, Osdorfer Born has received funding from different programs co-financed by the City of Hamburg and the German government since 1992 ("Revitalisierungsprogramm" 1992–1998; "Soziale Stadtteilentwicklung" 1999–2005; "Aktive Stadtteilentwicklung" 2005–2008; in parallel, from 2002 Bund-Länder-Programm "Soziale Stadt") [34]. Since 2014, Osdorfer Born, together with the adjoining neighborhood Lurup, has received funding from Hamburg's framework Programme for Integrated Urban District Development (Rahmenprogramm Integrierte Stadtteilentwicklung RISE) for neighborhood management programs and mainly social projects [34]. Additionally, SAGA has set up neighborhood management programs through its subsidiary company, Pro Quartier, to enhance social inclusion and to improve the built and natural environment [34].

According to the socio-spatial monitoring conducted by the City of Hamburg, Osdorfer Born is characterized by a significantly higher unemployment rate, a higher share of households receiving social welfare, and a higher school dropout rate than the Hamburg average [34]. Due to its characteristics, Osdorfer Born is considered to have a "low status" according to the latest social monitoring report published by Hamburg's Ministry of Urban Development and Housing, which is an improvement compared to the previous report where the area had a "very low status". Nevertheless, Osdorfer Born, together with Lurup, is again defined as a future funding area receiving support for urban regeneration [36].

4.3. The Stakeholders in the Project WiedergeBORN

The key stakeholders in the project WiedergeBORN were Stadtreinigung Hamburg (SRH), SAGA Unternehmensgruppe, and CHANCE and Pro Quartier (PQ), two subsidiary companies of SAGA.

Stadtreinigung Hamburg (SRH) is Hamburg's public waste management company. As stipulated by law [37], Stadtreinigung (SRH) manages the collection of household waste, street cleaning, winter maintenance services, and public toilets. Furthermore, it operates twelve recycling centers throughout the city, one of which is located next to Osdorfer Born. According to law [37], SRH is responsible for the collection and processing of household waste fractions [38]. These fractions are residual waste, which is collected and then incinerated to generate heat and electricity, and organic waste, which is brought to facilities for anaerobic digestion and subsequent composting to produce biogas and compost [39]. SRH is also responsible for collecting large or bulky household waste on demand, in exchange for a collection fee. Bulky waste is further processed at the recycling centers. Still usable items (i.e., furniture) can be brought to second-hand warehouses operated by Stilbruch, a subsidiary company of SRH. Other waste fractions, including packaging, glass, and paper, belong to the so-called dual system and are collected by private companies, including by the SRH subsidiary company WERT GmBH [38].

SAGA Unternehmensgruppe (SAGA) is a public housing company, fully owned by the City of Hamburg, and is Germany's largest municipal housing company. In 2022, SAGA rented out a total of 138,656 flats in Hamburg, of which 30,187 flats were publicly subsidized, and 108,500 were financed without public subsidies (of which 102,121 were without limited rents and 6348 had limited rents). Circa 270,000 persons live in a SAGA flat, representing 14% of Hamburg's population. As SAGA offers moderately priced flats, it is considered to have a balancing effect on the rental flat market, offering housing to lower- and middle-income households [40,41]. In Osdorfer Born, SAGA owns 3600 of the 5100 total flats. SAGA has conducted an energy-efficient renovation and modernization of its building stock in Osdorfer Born over the last few years [34]. In its sustainability strategy, SAGA addresses the reduction of energy and water usage linked with the refurbishment of buildings as its main ecological concerns. Regarding the topic of waste reduction, a reduction in the amount of residual waste was observed between 2011 and 2018, but the amount has increased again since then. The separated waste fraction of paper and organic waste has continuously increased. The collected amount of all fractions grew significantly in 2021 compared to 2020, which can be explained by the lockdowns during the COVID-19 pandemic. In the sustainability reports, SAGA mentions only an intelligent waste management system in cooperation with external partners since 2004 and holds contracts with SRH for waste collection, while no concrete measures on waste reduction are described [42–44].

CHANCE (CHANCE Beschäftigungsgesellschaft mbH) is a subsidiary non-profit company of SAGA that employs long-term unemployed persons to reintegrate them into the job market. The employees work as service and contact persons for the inhabitants, mainly in large housing estates of SAGA and some estates owned by other housing cooperatives. The employees are located in so-called lodges and serve as contact points for the inhabitants. They offer concierge services to the inhabitants and inform the facility management and caretakers about problems like technical issues, vandalism, littering, and security [45].

Pro Quartier (PQ) is a subsidiary company of SAGA that offers services like neighborhood management, mainly in SAGA estates, but also in other areas. In Osdorfer Born, PQ are responsible for the neighborhood management of the large housing estate owned by SAGA and other housing cooperatives.

Table 4 lists the further stakeholders that participated and their involvement in the project.

Table 4. Further stakeholders involved in the project WiedergeBORN.

Stakeholder Name and Description	Cooperation in WiedergeBORN
Nutzmüll e.V., (Nutzmüll means "useful trash" in German) is an association that has been active for several years in the area in the fields of waste prevention and sensitivity.	In cooperation with Nutzmüll, various collection activities were conducted, for example, bulky waste collection and shopping trolley collection.
Kinder Museum, Hamburg's museum for children, is located in the area	Cooperation during the festival and workshops with children
Bürgerhaus, a community centre	Cooperation in several workshops and information events
Born-Center, a shopping center	Cooperation during the festival; exhibitions on waste prevention and recycling
Citizens	Individually cooperated, for example collecting waste or informing about illegal waste dumping
Westwind, local newspaper	Several articles were published
Recycling center/yard Osdorf, one of 12 recycling centers operated by SRH in Hamburg	Cooperation on bulky waste and other hazardous or special waste fractions

Table 4. *Cont.*

Stakeholder Name and Description	Cooperation in WiedergeBORN
Churches	Cooperation with, they were supportive. Participation in swap meets (Tauschbörsen)
Daycare centers (Kita)	Cooperation on composting project
Bücherhalle (public library)	Organized and staged puppet theatre for children about waste separation
Hamburger Volkshochschule (adult education center)	Cooperation during the festival; exhibitions on waste prevention and recycling
Borner Runde (community roundtable in the neighborhood)	SRH and SAGA presented WiedergeBORN; the round table was used for feedback, broader outreach, and connection with further initiatives and associations in Osdorfer Born
Parent-Child-Centre	Practical waste separation training
Geschwister-Scholl-Stadtteilschule, a secondary school	School workshops on plastic recycling
Precious Plastic Hamburg, a non-profit company	School workshops on plastic recycling

4.4. Development of the Measures

The main measures and activities of WiedergeBORN were created based on outcomes of analyses and ideas for solutions from the projects REPAiR and FORCE. They then were further developed in a co-creative process between SRH and the other main stakeholders, PQ and CHANCE. In the next step, the concrete measures were elaborated in cooperation with further stakeholders (see Table 4), also using input from the inhabitants' survey.

4.5. Process of Cooperation

At the start of the project, a core group was established that consisted of two people from the research and sustainability unit SRH, a person from SRH who was hired as the local project manager to be present in Osdorfer Born, and three people from PQ (one from the PQ headquarters and two working in Osdorfer Born and who received additional funding from the project). PQ served as a contact to CHANCE (persons working in the lodges) and to the local SAGA office (See Table 4).

During the project, three workshops were held with the main stakeholders: one at the start to define measures and activities (see Table 3), one at mid-term to evaluate the progress, and one at the end to evaluate the project and its measures. Between these workshops, regular meetings with the core group of stakeholders were held to monitor the process of the measures. To gather broad input from citizens, a survey among inhabitants was conducted as a starting activity. In total, 354 residents were contacted, of whom 114 were interviewed at the door of their flat by employees of SRH with a semi-structured questionnaire. The survey had two aims: first, to generate citizens' interest in the subjects of waste and cleanliness and in future activities; and second, to learn from their experiences and understand which topics citizens want to be handled, to learn about their waste separation behavior, and to learn about their attitudes towards waste topics. Table 5 gives examples from the unpublished survey and how it contributed to the development of measures.

Table 5. SRH survey with inhabitants: Overview on outcomes with focus on organic waste and respective measures developed.

Question Category	Summarized Answers	Respective Measure Developed
What types of waste are generally separated at home?	Organic waste 19%, recyclables 46%, paper 87%	Measure 4: improving waste separation behavior, reaching different target groups
What would have to change to enable or facilitate the separate collection of organic waste?	Free bins and collection bags 9%, neighbors need to participate 8%, more information 9%, more space for separation bins in flat 7%, better labeling of tons 4%	Measure 1 Improving waste locations sites; Measure 3 Lodges as information points, distribution of free bins

5. The Main Measures and Activities of the Project—Description and Results

In this section, based on the empirical studies (document analysis, observation, interviews), the main measures in the WiedergeBORN project are presented, including the problem definition, the development and conduction of activities, and the evaluation of the measure.

During the co-creative process at the beginning of the project, the key stakeholders decided to conduct the following main activities. Each main activity was then further developed, divided into sub-activities, and additional stakeholders were involved (see Table 4).

Measure 1—Improving Waste Storage Locations: Site visits, workshops, and direct observations at the beginning of the project revealed that the neighborhood's waste storage areas faced significant issues. These included poor accessibility for people with disabilities, inadequate placement of bins, and an unclean environment that discouraged residents from using the sites properly. The organic waste bins were too small, difficult to reach, and placed in inconvenient areas. Additionally, labels on the bins were too small, not self-explanatory, and not accessible to all residents despite translation into multiple languages. As a result, many residents had difficulty understanding the waste separation requirements, leading to improper waste disposal.

In response, a set of physical improvements was implemented. These included constructing new, accessible structures at waste storage sites, providing larger and cleaner bins, and introducing more intuitive labeling that included larger pictograms in place of text. The new labels were designed to convey sorting instructions visually, minimizing language barriers. The project also adjusted bin sizes based on waste type, with larger bins for organic waste placed in easily accessible spots. Three main partners collaborated on these interventions: SAGA, the property owner responsible for providing appropriate waste storage locations (according to German law, waste must be stored on private property and the owner must provide a location where the bins are placed that is adequate and accessible) [38]; a service provider contracted by SAGA to maintain the cleanliness and order of bins; and SRH, which supplied bins and managed their regular emptying. Through close collaboration, SAGA undertook necessary construction, SRH provided improved bins and labels, and enhanced communication was established between SRH and the service provider to address issues like littering and vandalism more effectively.

The evaluation workshop indicated a positive reception from residents, caretakers, and visitors, who reported improved cleanliness and accessibility. Measurable improvements included an increase in organic waste volume collected and a decrease in incorrect waste sorting, both essential goals for SRH. According to stakeholders, a key success factor was the improved physical environment of the waste storage areas, which facilitated easier, more appealing, and more accessible use by residents. This change was seen as a foundation for fostering sustainable behavioral shifts among residents, as an accessible and well-maintained environment was more likely to encourage proper waste disposal practices. Furthermore, improved cooperation between service providers, SRH, and CHANCE, a local neighborhood support organization, allowed for a quicker response to issues such as vandalism and improper bin usage.

Measure 2—Cleanliness in the neighborhood: SAGA highlighted a lack of cleanliness, which could be observed, and subsequently, the causes of littering were further analyzed and discussed in the co-creative process. The littering comprised minor issues such as papers and cigarettes disposed of in public spaces and littering on a larger scale, namely the illegal dumping of bulky waste items. Several causes for the bulky waste problem were identified, including a lack of knowledge, "inhabitants do not know how and where to dispose of bulky waste", and the cost, as the collection and disposal of bulky waste must be paid individually by households and many inhabitants regard the fees as too high. As an alternative, the inhabitants can bring their bulky waste to a recycling center free of charge, but, although the closest recycling center is in only 1 km away from Osdorfer Born, according to PQ the transport is still problematic as many households in the large housing estate do not own a car. As a consequence, bulky waste was often left in front of the house. Bulky waste that is dumped illegally in the area must be collected by SRH in so-called 'sprint collections,' for which a fee is charged to SAGA and, therefore, is paid by all inhabitants. SAGA stated that the lack of cleanliness and littering are major concerns expressed by their tenants.

To improve the cleanliness in the neighborhood, more attractive public waste bins (colorful and "monster" bins to attract interest, especially of children) were provided. A more complex cooperation was needed to reduce the bulky waste problem. Before the project started, SAGA had already designated a garage where inhabitants could bring bulky waste. This offer was expanded by installing a container for bulky waste in front of another large building. The inhabitants could use the waste storage by contacting the concierges at the CHANCE lodges, who then could open the container or garage so that the bulky waste could be put in. A small vehicle was offered to assist with the transport from the apartments. When the container or garage was full, CHANCE informed SRH to collect the bulky waste and transport it to the recycling center. There, the usable bulky waste could be sorted as usual and brought to Stilbruch second-hand shops.

During the evaluation workshop and the interviews, the new colorful public waste bins were assessed positively, and minor littering visibly decreased. However, a need for more educational work on how to use the bins was expressed. The bulky waste measure was also regarded as successful. Illegal dumping of bulky household waste items in the area was significantly reduced, which translated into a considerable reduction of costs for SAGA, as the collection from the container and garage could be better planned. Furthermore, the bulky waste could be separated into different sub-fractions (especially furniture and WEEE—waste from electrical and electronic equipment) and consequently was easier to re-use or recycle. Additionally, the amount of separately collected WEEE increased because inhabitants were informed that they should dispose of WEEE as bulky waste and not as residual waste.

Measure 3—Lodges as information and communication points: The experience of SRH, and the analyses prior to the project, highlighted the lack of knowledge of the inhabitants about waste and recycling topics and the necessity to better inform residents to change their behaviors. To tackle this problem, at the project start, SRH and SAGA decided to use the existing lodges of the residential buildings and to involve CHANCE and its concierge employees as information and contact points for the project. The concierge employees of CHANCE were involved at an early stage and expressed high interest in the project, as many of them regarded waste and the lack of cleanliness as major problems in the area. They were trained at the beginning of the project on how to address the topics of waste prevention, recycling, and cleanliness.

The concrete measures were counselling sessions held during the lodges' opening hours. While residents' attendance at these sessions was lower than expected, they offered a good opportunity to exchange ideas with other local project partners. Craft and games afternoons were offered to inform about reuse and repair, but they met low responses due to comparable offers available in the neighborhood. Additionally, themed swap meets were held, intended to encourage resource-sharing among residents; the toy swap was well-received, while the clothes swap met with less enthusiasm, likely due to other competing events. It became clear that a longer period of time would be required to establish the idea and offers for such swap meets in the neighborhood. In the interviews, SRH and PQ stated that inhabitants expressed that used things have a negative stigma. Informational exhibits and displays on different waste-related topics, including showing previously used furniture from Stilbruch second-hand shops, were designed to increase local residents' use of various offers, but only met moderate interest. Although the impact was not directly measurable, these activities were still seen as valuable, as they increased general awareness and offered food for thought to the residents.

The distribution 300 small organic waste bins (Biomüllis) to separate organic waste and 10,000 compostable bags to the residents, free of charge, was regarded as successful by SRH. This measure was based on the findings from the analysis prior to the project and from the survey among residents that many households did not separate organic waste due, on one hand, to the lack of knowledge, but also to practical reasons like lack of space for several bins and the fear of a bad smell from organic waste. The small waste bins and the compostable bags could offer solutions to overcome these concerns.

The training on waste topics, which was offered by SRH to the CHANCE employees, was regarded as successful and enabled the CHANCE employees to serve as direct contact people and multipliers towards the residents. However, due to the COVID-19 pandemic, the trainings had to be reduced, and, also, the direct contact with the residents was limited.

Measure 4—Improving waste separation behavior: SRH regularly conducts studies on waste separation in different sample areas in Hamburg that examine the content of the different waste fraction bins to estimate correct or incorrect sorting of waste fractions. The latest study prior to the project showed that in Osdorfer Born, the separation practice is significantly worse than in other areas with different types of housing [32]. During the workshops, SRH employees confirmed that incorrect waste separation is a problem in Osdorfer Born (see Section 4.1). The amount of organic waste that is collected in the area was significantly lower than in other neighborhoods. SRH, in the past, hesitated to campaign for more organic waste collection in large housing estates like Osdorfer Born because SRH expected an increase of mis-filling/mis-disposal of non-organic waste into the organic waste bins. This corrupts the organic waste stream; thus, such impure organic waste cannot be further processed (in the organic waste treatment plant: fermentation, biogas production, composting of the rest material). Therefore, in the past, SRH preferred to collect less but pure organic waste.

To improve the waste separation behaviors of the inhabitants and to increase correct waste separation, several measures were developed and conducted. A four-week organic waste collection campaign was conducted to achieve a collection of qualitatively and quantitatively 'high-quality' organic waste through information and training of residents on the waste separation process. It included the distribution of 1100 free-of-charge organic waste bins for pre-sorting bins at home (Biomüllis) and flyers at an information stand in the neighborhood, join-in campaigns, and a raffle of attractive prizes. Circa 20% of the inhabitants in the focus area could be reached, with 254 visitors at the stand. An extra analysis of organic waste was conducted before and after the four-week campaign, with the results showing that the volume of organic waste had increased by 51%. The usage of compostable paper bags to dispose of organic waste increased to 70%, while the usage of plastic bags to dispose of organic waste was reduced from 53% to 18%. The most common contaminants in the organic waste were plastic bags and packaged food.

Further outreach measures included school workshops on plastic recycling with the non-profit company Precious Plastic Hamburg. According to the interviews with SRH and PQ, these workshops received good feedback from teachers and pupils and led to a sustainable change in behavior among pupils. SRH also conducted practical waste separation training in the parent-child center. The participants were interested in the topic. For SRH, it was a positive takeaway that different target groups can be reached through specific facilities in the neighborhood.

In the evaluation workshop, the measures were regarded as successful. It was stated that separation behavior had improved, but also that the residual waste bins still fill up very quickly and that there is potential for more separation. The stakeholders agreed that behavioral change would need time and requires continuous support in the neighborhood and repetition of the campaigns. Regarding the waste separation training and the school workshops, the importance of reaching different target groups via neighborhood facilities was stressed and the attempt to consider different levels of knowledge and the motivation of adults through their children were highlighted as positive outcomes.

Measure 5—WiedergeBORN Festival: A three-day neighborhood festival was conducted to address the themes of waste and sustainability in a playful, informative, and participatory way. The festival was organized in cooperation with various facilities like VHS, public library, initiatives like Precious Plastic, and the shopping center. It offered services like waste avoidance workshops and repair cafés, a puppet theatre on recycling, and games and challenges with prices. In total, 500 visitors across different age groups could be reached.

The festival was evaluated positively, as it reached many inhabitants who were not very familiar with the topics of waste and sustainability. Through the offers for children, their parents and relatives could be reached, and further participants could be targeted and mobilized via local stakeholders. The festival was regarded as a means to spread knowledge on the topic and to create a certain expertise among the citizens. It was important that the festival allowed dialogue with citizens rather than only one-way information. This enabled a further understanding of the views of inhabitants on waste and sustainability problems.

It can be stated that most of the measures were successful, and some still have a positive impact in the neighborhood. In contrast, the desired knowledge transfer of the lessons learnt from the project into the participating stakeholders' organizations and to other neighborhoods could not be implemented as desired by the core team of the project. Table 6 gives an overview on the measures and their innovative aspects compared to normal measures in this or other areas

Table 6. Measures and innovations introduced compared to other urban areas.

Measure Description	Innovative Aspects of Measure	Normal Measures in This or Other Areas
1—Improving Waste Storage Locations: physical improvements, accessibility, labelling, management of sites.	Measures developed in cooperation of different stakeholders (property owner, waste management, service provider).	Property owners provide standard waste storage locations, waste management company provides standard bins and labeling, not adapted to specific needs in areas. Limited cooperation between stakeholder
2—Cleanliness in the neighborhood: improved offers for bulky waste collection, and measures to improve cleanliness	Cooperation of concierge service of the property owner with waste management company; locally adapted measures against littering	Households are individually responsible for bulky waste disposal, no support from property owner; standard paper bins not addressing individual neighborhood
3—Lodges as information and communication points	Involvement of existing concierge employees, training as contact persons for waste and circularity topics	Concierge services normally are less interactive with inhabitants and other stakeholders. They are not responsible for circularity topics
4—Improving waste separation behavior	Four-week information campaign in the area as cooperation between various stakeholders. Educational workshops with different target groups (children, pupils, adults)	Normal waste separation behavior campaigns use classic media (posters, adverts) and address the citizens in general. Stakeholders like schools and childcare centers are less involved
5—WiedergeBORN Festival	Three-day neighborhood festival on waste and sustainability, organized with various local stakeholders	The topics waste and sustainability normally are addressed in classic information campaigns, not in neighborhood festivals

6. Discussion—Accounting of the Success and Failure Factors

As was described in Section 5, the stakeholders considered the project itself and most of its measures to be successful; however, the transfer of the lessons learned from the project was regarded as limited. To understand why, in this section, a categorization and discussion of the success and failure factors of the experimental project and its potential for transfer on a strategic planning level are conducted using the four pillars of accountability [11,27] as the analytical framework (see Section 2).

6.1. Responsibility: Success and Failure Factors

The guiding questions in the pillar 'responsibility' are: Which stakeholders (agents) were responsible to the citizens (principal)? Who was responsible for which actions? What were the responsibilities of stakeholders in regard to the actions monitored and in which form? As WiedergeBORN was an experimental strategic project involving several public, private, and civic stakeholders, the responsibility relations among these actors were also considered.

The main stakeholders of the project, SAGA, together with CHANCE, PQ, and SRH, are, as public actors, in general directly responsible to the government and the parliament of the City of Hamburg, and indirectly responsible to the citizens of Hamburg. Their specific main responsibilities are linked to their defined tasks and objectives, which are defined on a political level. In the case of SAGA, these are the offer of affordable housing and to care for the needs of its tenants. For SRH, the main tasks are waste management and cleanliness services in Hamburg (see Section 4.3).

According to the interviews, during the starting phase of the project, the responsibilities of the different public actors (SRH, SAGA with CHANCE and PQ) were not always clearly defined between the actors. As an example, the shared responsibilities for the waste storage locations (see measure 1) were mentioned in the interviews. SAGA owns the locations and is responsible for their construction and maintenance, but a sub-contractor is in charge of the cleanliness of the area. CHANCE can only collect complaints from inhabitants and is not allowed to intervene. Further, SRH owns the bins and oversees their contents and emptying them. The explanation and clear definition of the responsibilities, including the mandatory tasks and the objectives of the public actors, enabled the development of common goals for the project: to improve circularity in the area with a focus on waste management and cleanliness by developing and implementing place-based solutions with citizens and various stakeholders.

In the next step, the goals could be operationalized into measures to reach the agreed objectives. This would enable the joint development of new measures as well as the combination of different (existing) actions that normally would each have been under the responsibility of different actors (see five measures). The responsibilities for the implementation were clarified for both the new and the newly combined actions, which enabled their successful realization. In the interviews, the combination of actions was stated to have enabled synergies and made the project, as a whole, more effective and efficient. As described in Sections 4 and 5, many actions involved further public, private, and civic stakeholders. Again, the responsibilities and objectives of the different stakeholders were defined and then communicated between each other and the citizens, which supported the successful implementation of these actions.

The project enabled a clearer communication with the citizens regarding the responsibilities of each of the public stakeholders and the objectives of the project to improve waste management and cleanliness. This was possible through various information campaigns, exchanges, and direct contact because of the project's presence in the area and the way citizens were approached with different methods and by different actors.

In the interviews, it was stated that to successfully implement actions, it was good to combine the thematic responsibilities of the different public and other stakeholders and to commonly address the citizens in the area. This thematic and spatial focus was regarded as a key success factor, especially compared to customary actions of the different stakeholders that often are isolated and not adapted to specific spatial conditions. This corresponds with the findings of Williams [23] and Prendeville et al. [21], which show that, in order to achieve circularity objectives, spatially adapted solutions should involve stakeholders in their development and implementation and shared responsibility with citizens.

Continuation and transfer beyond the project: A major failure factor for the continuation of the activities of the project was, as stated by the interviewees, the lack of clear responsibilities beyond the project lifetime in Osdorfer Born. The different actors (CHANCE, SRH, PQ) continued some of the actions, but, with the closure of the project, the position of the local project manager also ended, and, therefore, a central responsible person was missing. Some of the lessons learned could be integrated into the work of the stakeholders and have been taken up as tasks in the respective units inside of the organizations: SAGA continues to use the developed recommendations for restructuring its waste storage locations; CHANCE was interested in using the training guideline for its lodge employees, but has not yet implemented a follow-up; SRH is implementing the concept of a local caretaker, but in an adapted way; and PQ is integrating the topic of circularity more in its work in other neighborhoods.

A negative aspect that was mentioned regarding the legacy of the project was the limited or missing involvement of higher tiers of the administration on the district level (integrated urban district development, climate mitigation) and the city level (Ministry for Urban Development and Housing, Ministry for the Environment, Climate, Energy, and Agriculture). As a consequence, a responsibility gap occurred at the end of the project. This is a major hinderance for the transferring of the project's lessons learned on a strategic level, due to missed potential for up-take into policy, as described by von Wirth et al. [9] and Bulkeley et al. [10].

6.2. Transparency: Success and Failure Factors

The guiding questions regarding transparency are: How transparent was the strategic project to stakeholders and citizens? Who among the stakeholders (agents) is communicating with citizens in general and with different target groups? What kind of information is offered to citizens, and in which forms and formats is information presented [11,27]?

In the interviews, the sharing of knowledge and information and making it understandable and transparent among the project partners was seen as a main positive factor. This was possible because of the close cooperation and trust-building between the partners during the project.

Information on the aims and activities of the project and on its effects on improved waste management and cleanliness were communicated to the inhabitants in a clear and transparent way by the project partners. The citizens were reached by using different channels to address diverse target groups (see measure 4 on waste separation behavior). The cooperation of the partners enabled the bundling of information towards citizens. This proved to be more efficient compared to the normally separated communication by each of the different public actors. The inhabitants could be better reached because the information was specific to the project and adapted to the citizens in the area. This was regarded as an important success factor for the project, as SRH conducts generalized information campaigns for the whole City of Hamburg and consequently does not reach specific target groups, while SAGA and PQ communicate with local inhabitants, but normally not about the topics of waste management, CE, or sustainability in general. This efficient form of communication with the inhabitants was regarded as important to involve citizens (see Section 6.4 participation) and change their behaviors over a longer term.

The idea to involve the CHANCE employees at the lodges to train them as direct contact persons to communicate with the inhabitants on the project and its topics was regarded as very successful. The CHANCE employees were considered as easily accessible for the inhabitants and were enabled to transmit information to them. However, because of COVID-19, some of the communication activities, especially in and with the CHANCE lodges, could not be fully implemented, and the training of the CHANCE employees had to be limited.

Continuation and transfer beyond the project: As a shortcoming, the interviewees highlighted that continuous communication with the citizens and a repetition of information on waste management, cleanliness, and circularity would be necessary: first, because new inhabitants need to be informed, and, second, because the information needs to be refreshed over time. With the end of the project, the communication efforts could not be continued with the same intensity.

6.3. Assessment: Success and Failure Factors

The guiding questions in the pillar assessment are how the measures and their outcomes are evaluated. Which data are collected, by whom, and with which methods, and how are the data evaluated?

The analyses of the socio-spatial situation of the project area and its inhabitants that were conducted before the start of the project were regarded as important for understanding the specific local problems and as crucial for the successful development and implementation of the measures that were adapted to the local situation. At the start of the project, the analysis of the regular waste separation assessment of SRH, which is normally used only internally, was shared and openly discussed with the other stakeholders. This was an important starting point for understanding the local problems and defining objectives and then measures in the project.

The assessment of waste separation (see measure 4) that was specially developed for the project and conducted before and after the waste separation awareness campaign made it possible for the project team to evaluate the effects of the measures immediately. Therefore, the significant improvement in organic waste separation due to the measures could be communicated to the citizens, which, according to SRH, had an additional motivation effect towards the inhabitants.

The survey among inhabitants (see Section 4.5) was seen as an important source of information to assess the status quo situation. Compared to the standard surveys that SRH conducts, this survey was regarded as more interactive and polled for deeper information about the behavior and attitude of the inhabitants on waste and circularity and on their personal perception of problems with waste management.

Overall, the intensive participation of the citizens in the project led to diverse and useful feedback from citizens on activities and outcomes of the measures and, according to the interviews, enabled a continuous assessment and consequent adaption of the actions.

The assessment of further measures (less bulky waste, increased re-use and recycling of bulky waste, improved cleanliness in the waste storage locations, less littering in public spaces) was primarily used for feedback among the stakeholders. In the interviews, it was highlighted that the positive results of the assessment were communicated internally to the management levels of SRH and SAGA. Based on the positively assessed measures, the members of the core group of the project could develop guidelines for the replication of some of the measures: the renovation and restructuring of the waste storage locations; the involvement of CHANCE lodges; and the involvement of day-care centers and schools.

Continuation and transfer beyond the project: A long-term assessment of the outcomes of the measures was not conducted after the project ended. According to the interviews, this information is necessary to evaluate the effects of the measures and the whole project, and it was suggested that a further assessment should be conducted a few years later. Furthermore, it was stated that a general evaluation of the project after its closure was missing. This was regarded as a hindering factor for the transfer of the project's learnings and the ability to integrate them in the strategies of the participating key stakeholders as well as on district and city level. Such a general evaluation serves to formalize learnings and experiences and permits an additional reflexive exchange among stakeholders [5,22,23].

6.4. Participation: Success and Failure Factors

In the fourth pillar, "participation", the involvement of citizens in the strategic project and of stakeholders in the decision-making process is examined. Guiding questions are: How are citizens and different groups involved? How are different stakeholders involved in the project? How are decisions made between different stakeholders?

As a main positive outcome of the project regarding participation, the direct and intensive contact with inhabitants via different channels was highlighted in the interviews. Through the bundling of activities and the cooperation between different stakeholders, many citizens who, according to SRH and PQ, normally are not interested in CE and waste management could be reached and directly engaged. Additionally, the cooperation with local experts and initiatives (see Table 4), who are closer to the inhabitants and have knowledge of how to approach and reach different groups of citizens, was also regarded as supportive when seeking to reach different groups of inhabitants. The intensive involvement of different groups of citizens enabled a more specific and better understanding of the local circumstances. As a consequence, the measures could be developed and implemented according to the problems and needs of different groups of inhabitants (e.g., considering language barriers and using children as multipliers in families). The combination of the activities of the different stakeholders to involve citizens was regarded as a success factor. This was only possible because of the project's strategic approach, as normally each of the stakeholders would approach citizens separately.

Some of the projects involving citizens were regarded as particularly successful. The waste separation contest with inhabitants in two streets enabled citizens to gain direct insight into the content of their waste bins and immediately learn about correct waste separation (see measure 4). The presence of the local project manager and of the CHANCE lodges as direct contact points were also considered as positive factors that lowered the barrier for citizens to get involved in the different activities.

Continuation and transfer beyond the project: The employees of CHANCE were trained as multipliers to involve citizens during the project, but the training had to be reduced due to the COVID-19 pandemic. This hindered the stabilization of the work of the CHANCE employees. Nevertheless, according to the interviews, the CHANCE lodges served as information points for the inhabitants beyond the project's lifetime, but without offering specific activities. The negative aspects mentioned in the interviews were that, despite the intensive involvement of citizens, it needs to be critically observed how far the situation (waste separation behavior, cleanliness) could be improved in the long term. Limiting factors were that not all citizens could be reached and that many inhabitants are facing other problems, such as financial or social issues, and therefore waste or recycling are not the main concerns.

Table 7 provides a summary of the success and failure factors in the project respectively beyond the project and assigns them to the four pillars of accountability.

Table 7. Summary of success and failure factures in the project and beyond the project.

Pillars of Accountability	Success and Failure Factors in Project	Beyond the Project
1—Responsibility	Understanding of responsibilities of stakeholders enabled development of common objectives, and consequently of integrated measures. Responsibilities of different stakeholders were made transparent towards citizens which facilitated their involvement.	Responsibility beyond the project was unclear which hindered continuation of activities and transfer.
2—Transparency	Sharing information and trust building between stakeholders supported their cooperation. This enabled integrated communication with citizens adapted to target groups.	Communication with citizens could not be continued in the same intensity.
3—Assessment	Special analysis of data and survey enabled the development of locally adapted solutions. Assessment of measures could be used to develop guidelines for future transfer.	Assessment of measures ended with the project. The experimental project itself was not evaluated.
4—Participation	Citizens and local stakeholders were involved with targeted activities. The bundling and connection of activities and the cooperation with local multipliers enabled them to reach more citizens.	The training of local multipliers had to be reduced due to COVID-19. This limited their work beyond the project.

7. Conclusions and Outlook

Regarding integrating CE into strategic urban planning, the lessons learned from the experimental project WiedergeBORN need to be regarded on the project level and beyond.

Considering the project level, based on the evaluation conducted in this study, I find that the strategic approach of the project can be rated as successful. The project was set-up based on an analysis of the situation (socio-economic, spatial). Subsequently, a cooperation between the main public stakeholders who are active in the area could be established. This enabled a common definition of objectives—in accordance with the sustainability strategies of the different stakeholders—to be reached with the project. Based on these objectives, measures and activities were developed and implemented. Further stakeholders from the public, private, and civic sectors were involved, which supported the successful implementation of most of the planned actions. The intensive exchange between the stakeholders enabled a collaborative learning process during the project. This was possible due to the exceptional situation of the project that gave the stakeholders a mandate to cooperate and act beyond their ordinary tasks. From a spatial development perspective, the concentration in a neighborhood proved to be key for the success of the measures because it enabled the bundling of activities in a functional area. The combination of social and educational measures that envisaged a behavioral change with physical measures that improved the built environment proved to be successful. This underlines the necessity of a socio-spatial approach. Another spatial aspect was the ability to overcome administrative micro-boundaries between involved stakeholders due to the cooperative project design.

Beyond the project, both in terms of timing and organizational levels, the legacy of the project was limited. Reasons can be found in the extraordinary project situation, which enabled extra funding notably to finance a project manager in the area. With the end of the project, the funding of the managing position ended, and with it the intensive exchange between stakeholders and the bundling of activities. Furthermore, all representatives of the key stakeholders had a mandate to work on the project. This enabled opportunities and the freedom to try things out as part of the project and collaboration between the stakeholders over short distances. With the end of the project, the work of the individual actors resumed according to the customary practice. Moreover, only a limited transfer of learnings from the working level to the decision-making level with the involved organizations, SRH and SAGA, was achieved. Consequently, SRH and SAGA/CHANCE only transferred the project results to a limited level into their strategies (sustainability strategies of SAGA, SRH, local development plans) and future work. Another shortcoming was the lack of involvement of strategic levels in the district and the city. Lastly, there was a lack of a strategy to maintain the legacy of the project. Such a strategy could enable the sustainability of project outcomes and enable transfer and replication of the results of the project.

I propose several recommendations for future projects that center place-based approaches to develop and implement solutions to improve waste management and circularity in collaboration with the relevant stakeholders and with local citizens.

One success factor was the local project manager, who enabled cooperation with stakeholders in the area and with citizens. This role is not transferable one-to-one in all cases, but its tasks could be integrated into the work of existing 'caretakers' like neighborhood or climate action managers and other intermediaries working at the interface with citizens.

The strategic approach led to accountable results whose measurable outcomes gained the interest of stakeholders. This could support replication in other projects, as the communication of tangible results is important both to citizens and within the organizations of stakeholders and to decision-makers [9,11].

The integration of environmental, social, and educational aspects in the measures and the cooperation with actors from these fields led to positive outcomes. For future projects on circularity, the involvement of actors representing social services and education can be recommended. This could be supported through strategies and programs that interlink different policy fields (e.g., local development plans).

The project placed citizens at the center of its objectives and measures, and it developed and implemented solutions with them. Furthermore, it directly impacted their wellbeing, improving cleanliness and the living environment through circularity objectives. This could be adopted by other projects to involve citizens in the sense of just transformation.

Another recommendation for future projects is to develop a plan for the legacy of the project and to better integrate higher decision-making levels from an early stage. This could support the integration of lessons learned into strategies of the participating stakeholders.

Lastly, comparable projects will need enabling framework conditions. A simple replication of a project like WiedergeBORN might cause unnecessarily high efforts and costs. To provide projects with funding and an organizational frame, circular economy aspects could be integrated into existing programs, e.g., social development programs (like RISE in Hamburg), local climate action plans, and urban development plans. These plans should consequently be adapted and used for future projects to enable the integration of social and environmental aspects to realize circularity [46,47]. An alternative could be the development of specific strategies on circularity, which for their part need to consider existing social, environmental, and spatial development strategies [5,22,23].

From a methodological perspective, the usage of accountability as a framework enabled a structured analysis of the project, especially regarding the relationship between the principal (citizens) and the agents (public actors), and in this case also between the different public actors. The four pillars of accountability are a basis for a practicable guideline to deconstruct a project in order to understand its potential contribution to strategic planning. However, in this case study, the four categories were partly overlapping and could not be always easily separated from each other.

To better understand the potential of experimental projects that establish CE and how they are or could be linked to strategic urban planning, an in-depth investigation of further examples would be required. An analysis of several experiments could then enable a comparative approach of a larger variety of cases.

Supplementary Materials: The following supporting information can be downloaded at: https://www.mdpi.com/article/10.3390/su162310643/s1, File S1: Guideline for interviews with key stakeholders.

Funding: This research received no external funding.

Institutional Review Board Statement: Not available.

Informed Consent Statement: Not available.

Data Availability Statement: The original contributions presented in the study are included in the article, further inquiries can be directed to the corresponding author.

Conflicts of Interest: The author declares no conflict of interest.

References

1. Vogt, R.; Derreza-Greeven, C.; Giegrich, J.; Dehoust, G.; Möck, A.; Merz, C. *The Climate Change Mitigation Potential of the Waste Sector*; UBA-Texte 56/2015; Umweltbundesamt: Dessau-Roßlau, Germany, 2015; Available online: https://www.umweltbundesamt.de/sites/default/files/medien/378/publikationen/texte_56_2015_the_climate_change_mitigation_potential_of_the_waste_sector.pdf (accessed on 10 October 2024).
2. Vogt, R.; Harju, N.; Auberger, A.; Bulach, W.; Merz, C.; Dehoust, G.; Stahl, H.; Gonser, J.; Küchen, V. *Ermittlung der Klimaschutzpotentiale in der Kreislaufwirtschaft für Deutschland und die EU. Teilbericht Deutschland*; UBA-Texte 83/2023; Umweltbundesamt: Dessau-Roßlau, Germany, 2023; Available online: https://www.umweltbundesamt.de/sites/default/files/medien/11740/publikationen/2023-06-07_texte_83_2023_klimkreis_teilbericht_deutschland.pdf (accessed on 10 October 2024).
3. Vogt, R.; Harju, N.; Auberger, A.; Bulach, W.; Merz, C.; Dehoust, G.; Stahl, H.; Gonser, J.; Küchen, V. *Ermittlung der Klimaschutzpotentiale in der Kreislaufwirtschaft für Deutschland und die EU. Teilbericht EU*; UBA-Texte 85/2023; Umweltbundesamt: Dessau-Roßlau, Germany, 2023; Available online: https://www.umweltbundesamt.de/sites/default/files/medien/11740/publikationen/2023-06-07_texte_85-2023_klimkreis_teilbericht_eu.pdf (accessed on 10 October 2024).
4. Superti, V.; Merino-Saum, A.; Baur, I.; Binder, C.R. Unraveling How the Concept of Circularity Relates to Sustainability: An Indicator-Based Meta-Analysis Applied at the Urban Scale. *J. Clean. Prod.* **2021**, *315*, 128070. [CrossRef]
5. Williams, J. *Circular Cities: A Revolution in Urban Sustainability*; Routledge: London, UK, 2021. [CrossRef]
6. Bulkeley, H.; Kern, K. Local Government and the Governing of Climate Change in Germany and the UK. *Urban Stud.* **2006**, *43*, 2237–2259. [CrossRef]
7. Bulkeley, H.; Castán Broto, V. Government by Experiment? Global Cities and the Governing of Climate Change. *Trans. Inst. Br. Geogr.* **2013**, *38*, 361–375. [CrossRef]

8. Castán Broto, V.; Bulkeley, H. A Survey of Urban Climate Change Experiments in 100 Cities. *Glob. Environ. Chang.* **2013**, *23*, 92–102. [CrossRef] [PubMed]
9. von Wirth, T.; Fuenfschilling, L.; Frantzeskaki, N.; Coenen, L. Impacts of Urban Living Labs on Sustainability Transitions: Mechanisms and Strategies for Systemic Change through Experimentation. *Eur. Plan. Stud.* **2019**, *27*, 229–257. [CrossRef]
10. Bulkeley, H.; Marvin, S.; Palgan, Y.V.; McCormick, K.; Breitfuss-Loidl, M.; Mai, L.; von Wirth, T.; Frantzeskaki, N. Urban Living Laboratories: Conducting the Experimental City? *Eur. Urban Reg. Stud.* **2018**, *26*, 317–335. [CrossRef]
11. Zengerling, C. Action on Climate Change Mitigation in German and Chinese Cities—A Search for Emerging Patterns of Accountability. *Habitat Int.* **2018**, *75*, 147–153. [CrossRef]
12. European Commission (EC). Closing the Loop—An EU Action Plan for the Circular Economy. Available online: https://eur-lex.europa.eu/resource.html?uri=cellar:8a8ef5e8-99a0-11e5-b3b7-01aa75ed71a1.0012.02/DOC_1&format=PDF (accessed on 11 June 2024).
13. Leipold, S.; Petit-Boix, A.; Luo, A.; Helander, H.; Simoens, M.; Ashton, W.S.; Babbitt, C.W.; Bala, A.; Bening, C.R.; Birkved, M.; et al. Lessons, Narratives, and Research Directions for a Sustainable Circular Economy. *J. Ind. Ecol.* **2023**, *27*, 6–18. [CrossRef]
14. Wolman, A. The Metabolism of Cities. *Sci. Am.* **1965**, *213*, 178–193. [CrossRef]
15. Kennedy, C.; Pincetl, S.; Bunje, P. The Study of Urban Metabolism and Its Applications to Urban Planning and Design. *Environ. Pollut.* **2011**, *159*, 1965–1973. [CrossRef] [PubMed]
16. Ghisellini, P.; Cialani, C.; Ulgiati, S. A Review on Circular Economy: The Expected Transition to a Balanced Interplay of Environmental and Economic Systems. *J. Clean. Prod.* **2016**, *14*, 11–32. [CrossRef]
17. Ellen MacArthur Foundation. *Towards the Circular Economy: Economic and Business Rationale for an Accelerated Transition*; Ellen MacArthur Foundation: Cowes, UK, 2013.
18. European Commission (EC). A New Circular Economy Action Plan For a Cleaner and More Competitive Europe. Available online: https://eur-lex.europa.eu/resource.html?uri=cellar:9903b325-6388-11ea-b735-01aa75ed71a1.0017.02/DOC_1&format=PDF (accessed on 11 June 2024).
19. Williams, J. Circular Cities. *Urban Stud.* **2019**, *56*, 2746–2762. [CrossRef]
20. Girardet, H. *Creating Regenerative Cities*; Routledge: Oxford, UK, 2015.
21. Prendeville, S.; Cherim, E.; Bocken, N. Circular Cities: Mapping Six Cities in Transition. *Environ. Innov. Soc. Trans.* **2018**, *26*, 171–194. [CrossRef]
22. Petit-Boix, A.; Leipold, S. Circular Economy in Cities: Reviewing How Environmental Research Aligns with Local Practices. *J. Clean. Prod.* **2018**, *195*, 1270–1281. [CrossRef]
23. Williams, J. Circular Cities: Planning for Circular Development in European Cities. *Eur. Plan. Stud.* **2023**, *31*, 14–35. [CrossRef]
24. Milligan, B.; O'Keeffe, M. Global Governance of Resources and Implications for Resource Efficiency in Europe. *Ecol. Econ.* **2019**, *155*, 46–58. [CrossRef]
25. Wandl, A.; Dąbrowski, M.; Berruti, G.; Acke, A.; Obersteg, A.; Varjú, V.; Taelman, S.; Arlati, A.; Grodzicka-Kowalczyk, M.; Kowalczyk, M. A Holistic Self-Assessment Tool for Circular Economy Transitions in Cities and Regions. *Europa XXI* **2023**, *44*, 10. [CrossRef]
26. Bäckstrand, K.; Zelli, F.; Schleifer, P. Legitimacy and Accountability in Polycentric Climate-Governance. In *Governing Climate Change. Polycentricity in Action*; Jordan, A., Ed.; Cambridge University Press: Cambridge, UK, 2018; pp. 338–356.
27. Zengerling, C. Governing the City of Flows: How Urban Metabolism Approaches May Strengthen Accountability in Strategic Planning. *Urban Plan.* **2019**, *4*, 187–199. [CrossRef]
28. Flick, U. *Qualitative Sozialforschung. Eine Einführung*, 7th ed.; Rowohlt: Reinbek bei Hamburg, Germany, 2016.
29. Lamnek, S.; Krell, C. *Qualitative Sozialforschung mit Online-Material*; Beltz: Weinheim, Germany, 2016.
30. Kuckartz, U.; Rädiker, S. *Fokussierte Interviewanalyse mit MAXQDA. Schritt für Schritt*; Springer: Wiesbaden, Germany, 2020. [CrossRef]
31. Mayring, P. *Qualitative Inhaltsanalyse. Grundlagen und Techniken*; Beltz: Weinheim, Germany, 2015.
32. U.E.C. (Oetjen-Dehne & Partner Umwelt- und Energie-Consult GmbH). *Sortieranalyse von PPK, Biomüll, Wertstoffen und Restmüll an ausgewählten Standplätzen, Jahr 2018*; U.E.C.: Berlin, Germany, 2018; Unpublished internal document.
33. Statistisches Amt für Hamburg und Schleswig-Holstein (Statistik Nord). Hamburger Stadtteil-Profile Berichtsjahr 2017; Hamburg, 2018. Available online: https://www.statistik-nord.de/fileadmin/Dokumente/NORD.regional/NR20_Statistik-Profile_HH-2018.pdf (accessed on 11 June 2024).
34. Bezirksamt Altona. Zweite Zwischenbilanzierung 2018–2021 und Fortschreibung 2022–2025 des Integrierten Entwicklungskonzepts. Available online: https://www.hamburg.de/resource/blob/52810/f90f500e1d87ee793b1711d770b89098/rise-osdorfer-born-zweite-zwischenbilanzierung-2018-bis-2021-und-fortschreibung-iek-2022-bis-2025-data.pdf (accessed on 11 June 2024).
35. Schubert, D. *Hamburger Wohnquartier: Ein Stadtführer durch 65 Siedlungen*; Reimer: Berlin, Germany, 2005.
36. Behörde für Stadtentwicklung und Wohnen (BSW). Sozialmonitoring Integrierte Stadtteilentwicklung–Bericht 2023. Available online: https://www.hamburg.de/resource/blob/286264/30117a83fe60f77cbd7344274fa5beb4/d-sozialmonitoring-bericht-2023-data.pdf (accessed on 14 June 2024).
37. Stadtreinigungsgesetz (SRG). Gesetz zur Errichtung der Anstalt Stadteinigung Hamburg (09.03.1994). Available online: https://www.landesrecht-hamburg.de/bsha/document/jlr-StrGHArahmen (accessed on 14 June 2024).

38. Behörde für Umwelt, Klima, Energie und Agrarwirtschaft (BUKEA). Abfallwirtschaftsplan Siedlungsabfälle 2023. Fortschreibung des Abfallwirtschaftsplans Siedlungsabfälle 2017. Available online: https://www.hamburg.de/resource/blob/159264/64686d5a99ee4e029419db63aa045034/d-awp-siedlungsabfaelle-2023--data.pdf (accessed on 14 June 2024).
39. Stadtreinigung Hamburg (SRH). *Konzern- und Nachhaltigkeitsbericht 2018*; Stadtreinigung Hamburg (SRH): Hamburg, Germany, 2019.
40. SAGA. Geschäftsbericht 2022. Available online: https://www.saga.hamburg/SAGA.Hamburg/%C3%9Cber-uns/Unternehmensgruppe/Gesch%C3%A4ftsbericht/2022/SAGA%20Gesch%C3%A4ftsbericht%202022.pdf (accessed on 11 June 2024).
41. Statista. Bestand an Mietwohnungen der SAGA Unternehmensgruppe in Hamburg von 2015 bis 2023. Available online: https://de.statista.com/statistik/daten/studie/1049836/umfrage/wohnungsbestand-der-saga-in-hamburg/ (accessed on 11 June 2024).
42. SAGA. Nachhaltigkeitsbericht 2019. Available online: https://www.saga.hamburg/Unternehmensgruppe/wir-fuer-hamburg/nachhaltigkeit (accessed on 11 June 2024).
43. SAGA. Nachhaltigkeitsbericht 2020. Available online: https://www.saga.hamburg/Unternehmensgruppe/wir-fuer-hamburg/nachhaltigkeit (accessed on 11 June 2024).
44. SAGA. Nachhaltigkeitsbericht 2022. Available online: https://www.saga.hamburg/SAGA.Hamburg/%C3%9Cber-uns/Unternehmensgruppe/Nachhaltigkeit/2022/SAGA%20Unternehmensgruppe%20Nachhaltigkeitsbericht%202022.pdf (accessed on 11 June 2024).
45. CHANCE. Über uns. Available online: https://www.chance.hamburg/ueber-uns (accessed on 11 June 2024).
46. Varjú, V.; Óvári, Á. Challenges of Applying Circular Economy Principles in Urban Planning Practice in Pécs (Hungary). *Plan. Pract. Res.* **2024**, *39*, 1–14. [CrossRef]
47. Bortolotti, A.; Verga, G.C.; Khan, A.Z. Which Circularity for Urban Design and Planning? A Compass to Navigate Circular Economy Research Knowledge and Methods. *Plan. Pract. Res.* **2023**, *38*, 1–20. [CrossRef]

sustainability

MDPI

Review

A Systematic Informetric Analysis and Literature Review of Food Waste Quantification Studies in the Food Service Industry

Ioannis Vardopoulos *, Konstadinos Abeliotis and Katia Lasaridi *

School of Environment, Geography and Applied Economics, Harokopio University of Athens (HUA), 17676 Kallithea, Greece
* Correspondence: ivardopoulos@post.com (I.V.); klasaridi@hua.gr (K.L.)

Abstract: Incorporating findings from 74 meticulously selected studies, this analysis delves into the complex challenge of measuring food waste (FW) in the food service industry (FSI). The objectives are to aggregate findings on FW quantities, scrutinize quantification methods, identify research gaps, and outline key areas for future exploration. The analysis uncovers an average FW of 167.2 g (min 4.7 g, max 1100 g, standard deviation 202) per meal (26.1%), with studies focusing on plate waste indicating an average of 125.0 g (min 4.7 g, max 292 g, standard deviation 97.3) per meal (29.0%). Variability stems from FW quantification from diverse terminology, inconsistent methodologies, and contextual factors, revealing a lack of standardized concepts. Geographic and sub-sector differences further contribute to the variation in reported FW quantities, affecting the generalizability and comparability of quantities across different FSI establishments and studies. Notably, the majority of the articles offer case-specific insights, limiting their relevance to broader FW generation estimates at the local, national, or sectoral level. Beyond the need for precise quantification, the study acknowledges that the primary challenge lies in addressing wasteful behavioral drivers and actions. Within this perspective, this study underscores the influential role of sociocultural factors and behavioral tendencies and advocates for the need for a deeper understanding of FW generation to formulate practical, sustainable FW prevention strategies. This study provides essential groundwork for harmonizing quantification techniques and guides policymakers and industry stakeholders for the development of practical policies and targeted measures and interventions to effectively reduce FW in the FSI.

Keywords: circular economy; sustainability; data; waste management; food waste; food service

1. Introduction

In recent years, global attention on food loss and waste (FLW), as understood by the Commission Delegated Decision (EU) 2019/1597 [1], has intensified due to its environmental, socioeconomic, and cultural risks [2]. Essential for pollution reduction and efficient resource use, Food Waste (FW) prevention is crucial [3]. FW negatively impacts profitability in food-related sectors [4], affecting producers, manufacturers, distributors, and the food service industry (FSI) [5]—which, for the scope of the current research, includes businesses, institutions, and companies preparing meals outside the home; restaurants, snack bar/cafeterias, schools, hospitals, prisons, hospitality establishments, and catering operations. Yet, beyond these economic and operational concerns, addressing FW is a sociomoral imperative due to widespread food insecurity and malnutrition [6].

 https://doi.org/10.3390/su17010103

FW is a significant sustainability concern, aligning with various Sustainable Development Goals (SDGs) [7]. SDGs, established by the United Nations (UN) in 2015, outline 17 global objectives by 2023 for a sustainable future [8]. FW prevention directly impacts SDG 12, promoting sustainable consumption and production, specifically SDG 12.3, urging a 50% international reduction in the *per capita* FW at the retail and consumer levels and significant FLW reduction across production and Food Supply Chains (FSCs) by 2030 [9].

Understanding and addressing FW generation in the FSC is crucial [10] and varies based on local settings and market dynamics [11]. Recent frameworks on FLW outline priorities: first, waste prevention, then the redistribution of surplus food, then utilization for animal feed, and finally, a range of approaches for upcycling and recycling food scraps [12]. Despite preferred hierarchical categories, resource allocation for each step of the hierarchy remains unclear.

Recent reports estimate 931 million tons of FW in 2019, with FSI contributing 26% [13]. However, along FSI, extreme variations arise due to consumer culture, lifestyle, and behavior factors, in addition to the local economic, regulatory, social, manufacturing, and technological *status quo* [14].

Numerous studies address FW to offer solutions for optimizing resource exploitation, preventing FW generation, and ultimately adapting the entire FW spectrum to a circular model. Scholars investigate FW concerning consumer nutrition [15] and behavior [16]. Some focus on the generated greenhouse gas emissions [17] or carbon footprint [18]. Others explore economic effects [19], including financial losses through economic–probabilistic cost (estimation) models [20]. Researchers investigate energy potentials [21] and discuss waste-to-fuel production opportunities [22]. Scientists look into FW prevention in the FSI to identify the environmental benefits [23], and others consider FW prevention's impact on global food safety [24]. Whether engaged at a global (macro level), regional (meso level), or local (micro level) point of view or associated with one or more FSC stages, there is unanimity on the need to improve prevention strategies and practices to reduce FW. Against this background, FW quantification emerges as a crucial first step. Yet, despite the richness of studies on FW, observations from quantification studies lack generalizability. This study aims to fill this gap by integrating fragmented findings to propel future research endeavors.

Despite the existing literature reviews on FW [25–35], there is still a lack of up-to-date reviews with a particular clear-cut focus on the entire FSI and FW quantification. This review aspires to cover this gap, providing a comprehensive reference for understanding FW in the FSI. It contributes to knowledge in three main ways:

1. Summarizing FW quantities and trends within the FSI uniformly;
2. Describing methods and techniques used to quantify FW in FSI settings, including challenges and limitations;
3. Identifying critical research gaps and concerns related to reducing FW in the FSI.

By providing a clear picture of the current state of FW in the FSI, this review can inform future research, policy, and practice aimed at FW reduction and sustainability.

2. Methodology

A bias-free systematic literature review on FSI FW was carried out (final check October 2024) by following a specific protocol. Records were identified using the 2020 Preferred Reporting Items for Systematic Reviews and Meta-Analyses (PRISMA) Statement.

Scopus database was used to conduct the analysis, provided that it entails credible, high-quality scientific publications with a strong reputation [36]. Using Scopus as the sole database ensures reliable data and consistency, all while offering advanced search capabilities. Alternative datasets like Web of Science or PubMed were not considered, as their narrower scopes are less suited to the interdisciplinary objectives of this study, and their indexed content is most likely already included in Scopus.

The terms used in search fields of title, abstract, and keywords were "food waste*" OR "food loss*" OR "food loss and waste*" OR "plate waste*" AND "quantification*" OR "quantify*" OR "measure*" OR "scale*" OR "estimat*" OR "assessment*" OR "assess*" OR "track*" OR "manage" OR "volume*" OR "method*" AND "food service*" OR "cater*" or "hospitalit*" OR "touris*" OR "hospital*" OR "school*" OR "universit*" OR "canteen*" OR "airline*" OR "foodservice*" OR "flight*" OR "restaurant*" OR "hotel*". Synonyms and alternative terms were incorporated to account for variations in terminology across studies, enhancing the breadth and inclusivity of the search.

The keywords were selected to align with the research objectives, focusing on core concepts such as FW quantification, methodological variations, and specific FSI sub-sectors. An iterative testing process in the Scopus database was conducted to refine the keywords, ensuring they captured the most relevant and high-quality studies. Additionally, the keywords were reviewed and validated through discussions among the authors and collaborating field experts to ensure their relevance and comprehensiveness.

The search of published records was conducted with no date or language restrictions.

All records (n = 1956) were extracted to an Excel. Duplicate, non-English titles, non-journal articles, and records beyond 1990–2024 were removed. A first screening of unrelated records was carried out by reading titles, leaving 677 records. A second screening of abstracts and full texts identified 60 records providing new empirical evidence on FSI FW amounts. Study design type was not a limiting factor; however, records without English full texts, review studies, and studies lacking recorded quantification observations on a meal (or similar) basis were excluded. Observations not extrapolated to the number of guests or on a meal basis per day, week, or year were also excluded. Studies from all geographic regions were included. A total of 14 records were additionally found in reference lists. Applying these criteria, a total of 74 publications were considered eligible for review (See Supplementary Materials) (Figure 1).

To ensure data consistency and reliability, extracted data were independently reviewed by each member of the group of authors. Each record was cross-verified to confirm alignment with the inclusion and exclusion criteria. Any discrepancies were resolved through group discussions, ensuring consensus on key data points, such as FW quantities, measurement methods, and reporting units. The data were further validated by comparing extracted values against other reviews when such occasions arose, providing an additional layer of verification and confidence in the findings. This multi-step process minimized the risk of errors and enhanced the robustness of the review.

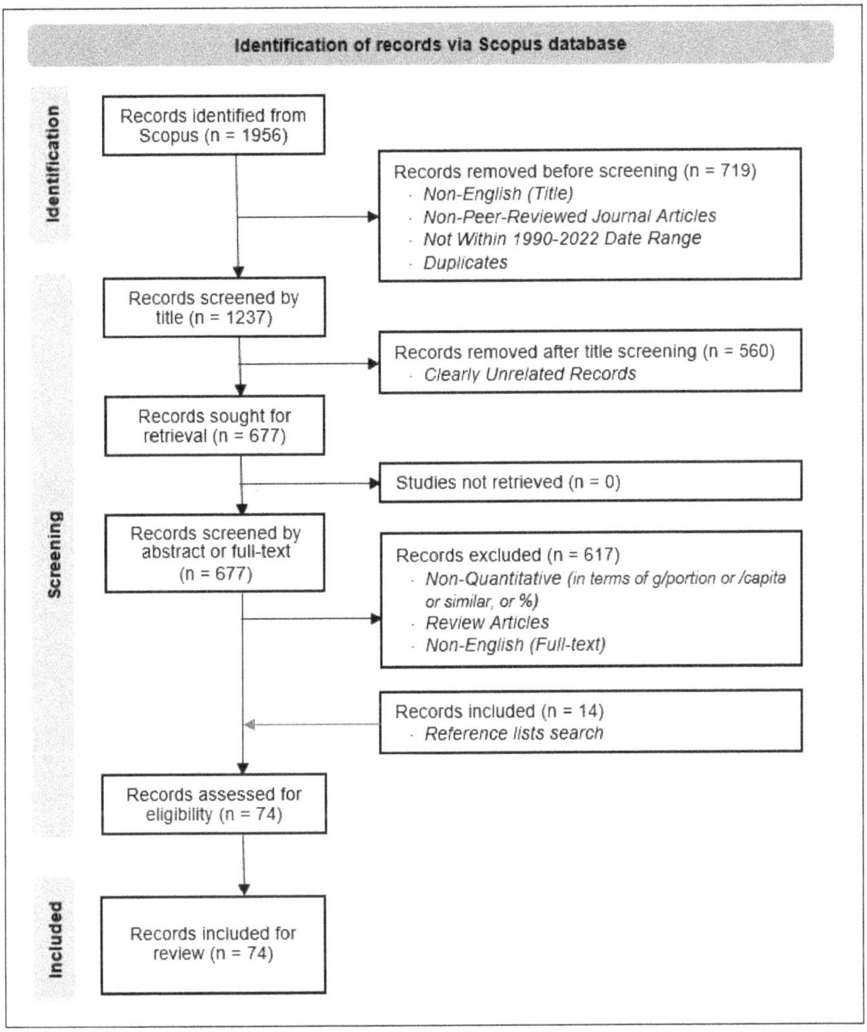

Figure 1. The PRISMA flow diagram.

3. Results and Discussion

3.1. General Characteristics

Over the span of more than three decades, a consistently growing trend in the number of articles has been noted (Figure 2), while 2019 is the year that most of the articles were published. This significant increase in research interest in FW quantification suggests that there is a growing recognition of the importance of this topic, as well as a need for further exploration and investigation. Potential reasons behind this growing trend perhaps are also associated with the diversity of research approaches and methods used in the field, which reflect different disciplinary backgrounds, research questions, or epistemological assumptions, as well as the practical implications of the research for policy or practice, and how can be elevated into actionable recommendations or interventions in different places, people, and customs around the globe.

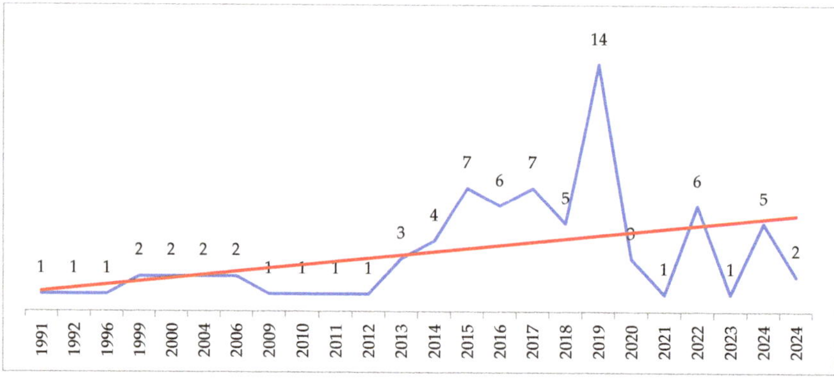

Figure 2. Distribution of reviewed articles on food waste quantification by publication year, highlighting the increasing research interest over three decades, with a peak in 2019.

Geographic-wise, the current list of studies under examination offers ample geographical distribution. Figure 3 illustrates geographical areas on a country basis covered by the selected for review studies.

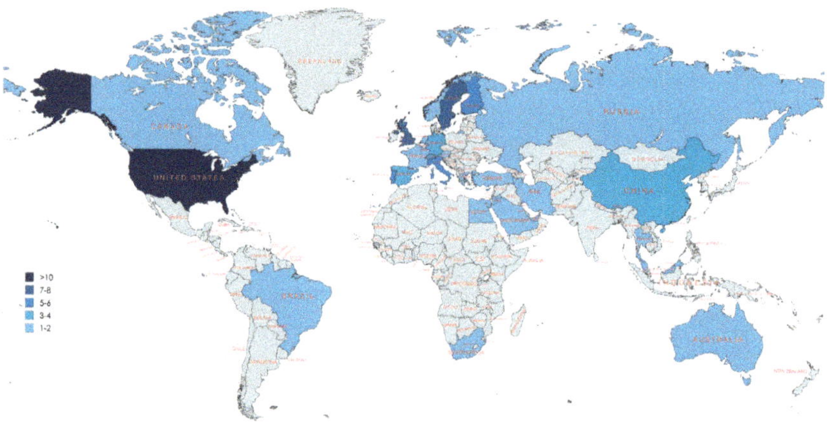

Figure 3. Countries referred to or studied by the selected records of studies for review. Color palette denotes the times the country is referred to or studied. Authors' personal elaboration using www.mapchart.net.

From a scientometric perspective, Eriksson, M., and Malefors, C. emerge as the leading contributors to the field of food waste quantification, showcasing a consistent and systematic body of work. Similarly, *Sustainability*, *Resources, Conservation and Recycling*, and *Waste Management* stand out as the top three journals consistently publishing research in this domain.

3.2. Food Waste Quantities

The average FW portion reported by the reviewed studies is 167.2 g (min 4.7 g, max 1100 g, SD 201.5 g) per meal, coming from an average meal size of 450 g (min 134 g, max 940 g) (Table 1). This calculation completely obliterates all delimitations set by the studies reviewed, such as type of establishment(s), kitchen type, measurement method(s), FW stages, and time involved, among others. It instead represents the mean value of the (total or average) reported FW per meal in grams (or equivalent unit) or percentage, which is used here as an indicative order of magnitude.

Table 1. Results categorization per establishment type, country, year, and quantification methodology.

	g/meal	%	
Average (N = 74)	167.2	26.1	# ≠
Average—Excluding Outliers	127.2	25.9	
Average—PL	125.0	29.0	≠
Average Schools (N = 19)	95.9	32.4	
Average Universities (N = 11)	191.8	21.1	
Average HoReCa [a] (N = 19)	209.3	21.6	#
Average Hospitals (N = 14)	191.3	32.7	
Average Airlines (N = 3)	33.6	17.9	≠
Developing Countries (N = 11)	386.8	25.3	#
Developed Countries (N = 63)	122.1	26.2	≠
America (N = 18)	107.6	33.2	
Europe (N = 38)	109.5	24.0	≠
Asia (N = 13)	313.1	20.3	#
Africa (N = 5)	282.1	26.3	
1990–1999 (N = 5)	176.6	34.0	
2000–2009 (N = 7)	116.2	32.9	
2010–2019 (N = 51)	181.0	25.1	# ≠
2020–2024 (N = 11)	143.2	-	
Direct Measurement (N = 50)	144.8	26.1	
Indirect Measurement (N = 15)	100.2	25.6	
Mixed Measurement (N = 9)	408.4	27.0	#
Evidence generated may not be reliable for further extrapolation or to warrant assertions. Different FW categories and potentially different FW definitions, as well as potential variation among FSI units and meal and food types (due to the lack of a common standard), were not reckoned with for the purposes of the present systematic meta-synthesis. Additionally, with the smaller sample size, the representation becomes less reflective.			

[a] HoReCa stands for hotels, restaurants (including hotel restaurants), caterings, canteens, and cafes; # Outlier with unusually high value included; ≠ Outlier with unusually low value included.

Additionally, although the basis of estimation among the studies reviewed varies—such as "per portion", "per person per meal", or "per student"—the nuanced differences behind the reading of the aforementioned bases of estimation are considered to be extremely thin and perforated. Therefore, for the scope of this review, "per meal" is used as a standard basis to adequately enclose the intended meaning across all cases.

The average FW reported, expressed as a percentage, amounts to 26.1% (min 5%, max 58.5%). Disregarding noteworthy outliers, such as the 1100 g per person FW from all levels of a luxury hotel-restaurant food service supply chain in Malaysia [37], or the reported 4.7 g per meal pertains to flight snacks [38], and the 6 g referencing pre-consumer quantities [39], the adjusted average FW is set to 127.2 g per meal, with a percentage of 25.9%.

Studies exclusively measuring plate waste (PL) report an average of 125.0 g (min 4.7 g, max 292 g, SD 97.4) per meal, equivalent to 29.0%. These findings align with previous research, which consistently highlights that approximately one-third of food is wasted [40,41]. Some studies refer to only pre-consumer or post-preparation FW amounts, but their number is not adequate to draw meaningful averages or generalizations. Attempts to categorize studies based on edible, inedible, or mixed, and avoidable, unavoidable, or overall FW amounts were hindered by significant definition(s) variations. Consequently, grouping a reliable number of studies for conclusive estimates proved unfeasible.

As evident, reported figures vary significantly in terms of representing different establishment types. Therefore, a necessary categorization per establishment type was conducted, including restaurants, hotels, hospitals, schools, universities, and airlines' in-flight catering, among other types of the FSI (Figure 4).

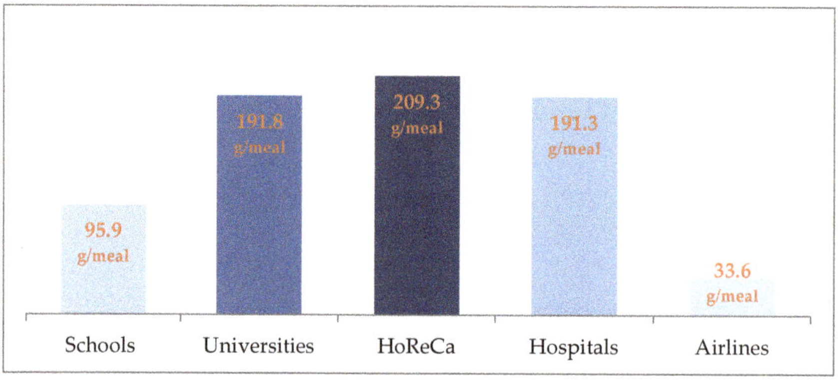

Figure 4. Average FW across various sectors.

This unintended comparison reveals significant sectoral variations in FW generation, with the highest average waste observed in the HoReCa sector (209.3 g/meal) and the lowest in airlines (33.6 g/meal). These findings align with expectations, as the HoReCa sector's high waste reflects its operational nature, while airlines' tightly controlled portion sizes and pre-packaged meals limit the opportunity for excessive waste. Similarly, the relatively high waste in universities (191.8 g/meal) and hospitals (191.3 g/meal) corresponds to their large and diverse populations with varying dietary needs and preferences. Finally, schools, with a moderate average of 95.9 g/meal, reflect demographic factors and operational practices that balance nutrition and waste management. These results underscore the influence of operational practices and consumer behavior on FW, suggesting that tailored, sector-specific interventions could significantly reduce waste.

3.3. Food Waste Quantification Methods and Techniques

In measuring methods, some articles used single methods. Others employed a combination. Generally, FW quantification methods fall into two main groups: direct (physical) measurement (weighing) and structured methods of calculation or estimation.

Direct measurement embraces methods including direct weighting and volumetric evaluation for straightforward FLW measurement. Individual measurements record FW per meal/tray/person or weight of each food component, whereas aggregate measures include gathering and weighing the entire population's (sample) FW [35,42]. Despite providing precise data, it demands significant costs, time, energy, expertise coherent with the FSC stage, and direct access to FW sources. Among its benefits is the ability to track FW effectively and provide measurable progress updates over time. Most articles in this review employ direct measurement (ω = 50), including waste composition analysis (WCA). WCA involves manually sorting, weighting, and classifying FW streams from other components that are not FW (such as packaging) [43]. WCA takes quantification techniques one step further, considering that both the FSI units and local authorities will be able to develop realistic alternatives and creative solutions, such as animal feeding or (bio)composting for FW management [44]. By separating edible and inedible food parts, WCA calculates FW amounts and quality in terms of food typology [45], providing insights into the FW amounts that can actually be prevented if suitable actions are adopted. WCA provides information for monetary, environmental, and nutritional analyses. Even as such, WCA requires advanced expertise, large samples, time, and substantial financial resources. Limitations also include focusing only on waste set out for collection, and not waste disposed of or reused (even for other purposes), and lacking details on the causes of the FW generated [39,46]. Table 2 lists the relevant articles and their main features.

Table 2. Overview and key outcomes of the articles employing direct measurement quantification methods.

N/N	Source	Establishment Type	Study Period (Days)	Inventory	Key Measurement Findings	Country	Summary (Authors' Elaboration)
1	[47]	University	24	· PL	11.3%	CA	Portion size and plate composition determine PL.
2	[48]	Restaurant	28	· Pre-Consumer	123 g/customer/day	CA	Quality assurance standards consistently influenced FW across eateries.
3	[49]	Restaurant	-	· Edible · Avoidable	98 g FW/cap/meal	CN	Tourists accounted for more than half of HoReCa FW.
4	[50]	Restaurant	-	· PL	93 g/cap/meal	CN	FW per capita per meal varies greatly by city, consumer type, restaurant category, and meal purpose.
5	[51]	School	-	· PL	130 g/cap/meal	CN	Buffet meals produce less PL than boxed dinners. Key determining variables are food supply patterns, service quality, dietary habits, and students' understanding of food production.
6	[52]	University	90	· All Stages · +Inedible	16.3%	USA	In campus dining facilities, fruits and vegetables are the greatest FW sources, followed by grains.
7	[38]	Airline	-	· PL · +Liquid	4.7%	CY	Passengers' habits and ethics, and the policies of each airline, generate various types of waste during flights.
8	[53]	Restaurant, Canteen	7	· PL · +Liquid	19%	FI	In self-service restaurants, service waste is the prevailing FW source. Leftovers are the primary source of FW in sit-down-and-order restaurants. Leftovers vary based on the company strategy, kind, portion sizes, and menu.
9	[54]	School, Day-care Center, Canteen, Restaurant, Cafe	7	· All Stages · +Liquid	122 g/portion or 21.8%	FI	Primary causes of FW in the FSI are buffet services and overproduction.

Table 2. *Cont.*

N/N	Source	Establishment Type	Study Period (Days)	Inventory	Key Measurement Findings	Country	Summary (Authors' Elaboration)
10	[55]	Hospital	14	· PL · +Serving Losses	117.1 g/person/day	DE	FW is reduced via an effective communication structure and compliance of meal quality and quantity with customer expectations, needs, and preferences.
11	[56]	Restaurant	164	· PL · +Liquid	200 g/meal	IT	Restaurants having seasonal operation, characteristic of tourist locations, are predicted to generate more waste per meal as the season comes to an end.
12	[57]	School	-	· Unserved	15.4%	IT	Inefficiency in school catering services is assessed by the volume of food that is prepared but not served during meal time.
13	[58]	School	-	· PL · +Inedible	22%	IT	Large-sample studies are essential for providing national and comparative FW statistics.
14	[59]	University	-	· PL	13%	JO	Basic cost of food and students' modest income are significant in limiting PL quantity.
15	[60]	University	45	· PL · +Liquid	76.5 g/cap	PT	Simple prevention initiatives for improving habits and increasing university canteen sustainability may be beneficial.
16	[61]	Hospital	56	· PL	292 g/meal	PT	FW is unavoidable, but it is possible to lessen it.
17	[5]	University	-	· PL · Leftovers · +Inedible	280 g/person	PT	FW amount is a quality sign.
18	[62]	University	16	· All Stages · +Unavoidable	867 g/sale	QA	FW production exceeds the highest globally reported levels. Overproduction is the primary cause of FW caused by food display.
19	[63]	University	21	· All Stages · +Unavoidable	185 g/student/meal/day	ZA	FW was found to be greater than other studies. FW was found to be affected by distance to the dining hall, gender, meal times, and meal options available.
20	[64]	School	5	· All Stages · +Potentially Avoidable	80 g/person/day	ES	The most important aspect in decreasing FW is the human factor in terms of top management and consumers.
21	[65]	Schools, Restaurant	14	· All Stages	92.2 g/portion	SE	1/5 of the food supplied to FSI establishments is lost. PL is the single most significant cause of loss.
22	[66]	School, Elderly Care	28	· PL · +Unavoidable	75 g/portion	SE	FW calculated with extreme variations among kitchen types. FW was 64% serving waste and 33% PL.
23	[67]	School, Hospital, Elderly Care, Canteen, Hotel, Restaurant	-	· All Stages	75.3 g/portion	SE, NO, FI, DE	Almost every institution wastes around 20% of food served.

Table 2. *Cont.*

N/N	Source	Establishment Type	Study Period (Days)	Inventory	Key Measurement Findings	Country	Summary (Authors' Elaboration)
24	[68]	Restaurant	14	· All Stages	138 g/guest	U.K.	Several constraints hinder FW control in ethnic food establishments.
25	[69]	School	4	· PL	46%	USA	The choice and consumption of school meals have been positively impacted by the revised standards.
26	[70]	School	7	· PL	116 g/student	USA	School lunch FW may be decreased by researching and adopting prevention methods.
27	[71]	School	8	· PL	46%	USA	Annually, schoolchildren waste an estimated 26.1% of the entire food expenditure at lunch. For the majority of meal components, less than 85% was eaten.
28	[72]	University	6	· PL · +Inedible	93.5 g/meal	USA	Tray removal is an easy solution for universities and other dining facilities to lessen their environmental effect while saving money.
29	[73]	University	14	· PL · Leftovers · +Inedible	181 g/meal	USA	Tray use in an all-you-can-eat university dining facility generates substantially more FW than not using trays.
30	[74]	School	14	· PL · +Liquid	148 g/meal	USA	When schoolchildren take recess before lunch, FW is decreased.
31	[75]	School	24	· PL · +Liquid	23.5%	USA	Lunch PL can be reduced by changing food preparation methods and limiting competitive food purchases.
32	[76]	School	1	· PL · Liquid	38.1%/meal	USA	Results indicate that revised nutrition standards and policies have led to more nutritious meals and less PL.
33	[77]	School	6	· PL	52.2 g/student/day	USA	FW, paper products, and plastics were found to be the greatest waste fractions in the schools' cafeterias.
34	[78]	Hospital	2	· PL	87 g/patient	U.K.	The hospital feeding system is responsible for FW amounts that by far surpass official averages.
35	[79]	University	2	· PL	29.4 g/student	USA	Several strategies may be used by university dining facilities to boost sustainability.
36	[43]	Airline	-	· PL	62.5 g/meal	EG	Passengers' in-flight FW varies depending on the meal type and content
37	[80]	Hotel, Restaurant, Canteen	-	· All Stages	154.8 g/portion	CY, DE, GR, PT, NO, ES, SE	Prioritizing catering units with the highest FW volume might be a cost-effective method for minimizing FW.
38	[81]	University	15	· All Stages	26.7%	ZA	With emphasis on reducing FW when developing meal plans and menus, student preferences should be taken into account.

Table 2. *Cont.*

N/N	Source	Establishment Type	Study Period (Days)	Inventory	Key Measurement Findings	Country	Summary (Authors' Elaboration)
39	[82]	Public Catering for School and Care Homes	-	· All Stages	47 g/portion	SE	Achieving FW objectives would require a thorough understanding of circumstances, in addition to the establishment of appropriate tools and ways to assist FSI units in reducing FW and reporting progress.
40	[83]	Hospital	-	· PL	129 g/meal	U.K.	FW was greater with the plated meal service.
41	[84]	Hospital	2	· PL	251.9 g/meal	SA	Average PL accounted for 40% of the meal cost per person per day.
42	[85]	Hospital	-	· PL · Trolley Waste	36.3%/meal	U.K.	FW was lower at breakfast than at lunch and dinner.
43	[86]	Hospital	28	· PL	35.3%/meal	U.K.	More than 35% of hospital food was wasted.
44	[87]	Hospital	1	· PL	30.4%/meal	U.K.	The implementation of a bulk meal provision system has resulted in a reduction in FW, but the amount is still high.
45	[88]	Hospital	365	· PL	34,05%/meal	Iran	High levels of plate waste contribute to malnutrition-related complications in hospitalized patients.
46	[89]	Hospital	21	· PL · Tray Waste	412 g/patient/day	Saudi Arabia	Expanding the concept of "plate waste" to "tray waste" reveals the need for a more comprehensive, sustainable approach to hospital food services.
47	[90]	School	134	· PL	44 g/meal	Sweden	Automation can accurately quantify plate waste. Study reveals that a few students produce the majority of school canteen plate waste.
48	[91]	Restaurant	17	· PL	252 g/person	China	A scarcity mindset reduces food waste, regardless of self-construal.
49	[92]	Hotel Restaurant	7	· PL	46 g/meal/person	Turkey	Tourists produce more food waste on vacation than in their everyday lives.
50	[93]	School	364	· PL · +Serving Losses	45 g/guest	Sweden	Popular and vegetarian school meals generate less food waste than unpopular meals, challenging the preconception that vegetarian options increase waste.

Note 1: Studies are listed in no particular order. Note 2: The listed values lack the complete context in which each study was conducted. To gain a deeper understanding, it is advised to refer to the specifics of each individual study. Note 3: Symbol "+" is used here to indicate that a certain fraction is included (e.g., inedible, liquid, etc.).

Non-direct measuring methods involve various strategies, with the most commonly used discussed below:

1. Questionnaires (also known as surveys or interviews) are sound methods for capturing quantitative and qualitative data on FW amounts. They broadly fall into two categories: (i) collecting available data and (ii) calculating (via estimations) FW amounts. While resourceful, sufficiently consistent, repeatable, and able to build up large samples (snowball sampling). Data will likely not be free of inherent inconsistencies and inaccuracies [50];

2. The mass balance measurement method uses input–output analysis to determine FLW, which is the ratio of raw materials entering production to finished goods served or sold. This technique can affordably produce estimates on FLW when input and output data are available that would otherwise be unattainable [94]. Material Flow Analysis (MFA), a key technique, provides an in-depth, comparable view of the food industry, revealing FLW patterns and trouble spots. MFA serves as a basic and clear framework for decision-making [95] together with a well-supported inventory of additional tools and techniques (e.g., Life Cycle Assessment), key performance environmental indicators (e.g., Shrink Ratio, or Waste Percentage per Customer), and environmental performance management frameworks (e.g., ISO 14001 [96]). Nevertheless, MFA has limitations, including data gaps and uncertainty estimation [97];
3. Food (/kitchen) diaries refers to (group of) subjects (e.g., families, students, etc.) estimating (visual estimation included) and self-reporting FW quantity in a given day/period. Subjects are encouraged to identify waste-generation stages and disposal methods and point to key reasons for generating FW. Despite consistent use, they come with drawbacks, including challenges in subject enlistment, potential sample bias, and data quality concerns [46,98].

Food diaries is one of the FW measurement methods accepted by the Commission Delegated Decision (EU) 2019/1597, specifically for the FSI, the others being direct measurement, WCA, and counting/scanning. Questionnaires and interviews, as well as mass balance, coefficients, and production statistics, are not considered valid methods for FSI FW measurement, at least for the mandatory national reporting to the Eurostat. Nevertheless, this review's findings indicate no significant differences in reported results compared to the direct measurement studies. The records using non-direct measurement methods are summarized in Table 3.

Table 3. Overview and key outcomes of the articles using questionnaires, surveys, and interviews to estimate FW amounts.

N/N	Source	Establishment Type	Study Period (Days)	Inventory	Key Measurement Findings	Country	Summary (Authors' Elaboration)
1	[99]	Hospital	·	All Stages	111 g/guest/meal	SE	FW in hospitals is higher compared to other industries, yet there are substantial variations among them. PL accounts for 42%, serving waste for 36%, and kitchen waste for 22%.
2	[100]	Big Revenue Food Service Companies	·	All Stages	13.5%	ES	Businesses work to minimize FW based on economic considerations. The biggest losses come from fruits and cooked dishes. PL accounted for 30% of overall FW.
3	[101]	Restaurant, Café, Canteen, Catering	·	All Stages	10.6%	IT	70% of the FW consists of food that remains unsold and possibly exploitable. Three primary FW generators were identified: raw materials or packaged products nearing their expiration date; unsold unpacked products; and leftovers.
4	[102]	Restaurant	·	All Stages	28.7%	IT	Restaurant managers' attitudes, menu types, and restaurant size are all variables that contribute to FW. Restaurants that serve meat tend to waste more.

Table 3. *Cont.*

N/N	Source	Establishment Type	Study Period (Days)	Inventory	Key Measurement Findings	Country	Summary (Authors' Elaboration)
5	[103]	Restaurant		· All Stages	20%	CY	Certain variations in FW were identified for insular island states, such as the large tourist flows, particularly during the summer months, as well as the contribution of imports and exports to the final FW quantity.
6	[104]	Hospital	7	· PL	58.5%	USA	A visual PL study benefits in measuring performance, menu planning, and effectiveness evaluation.
7	[105]	School, Day-care, Workplace	14	· All Stages · + Liquid	78.6 g/portion	FI	FW case studies and monitoring systems can be used to decrease FW in the FSI.
8	[106]	Bars, Warehouse, Snack bar, Restaurant, Wiskeria	-	· All Stages	118 g/meal	BR	The results indicate an average of 339 kg/month (or 13 kg/day) FW per restaurant, with a EUR 511 monthly revenue.
9	[107]	School	5	· PL	40.7%	FR	There is a strong correlation between food liking and food consumption, and consequently, FW.
10	[42]	School	11	· All Stages · + Potentially Avoidable	75 g/meal/cap	ES	Employees and administrators are very interested in the issue and would be prepared to adopt audits and reduction measures despite the fact that there is little awareness of the quantity of FW at school canteens.
11	[108]	Hospital	60	· PL	130 g/plate	NL	PL amount correlates with movements made during mealtime by laboratory and housekeeping employees, guests, and doctors.
12	[109]	Hospital	4	· PL	23.5%	AU	Room service in a public hospital improves nutritional intake and increases patient satisfaction while decreasing FW and cost.
13	[94]	School, Catering		· All Stages	88.5 g/meal	CH	The most common reasons for PL were the too-large portions served by staff and customers' lack of appetite.
14	[110]	Service Sector		· All Stages	15%	EU	FW case studies and monitoring systems can be used to decrease FW in the FSI.
15	[111]	Service Sector		· All Stages	20%	CH	The FW amount is not a reliable predictor of the ability to prevent or minimize FW. An extensive examination distinguishing avoidable and unavoidable, kitchen and PL, etc., toward understanding causalities enables one to derive strategies to decrease FW.

Note 1: Studies are listed in no particular order. Note 2: All stages indication includes preparation and storage FLW on the pre-consumer side, as well as serving and plate scrapings FLW from the post-consumer group. Note 3: 1–5 are articles using questionnaires, surveys, and interviews. Note 4: 6–12 are studies referring to food diaries. Note 5: 13–15 are publications working on mass balance. Note 6: Symbol "+" is used here to indicate that a certain fraction is included (e.g., inedible, liquid, etc.).

Following the guidelines of the Commission Delegated Decision (EU) 2019/1597, several studies combine methods. Smith et al. (2014) [112] primarily used a visual estimation model, validated by direct measurement in at least 20% of cases, reporting a 92% agreement between weighted and visually assessed FW. Liz Martins et al. (2014) [113] indicate the visual estimation technique overstating the FW value by an average (bias) of

8 g. Thamagasorn et al. (2019) [114] and Paparguropoulou et al. (2016; 2019) [37,115] all employed a complex methodology, including individual interviewing, group discussions, in situ FW estimation based on visual observations, and quantitative data collection. Pirani et al. (2016) [116] quantification methods involved MFA and physical measurements. Beretta and Hellweg (2019) [23] collected datasets of FW physical measurement coupled with interviews and filled data gaps using estimations. Prescott et al. (2019) [117] used a mixed-method approach consisting of key informants' interview sessions, systematic kitchen observations, and direct measurement. Filomonau et al. (2022) [118] combined approaches, validating the FW estimates with on-site observation and direct measurements.

Be that as it may, and provided that the aim is to overcome barriers to achieving the most accurate mass FW measurement to enrich future approaches, the findings indicate that combining direct and structured methods could effectively address the limitations of each while leveraging their strengths [119]. For example, structured methods like questionnaires or food diaries could be validated against direct measurements in a subset of cases to enhance accuracy while minimizing resource intensity. Additionally, integrating techniques such as WCA and MFA offers a comprehensive framework for addressing FW challenges. Combined approaches can indeed pave the way for standardized protocols, improving data comparability across contexts and supporting more targeted and efficient (food) waste reduction strategies [120].

3.4. Factors Affecting Food Waste Quantification

3.4.1. Conceptual and Terminological Confusion

The lack of a globally agreed definition for FLW complicates scientific discourse and hinders progress in the field. FUSIONS considers no distinction between the terms FW and Food Loss (FL) and includes both edible and inedible parts of food. In contrast, the FAO and the FLW Protocol understand FL as edible (excluding inedible) for human consumption food, mass throughout the primary production, processing, and manufacturing FSC stages, clarifying that FL occurs at retail and distribution, and consumption parts of the FSC is FW. This complex confusion has been carried over to academia. Some researchers consider FW as waste occurring at the end of the FSC (retail and final consumption), caused only by wasteful behavior [121]. Others differentiate, including unavoidable FL, arguing that it is (still) challenging to distinguish between wasteful behavior and other causes of FL [111]. Similarly, the term "Food Surplus" also creates confusion, referring either to unconsumed but suitable food [122] or the excess of nutrients in food that is appropriate for human consumption [123]. Categorizations like avoidable, possibly avoidable, and non-avoidable add further complexity. Falasconi et al. (2015) [57] introduce the term "physiological" non-avoidable, referring to unserved food intentionally prepared in excess to ensure a sufficient supply for potential additional requests. These classifications prompt critical examination of what should be defined as FW and what qualifies as avoidable, specifically in terms of edibility. Although food contains inedible components, making de facto distinction between edible and inedible parts seems challenging and subjective in terms of variation among cultures and customer tastes and preferences [124]. For instance, lamp tripe or testicles, commonly served in a variety of ethnic cuisines (e.g., Greek), are rarely seen in other parts of the world (e.g., USA). Similarly, watermelon rinds may be considered edible in some cultures but inedible in others. Therefore, current terms, albeit widely used in studies, pose significant challenges in comparing findings. This complexity hampers the ability to generalize results and impedes the development of effective waste prevention strategies and policies across diverse contexts. On that account, harmonized definitions and standardized frameworks, combined with culturally specific strategies, are essential to improving comparability, shaping policy, and implementing effective food waste reduction solutions.

3.4.2. Sector's Sections, Types, and Segments

FSI FW is the sum of certain waste-generation phases in and around preparation and consumption. Some scholars [106] consider three phases, namely FW occurring during preparation, consumption, and post-preparation. Others adopt pre-consumer and post-consumer phases. Nonetheless, FW during preparation may result from various factors, including staff inexperience, inadequate infrastructure, or human errors. Post-preparation involves two. Service, involving staff taking orders and serving, with factors like lack of training, failure to meet basic aesthetic standards, human error incidents, unsold or uneaten food, etc., contribute to FW. PL, involving cooking and management aspects (e.g., on portion size [125] or food type) and customer demographics (e.g., gender [54], age [42], nationality [126], income [127], education [126], and marital status [128]). Factors like season [57], day of the week [42], time of day, length of mealtime [129], mood, dietary preferences [54], palatability [94], raised awareness [64], eating alone or with others [130], and overall food experience (food options, atmosphere, décor, comfort, location, service speed, service quality, price, etc.) [128] can also contribute to FW amount and complexities.

Distinct cookery configurations also influence FW production. A first breakdown includes four types: a) kitchens preparing food on-site to serve in-house; b) kitchens preparing food on-site to be served out-of-house; c) kitchens preparing food on-site to be sent to other (satellite) kitchens for further processing or handling, and/or serving; and d) (satellite) kitchens that receive cooked food from central production units for further processing, or handling, and/or serving [66]. A second segmentation could concern hot kitchens, cold kitchens, confectionery rooms, and meal portioning areas [114]. Different cookery types involve different waste-generation phases and different equipment. Therefore, FW prevention measures can hardly concern horizontal cookery production units (e.g., some kitchens cannot save leftover food for later use or cannot handle variation in demand, etc.). In addition to the cookery type, the food service style can also significantly influence FW amounts. An all-you-can-eat buffet dinner typically overproduces food (due to difficulty in estimating consumption as well as presentation reasons), while a self-service fast food restaurant is quite the contrary [54].

Food service provision includes the following: a) for-profit enterprises that provide food and beverages as a core activity, such as restaurants; b) for-profit establishments that bring nourishment as a secondary (or auxiliary) activity, such as hotels; and c) public or institutional entities such as schools, nurseries, nursing houses, hospitals, or penitentiaries. The diverse scope, size, and location of FSI establishments introduce uncertainty in FW quantification. Considerations also include phenomena like the "Freshman 15" in universities [73] or the increased FW by travelers in tourist destinations [103]. Thus, understandably, a thorough study is necessary to fully comprehend the FW dynamics throughout the numerous sectors and sub-sectors, as well as other characteristics and parameters associated with the FSI (e.g., [131]). Indeed, [118] correlated FW with facility size and business strategy, aligning with [102].

3.4.3. Moisture Content Considerations

A potential bias in FW quantification procedures is the varying water content in food items [132]. FW inherent moisture content [133], fluctuating influenced by cooking methods, menu, temperature, and time factors presents significant impact measurements [29]. According to EC Regulation No 178/2002 (2002), water purposely added to food during its production, preparation, or treatment is regarded as food and must be included in the FW calculation. While FUSIONS offers instructions on water flows, research hardly refers to how water content and moisture fluctuations were addressed, or if addressed at all. For example, hospital patients on liquid or pureed diets generate more putrescible FW, and handling, processing, and measuring can be challenging. Limited exceptions dwell into moisture [43,62] or clearly refer to liquid FW [117], yet further research is needed to (i) determine the degree of moisture content fluctuations, (ii) develop an unbiased moisture-wise methodology, and (iii) formulate coefficients to estimate water content in FW.

3.4.4. Waste No More Time

Methodological challenges in FW measurement include considerations of time, encompassing the duration of measurement procedures and the impact of seasonality. Regarding duration, there is a lack of guidance on the recommended time and frequency for measuring FW in FSI establishments, except [134] reference to catering units. Questions arise about whether the measurement should be continuous or occasional and for what duration or frequency. Additionally, the timing of FW collection and weighing is crucial across methodologies, as it can significantly affect results, for instance, if conducted when a wasted piece of tomato has dried up.

While some studies suggest variation in FW amount on different week days, the season (calendar year) and period (work circle) significantly affect FW generation [116,135]. Shifting menu cycles may also be a factor influencing FW quantities, as it leads to the disposal of the one menu cycle food stock in order to welcome the starting of the new one. Additionally, meal-order and serving wait time (perceived as the standard time necessary or as time wasted by possible maladministration or accident) also affect FW production [136].

Malefons et al. (2022) [82] attempted to longitudinally monitor FW to identify trends and assess the efficacy of existing prevention measures. While this may be an initial effort, the authors advocate for continuous FW tracking in the future, emphasizing the necessity to establish a common understanding and foundation.

3.4.5. Behavioral and Cultural Factors

FW is the result of in-action activities, as well as behavioral and cultural factors [30]. State-funded school meals in Sweden, Finland, and Estonia, known as "pedagogic meals", are ingrained in the social culture, considered teaching opportunities, promoting a balanced, nutrient-rich diet, home economics, sustainable resource management, and FW reduction [54]. This sociocultural phenomenon offers insights into the qualitative and quantitative characteristics of food consumed and FW produced. Contrarily, this welfare state model, affecting food culture and fostering a reduced amount of FW at schools [42], stands far from the liberal welfare regimes seen in the USA, Australia, or other parts of the world, where food culture widely endorses convenience or pre-cooked food options.

Cultural peculiarities, such as Brazil's "comida por quilo" where food is priced by weight [106], China's "miànzi" concept emphasizing over-ordering as a sign of hospitality or social status [137], Africa's Ubuntu-inspired communal care expressed through food [138], and religious principles (e.g., Judaism, Islam, or Hinduism) that discourage wastefulness [139], collectively highlight how cultural attitudes and traditions can shape FW patterns [140].

Other socio-culturally influenced practices, like taking restaurant leftovers home, often linked with societal norms and environmental awareness, can affect FW generation. Similarly, ethnic food, often characterized by complex preparation cooking techniques or beyond the ordinary cooking recipes and unique ingredients, which may sometimes carry overtones of excessive refinement, is also strongly linked to high percentages of FW.

For example, (exotic) fruits, often used in ethnic cuisines, are associated with substantial inedible components [111].

3.5. Study Limitations, Implications, and Future Research Directions

Specific (de)limitations were adopted from the outset of this research, while a few weak points were additionally identified in the process. Future research should, therefore, further elaborate and potentially improve the developed research protocol and consequently validate or supplement current findings.

Largely depending on the authors' subjective views on the field and topic, inclusion and exclusion criteria were established. Hence, it may be worthwhile to consider making this component of the study protocol more expansive or more restricted.

Additionally, keywords, while chosen diligently, may not be all-encompassing, potentially overlooking relevant research. The focus on English publications in Scopus-indexed journals excludes studies in other languages or from diverse sources, warranting consideration for a more comprehensive global scope.

Moreover, the articles reviewed herein used different FW data collection methods, criteria, units, etc., which, in addition to the imperfection in light of the human involvement factor, creates additional bias and could be considered a drawback, urging future efforts to establish common functional units and measurement methods in the entire FSI.

Furthermore, while this study touches on aspects related to product quality and quantity, such as overproduction, portion sizes, and mismatches between food quality and consumer expectations, a deeper analysis of how these factors influence FW was beyond the scope of this review. Future research should explore these dimensions in greater detail, considering geographical, cultural, and policy-related/legal contexts, to better understand their impact on FW generation and to develop more effective FW prevention strategies.

Limitations also include the absence of studies meeting criteria for certain sectors, like penitentiaries, or emerging areas like online food ordering and delivery platforms, which could emerge as a research topic in its own right.

Finally, this study focused on FW in a particular sector. To demonstrate the whole impact, comprehend the causes, and ascertain the magnitude of the FW phenomenon on the economy, society, and the environment, all levels of the FSC must be examined to formulate feasible and efficient solutions.

In future studies of gap areas, the current research recognized that emerging technologies, such as AI-driven waste tracking, IoT-enabled smart scales, and advanced data analytics platforms, hold significant potential for FW reduction [141]. These tools provide automated insights into waste patterns and support the design of effective interventions, making them a promising avenue for future research [142]. Additionally, exploring the integration of consumer-facing apps with AI-driven systems could further enhance waste prevention and management strategies in the FSI (and beyond).

4. Conclusions

The study systematically reviewed FW quantification studies in the FSI. Findings emphasize the crucial necessity for universally applicable terminology and methods in FW quantification. Despite broad geographical coverage and variations in scope, size, scale, and quality, a conspicuous absence of well-defined, standardized concepts and methodologies exists. This is further compounded by the multiple contextual and sociocultural factors involved, as well as the diverse (de)limitations and biases in data collection and analyses, including estimations. This complexity affects outcomes in waste quantity range, generalizability, and comparability.

This review contends that, while scientifically sound, the majority of studies can only draw inferences in specific case contexts, limiting their support for horizontal FW prevention measures and management strategies. The actual challenge, however, lies not in FW measurement but in behaviors causing FW. Behaviors and actions related to FW generation, and consequently prevention, are intricate.

This analysis emphasizes the need for experience and expertise to strike the right balance between FW measurement precision and completeness, as well as the resources allocated for prevention and reduction.

Supplementary Materials: The following supporting information can be downloaded at: https://www.mdpi.com/article/10.3390/su17010103/s1.

Author Contributions: This research represents a collaborative effort among all authors. Conceptualization was undertaken by I.V. and K.L., with data curation managed by I.V. and K.L. Formal analysis and investigation were conducted by I.V., while methodology was developed collaboratively by I.V. and K.L. Funding acquisition and supervision were carried out by K.L. Project administration was shared among I.V., K.A. and K.L. Validation was performed by K.A. and K.L., and visualization was handled by I.V. The original draft was written by I.V., with review and editing also carried out by I.V. All authors have read and agreed to the published version of the manuscript.

Funding: This research is partly funded by the LIFE-IP CEI Greece project "Circular Greece-Circular economy implementation in Greece" (LIFE18 IPE/GR/000013) which is cofinanced by the EU and the Green Fund public-law entity reporting to the Hellenic Ministry of Environment and Energy.

Acknowledgments: The authors would like to thank all who directly or indirectly contributed to the completion of this study. Particular acknowledgment is given to the delegates of the 2nd International Conference RETASTE: Rethink Food Waste, held on 20–21 October 2022 in Heraklion, Greece, whose discussions and feedback on earlier drafts were instrumental in shaping this work. Appreciation is also expressed to the anonymous reviewers for their constructive critiques and valuable insights, which significantly improved the quality of the study.

Conflicts of Interest: The authors declare no conflict of interest.

References

1. European Commission. Commission Delegated Decision (EU) 2019/1597 of 3 May 2019 Supplementing Directive 2008/98/EC of the European Parliament and of the Council as Regards a Common Methodology and Minimum Quality Requirements for the Uniform Measurement of Levels of Food Wast. *Off. J. Eur. Union* **2019**, 77–85.
2. Sigala, E.G.; Chroni, C.; Boikou, K.; Abeliotis, K.; Panagiotakos, D.; Lasaridi, K. Quantification of Household Food Waste in Greece to Establish the 2021 National Baseline and Methodological Implications. *Waste Manag.* **2024**, *190*, 102–112. [CrossRef] [PubMed]
3. Abeliotis, K.; Lasaridi, K. Food Waste Prevention: Reduction, Reuse, and Recycling. *Resources* **2022**, *12*, 3. [CrossRef]
4. Chroni, C.; Vardopoulos, I.; Abeliotis, K.; Lasaridi, K. Food Waste Generation in Hospitality and Food Services Sector in Greece. In Proceedings of the 10th International Conference on Environmental Management, Engineering, Planning and Economics (CEMEPE 2023) and SECOTOX Conference, Skiathos, Greece, 5–9 June 2023; Kungolos, A., Laspidou, C., Aravossis, K., Samaras, P., Marnellos, D., Melidis, P., Papapolymerou, G., Zafeirakou, A., Emmanouil, C., Manakou, V., Eds.; Grafima: Thessaloniki, Greece, 2023; p. 135, ISBN 9786185710255.
5. Ferreira, M.; Liz Martins, M.; Rocha, A. Food Waste as an Index of Foodservice Quality. *Br. Food J.* **2013**, *115*, 1628–1637. [CrossRef]
6. Talwar, S.; Kaur, P.; Kumar, S.; Salo, J.; Dhir, A. The Balancing Act: How Do Moral Norms and Anticipated Pride Drive Food Waste/Reduction Behaviour? *J. Retail. Consum. Serv.* **2022**, *66*, 102901. [CrossRef]
7. Kaur, G.; Lasaridi, K.; Wong, J. Sustainable Food Waste Management: An Introduction. In *Current Developments in Biotechnology and Bioengineering. Sustainable Food Waste Management: Resource Recovery and Treatment;* Wong, J., Kaur, G., Taherzadeh, M., Pandey, A., Lasaridi, K., Eds.; Elsevier: Amsterdam, The Netherlands, 2021; pp. 1–10.
8. Loizia, P.; Voukkali, I.; Zorpas, A.A.; Navarro Pedreño, J.; Chatziparaskeva, G.; Inglezakis, V.J.; Vardopoulos, I.; Doula, M. Measuring the Level of Environmental Performance in Insular Areas, through Key Performed Indicators, in the Framework of Waste Strategy Development. *Sci. Total Environ.* **2021**, *753*, 141974. [CrossRef]

9. Chroni, C.; Lasaridi, K.; Abeliotis, K.; Manios, T. Closing the Food Chain Loop Through Waste Prevention. In *Current Developments in Biotechnology and Bioengineering. Sustainable Food Waste Management: Resource Recovery and Treatment*; Wong, J., Kaur, G., Taherzadeh, M., Pandey, A., Lasaridi, K., Eds.; Elsevier: Amsterdam, The Netherlands, 2021; pp. 107–135.
10. Vardopoulos, I.; Chroni, C.; Abeliotis, K.; Lasaridi, K. Assessing Food Waste Generation in the Food Service Sector Mission Impossible? In Proceedings of the 2nd International Conference RETASTE: Rethink Food Waste, Heraklion, Greece, 20–21 October 2022; Manios, T., Lasaridi, K., Daliakopoulos, I., Eds.; Book of Abstracts. Hellenic Mediterranean University: Heraklion, Greece, 2022; pp. 50–51, ISBN 9786188477452.
11. Economou, F.; Chatziparaskeva, G.; Papamichael, I.; Loizia, P.; Voukkali, I.; Navarro-Pedreño, J.; Klontza, E.; Lekkas, D.F.; Naddeo, V.; Zorpas, A.A. The Concept of Food Waste and Food Loss Prevention and Measuring Tools. *Waste Manag. Res. J. Sustain. Circ. Econ.* **2024**, *42*, 651–669. [CrossRef] [PubMed]
12. Zorpas, A.A.; Doula, M.K.; Jeguirim, M. Waste Strategies Development in the Framework of Circular Economy. *Sustainability* **2021**, *13*, 13467. [CrossRef]
13. Forbes, H.; Quested, T.; O'Connor, C. *Food Waste Index Report 2021*; United Nations Environment Programme: Nairobi, Kenya; University of Oxford: Oxford, UK; United Nations World Tourism Organisation: Madrid, Spain; World Meteorological Organization: Geneva, Switzerland, 2021.
14. Fiore, M.A.; Pellegrini, G.; Sala, P.L.; Conte, A.; Liu, B. Attitude toward Food Waste Reduction: The Case of Italian Consumers. *Int. J. Glob. Small Bus.* **2017**, *9*, 185. [CrossRef]
15. Ahmed, M.; Mandic, I.; Lou, W.; Goodman, L.; Jacobs, I.; L'Abbé, M. Validation of a Tablet Application for Assessing Dietary Intakes Compared with the Measured Food Intake/Food Waste Method in Military Personnel Consuming Field Rations. *Nutrients* **2017**, *9*, 200. [CrossRef] [PubMed]
16. Pocol, C.B.; Pinoteau, M.; Amuza, A.; Burlea-Schiopoiu, A.; Glogovețan, A.-I. Food Waste Behavior among Romanian Consumers: A Cluster Analysis. *Sustainability* **2020**, *12*, 9708. [CrossRef]
17. Porter, S.D.; Reay, D.S.; Higgins, P.; Bomberg, E. A Half-Century of Production-Phase Greenhouse Gas Emissions from Food Loss & Waste in the Global Food Supply Chain. *Sci. Total Environ.* **2016**, *571*, 721–729. [CrossRef] [PubMed]
18. Eriksson, M.; Strid, I.; Hansson, P.-A. Carbon Footprint of Food Waste Management Options in the Waste Hierarchy—A Swedish Case Study. *J. Clean. Prod.* **2015**, *93*, 115–125. [CrossRef]
19. Garske, B.; Heyl, K.; Ekardt, F.; Weber, L.; Gradzka, W. Challenges of Food Waste Governance: An Assessment of European Legislation on Food Waste and Recommendations for Improvement by Economic Instruments. *Land* **2020**, *9*, 231. [CrossRef]
20. Fiore, M.A.; Contò, F.; Pellegrini, G. Reducing Food Losses: A (Dis)-Opportunity Cost Model. *Riv. Stud. Sulla Sostenibilita* **2015**, *1*, 151–166. [CrossRef]
21. Melikoglu, M. Appraising Food Waste Generation and Forecasting Food Waste to Energy Potentials of Hospitals in Turkey: A Global to Local Analysis. *Sustain. Prod. Consum.* **2020**, *24*, 292–297. [CrossRef]
22. Abad, V.A.; Cherrett, T.; Holdsworth, P. Waste-to-Fuel Opportunities for British Quick Service Restaurants: A Case Study. *Resour. Conserv. Recycl.* **2015**, *104*, 239–253. [CrossRef]
23. Beretta, C.; Hellweg, S. Potential Environmental Benefits from Food Waste Prevention in the Food Service Sector. *Resour. Conserv. Recycl.* **2019**, *147*, 169–178. [CrossRef]
24. Kasza, G.; Szabó-Bódi, B.; Lakner, Z.; Izsó, T. Balancing the Desire to Decrease Food Waste with Requirements of Food Safety. *Trends Food Sci. Technol.* **2019**, *84*, 74–76. [CrossRef]
25. Hoteit, M.; Antar, E.; Malli, D.; Fattouh, F.; Khattar, M.; Baderddine, N.; El Cheikh, J.; Chalak, A.; Abiad, M.G.; Hassan, H.F. A Review on Hospital Food Waste Quantification, Management and Assessment Strategies in the Eastern Mediterranean Region. *J. Agric. Food Res.* **2024**, *15*, 100959. [CrossRef]
26. Dhir, A.; Talwar, S.; Kaur, P.; Malibari, A. Food Waste in Hospitality and Food Services: A Systematic Literature Review and Framework Development Approach. *J. Clean. Prod.* **2020**, *270*, 122861. [CrossRef]
27. Hoehn, D.; Vázquez-Rowe, I.; Kahhat, R.; Margallo, M.; Laso, J.; Fernández-Ríos, A.; Ruiz-Salmón, I.; Aldaco, R. A Critical Review on Food Loss and Waste Quantification Approaches: Is There a Need to Develop Alternatives beyond the Currently Widespread Pathways? *Resour. Conserv. Recycl.* **2023**, *188*, 106671. [CrossRef]
28. Filimonau, V.; De Coteau, D.A. Food Waste Management in Hospitality Operations: A Critical Review. *Tour. Manag.* **2019**, *71*, 234–245. [CrossRef]
29. Corrado, S.; Sala, S. Food Waste Accounting along Global and European Food Supply Chains: State of the Art and Outlook. *Waste Manag.* **2018**, *79*, 120–131. [CrossRef]
30. Özbük, R.M.Y.; Coşkun, A. Factors Affecting Food Waste at the Downstream Entities of the Supply Chain: A Critical Review. *J. Clean. Prod.* **2020**, *244*, 118628. [CrossRef]
31. Pirani, S.I.; Arafat, H.A. Solid Waste Management in the Hospitality Industry: A Review. *J. Environ. Manag.* **2014**, *146*, 320–336. [CrossRef] [PubMed]

32. Shanks, C.B.; Banna, J.; Serrano, E.L. Food Waste in the National School Lunch Program 1978-2015: A Systematic Review. *J. Acad. Nutr. Diet.* **2017**, *117*, 1792–1807. [CrossRef] [PubMed]
33. Amicarelli, V.; Bux, C. Food Waste Measurement toward a Fair, Healthy and Environmental-Friendly Food System: A Critical Review. *Br. Food J.* **2021**, *123*, 2907–2935. [CrossRef]
34. Diana, R.; Martianto, D.; Baliwati, Y.F.; Sukandar, D.; Hendriadi, A. Food Waste in Indonesian Hospitals: A Systematic Review. *Nutr. Food Sci.* **2022**, *53*, 881–900. [CrossRef]
35. Williams, P.; Walton, K. Plate Waste in Hospitals and Strategies for Change. *E-Spen Eur. e-J. Clin. Nutr. Metab.* **2011**, *6*, e235–e241. [CrossRef]
36. Pranckutė, R. Web of Science (WoS) and Scopus: The Titans of Bibliographic Information in Today's Academic World. *Publications* **2021**, *9*, 12. [CrossRef]
37. Papargyropoulou, E.; Wright, N.; Lozano, R.; Steinberger, J.; Padfield, R.; Ujang, Z. Conceptual Framework for the Study of Food Waste Generation and Prevention in the Hospitality Sector. *Waste Manag.* **2016**, *49*, 326–336. [CrossRef]
38. Tofalli, N.; Loizia, P.; Zorpas, A.A. Passengers Waste Production during Flights. *Environ. Sci. Pollut. Res.* **2018**, *25*, 35764–35775. [CrossRef]
39. Jayasekara, A.P.; McGrath, D.; Kravchuk, O.; Zhou, S.J.; Morris, H. Mapping Pre-Consumer Food Waste in Quick Service Restaurants on a University Campus: Two Australian Case Studies. *J. Clean. Prod.* **2024**, *434*, 139978. [CrossRef]
40. Ishangulyyev, R.; Kim, S.; Lee, S. Understanding Food Loss and Waste—Why Are We Losing and Wasting Food? *Foods* **2019**, *8*, 297. [CrossRef]
41. Ioannou, T.; Bazigou, K.; Katsigianni, A.; Fotiadis, M.; Chroni, C.; Manios, T.; Daliakopoulos, I.; Tsompanidis, C.; Michalodimitraki, E.; Lasaridi, K. The "A2UFood Training Kit": Participatory Workshops to Minimize Food Loss and Waste. *Sustainability* **2022**, *14*, 2446. [CrossRef]
42. Derqui, B.; Fernandez, V. The Opportunity of Tracking Food Waste in School Canteens: Guidelines for Self-Assessment. *Waste Manag.* **2017**, *69*, 431–444. [CrossRef]
43. El-Mobaidh, A.M.; Razek Taha, M.A.; Lassheen, N.K. Classification of In-Flight Catering Wastes in Egypt Air Flights and Its Potential as Energy Source (Chemical Approach). *Waste Manag.* **2006**, *26*, 587–591. [CrossRef]
44. Giamouri, E.; Pappas, A.C.; Papadomichelakis, G.; Simitzis, P.E.; Manios, T.; Zentek, J.; Lasaridi, K.; Tsiplakou, E.; Zervas, G. The Food for Feed Concept: Redefining the Use of Hotel Food Residues in Broiler Diets. *Sustainability* **2022**, *14*, 3659. [CrossRef]
45. Rajan, J.; Fredeen, A.L.; Booth, A.L.; Watson, M. Measuring Food Waste and Creating Diversion Opportunities at Canada's Green University TM. *J. Hunger Environ. Nutr.* **2018**, *13*, 573–586. [CrossRef]
46. Quested, T.E.; Palmer, G.; Moreno, L.C.; McDermott, C.; Schumacher, K. Comparing Diaries and Waste Compositional Analysis for Measuring Food Waste in the Home. *J. Clean. Prod.* **2020**, *262*, 121263. [CrossRef]
47. Von Massow, M.; McAdams, B. Table Scraps: An Evaluation of Plate Waste in Restaurants. *J. Foodserv. Bus. Res.* **2015**, *18*, 437–453. [CrossRef]
48. McAdams, B.; von Massow, M.; Gallant, M.; Hayhoe, M.-A. A Cross Industry Evaluation of Food Waste in Restaurants. *J. Foodserv. Bus. Res.* **2019**, *22*, 449–466. [CrossRef]
49. Wang, L.; Xue, L.; Li, Y.; Liu, X.; Cheng, S.; Liu, G. Horeca Food Waste and Its Ecological Footprint in Lhasa, Tibet, China. *Resour. Conserv. Recycl.* **2018**, *136*, 1–8. [CrossRef]
50. Wang, L.; Liu, G.; Liu, X.; Liu, Y.; Gao, J.; Zhou, B.; Gao, S.; Cheng, S. The Weight of Unfinished Plate: A Survey Based Characterization of Restaurant Food Waste in Chinese Cities. *Waste Manag.* **2017**, *66*, 3–12. [CrossRef]
51. Liu, Y.; Cheng, S.; Liu, X.; Cao, X.; Xue, L.; Liu, G. Plate Waste in School Lunch Programs in Beijing, China. *Sustainability* **2016**, *8*, 1288. [CrossRef]
52. Costello, C.; Birisci, E.; McGarvey, R.G. Food Waste in Campus Dining Operations: Inventory of Pre- and Post-Consumer Mass by Food Category, and Estimation of Embodied Greenhouse Gas Emissions. *Renew. Agric. Food Syst.* **2016**, *31*, 191–201. [CrossRef]
53. Katajajuuri, J.-M.; Silvennoinen, K.; Hartikainen, H.; Heikkilä, L.; Reinikainen, A. Food Waste in the Finnish Food Chain. *J. Clean. Prod.* **2014**, *73*, 322–329. [CrossRef]
54. Silvennoinen, K.; Heikkilä, L.; Katajajuuri, J.-M.; Reinikainen, A. Food Waste Volume and Origin: Case Studies in the Finnish Food Service Sector. *Waste Manag.* **2015**, *46*, 140–145. [CrossRef]
55. Strotmann, C.; Friedrich, S.; Kreyenschmidt, J.; Teitscheid, P.; Ritter, G. Comparing Food Provided and Wasted before and after Implementing Measures against Food Waste in Three Healthcare Food Service Facilities. *Sustainability* **2017**, *9*, 1409. [CrossRef]
56. Tatàno, F.; Caramiello, C.; Paolini, T.; Tripolone, L. Generation and Collection of Restaurant Waste: Characterization and Evaluation at a Case Study in Italy. *Waste Manag.* **2017**, *61*, 423–442. [CrossRef] [PubMed]
57. Falasconi, L.; Vittuari, M.; Politano, A.; Segrè, A. Food Waste in School Catering: An Italian Case Study. *Sustainability* **2015**, *7*, 14745–14760. [CrossRef]
58. Boschini, M.; Falasconi, L.; Giordano, C.; Alboni, F. Food Waste in School Canteens: A Reference Methodology for Large-Scale Studies. *J. Clean. Prod.* **2018**, *182*, 1024–1032. [CrossRef]

59. Al-Domi, H.; Al-Rawajfe, H.; Aboyousif, F.; Yaghi, S.; Mashal, R.; Fakhoury, J. Determining and Addressing Food Plate Waste in a Group of Students at the University of Jordan. *Pakistan J. Nutr.* **2011**, *10*, 871–878. [CrossRef]
60. Pinto, R.S.; dos Santos Pinto, R.M.; Melo, F.F.S.; Campos, S.S.; Cordovil, C.M.-S. A Simple Awareness Campaign to Promote Food Waste Reduction in a University Canteen. *Waste Manag.* **2018**, *76*, 28–38. [CrossRef] [PubMed]
61. Dias-Ferreira, C.; Santos, T.; Oliveira, V. Hospital Food Waste and Environmental and Economic Indicators—A Portuguese Case Study. *Waste Manag.* **2015**, *46*, 146–154. [CrossRef]
62. Abdelaal, A.H.; McKay, G.; Mackey, H.R. Food Waste from a University Campus in the Middle East: Drivers, Composition, and Resource Recovery Potential. *Waste Manag.* **2019**, *98*, 14–20. [CrossRef]
63. Painter, K.; Thondhlana, G.; Kua, H.W. Food Waste Generation and Potential Interventions at Rhodes University, South Africa. *Waste Manag.* **2016**, *56*, 491–497. [CrossRef]
64. Derqui, B.; Fernandez, V.; Fayos, T. Towards More Sustainable Food Systems. Addressing Food Waste at School Canteens. *Appetite* **2018**, *129*, 1–11. [CrossRef]
65. Engström, R.; Carlsson-Kanyama, A. Food Losses in Food Service Institutions Examples from Sweden. *Food Policy* **2004**, *29*, 203–213. [CrossRef]
66. Eriksson, M.; Persson Osowski, C.; Malefors, C.; Björkman, J.; Eriksson, E. Quantification of Food Waste in Public Catering Services—A Case Study from a Swedish Municipality. *Waste Manag.* **2017**, *61*, 415–422. [CrossRef]
67. Malefors, C.; Callewaert, P.; Hansson, P.-A.; Hartikainen, H.; Pietiläinen, O.; Strid, I.; Strotmann, C.; Eriksson, M. Towards a Baseline for Food-Waste Quantification in the Hospitality Sector—Quantities and Data Processing Criteria. *Sustainability* **2019**, *11*, 3541. [CrossRef]
68. Filimonau, V.; Nghiem, V.N.; Wang, L. Food Waste Management in Ethnic Food Restaurants. *Int. J. Hosp. Manag.* **2021**, *92*, 102731. [CrossRef]
69. Cohen, J.F.W.; Richardson, S.; Parker, E.; Catalano, P.J.; Rimm, E.B. Impact of the New U.S. Department of Agriculture School Meal Standards on Food Selection, Consumption, and Waste. *Am. J. Prev. Med.* **2014**, *46*, 388–394. [CrossRef] [PubMed]
70. Byker, C.J.; Farris, A.R.; Marcenelle, M.; Davis, G.C.; Serrano, E.L. Food Waste in a School Nutrition Program After Implementation of New Lunch Program Guidelines. *J. Nutr. Educ. Behav.* **2014**, *46*, 406–411. [CrossRef]
71. Cohen, J.F.W.; Richardson, S.; Austin, S.B.; Economos, C.D.; Rimm, E.B. School Lunch Waste Among Middle School Students. *Am. J. Prev. Med.* **2013**, *44*, 114–121. [CrossRef]
72. Kim, K.; Morawski, S. Quantifying the Impact of Going Trayless in a University Dining Hall. *J. Hunger Environ. Nutr.* **2012**, *7*, 482–486. [CrossRef]
73. Sarjahani, A.; Serrano, E.L.; Johnson, R. Food and Non-Edible, Compostable Waste in a University Dining Facility. *J. Hunger Environ. Nutr.* **2009**, *4*, 95–102. [CrossRef]
74. Getlinger, M.J.; Laughlin, C.V.; Bell, E.; Akre, C.; Arjmandi, B.H. Food Waste Is Reduced when Elementary-School Children Have Recess before Lunch. *J. Am. Diet. Assoc.* **1996**, *96*, 906–908. [CrossRef] [PubMed]
75. Marlette, M.A.; Templeton, S.B.; Panemangalore, M. Food Type, Food Preparation, and Competitive Food Purchases Impact School Lunch Plate Waste by Sixth-Grade Students. *J. Am. Diet. Assoc.* **2005**, *105*, 1779–1782. [CrossRef] [PubMed]
76. Schwartz, M.B.; Henderson, K.E.; Read, M.; Danna, N.; Ickovics, J.R. New School Meal Regulations Increase Fruit Consumption and Do Not Increase Total Plate Waste. *Child. Obes.* **2015**, *11*, 242–247. [CrossRef] [PubMed]
77. Wilkie, A.; Graunke, R.; Cornejo, C. Food Waste Auditing at Three Florida Schools. *Sustainability* **2015**, *7*, 1370–1387. [CrossRef]
78. Sonnino, R.; McWilliam, S. Food Waste, Catering Practices and Public Procurement: A Case Study of Hospital Food Systems in Wales. *Food Policy* **2011**, *36*, 823–829. [CrossRef]
79. Babich, R.; Smith, S. "Cradle to Grave": An Analysis of Sustainable Food Systems in a University Setting. *J. Culin. Sci. Technol.* **2010**, *8*, 180–190. [CrossRef]
80. Eriksson, M.; Malefors, C.; Callewaert, P.; Hartikainen, H.; Pietiläinen, O.; Strid, I. What Gets Measured Gets Managed—Or Does It? Connection between Food Waste Quantification and Food Waste Reduction in the Hospitality Sector. *Resour. Conserv. Recycl. X* **2019**, *4*, 100021. [CrossRef]
81. Marais, M.; Smit, Y.; Koen, N.; Lötze, E. Are the Attitudes and Practices of Foodservice Managers, Catering Personnel and Students Contributing to Excessive Food Wastage at Stellenbosch University? *S. Afr. J. Clin. Nutr.* **2017**, *30*, 60–67. [CrossRef]
82. Malefors, C.; Strid, I.; Eriksson, M. Food Waste Changes in the Swedish Public Catering Sector in Relation to Global Reduction Targets. *Resour. Conserv. Recycl.* **2022**, *185*, 106463. [CrossRef]
83. Wilson, A.; Evans, S.; Frost, G. A Comparison of the Amount of Food Served and Consumed According to Meal Service System. *J. Hum. Nutr. Diet.* **2000**, *13*, 271–275. [CrossRef]
84. Al-shoshan, A.A. Study of the Regular Diet of Selected Hospitals of the Ministry of Health in Saudi Arabia: Edible Plate Waste and Its Monetary Value. *J. R. Soc. Health* **1992**, *112*, 7–11. [CrossRef] [PubMed]
85. Edwards, J.S.A.; Nash, A.H.M. The Nutritional Implications of Food Wastage in Hospital Food Service Management. *Nutr. Food Sci.* **1999**, *99*, 89–98. [CrossRef]

86. Barton, A.D.; Beigg, C.L.; MacDonald, I.A.; Allison, S.P. High Food Wastage and Low Nutritional Intakes in Hospital Patients. *Clin. Nutr.* **2000**, *19*, 445–449. [CrossRef]
87. Kelly, L. Audit of Food Wastage: Differences between a Plated and Bulk System of Meal Provision. *J. Hum. Nutr. Diet.* **1999**, *12*, 415–424. [CrossRef]
88. Vahabzadeh, D. Food Intake, Plate Waste and Its Association with Malnutrition in Hospitalized Patients. *Nutr. Hosp.* **2017**, *34*, 1376–1381. [CrossRef] [PubMed]
89. Sahal Alharbi, N.; Yahia Qattan, M.; Haji Alhaji, J. Towards Sustainable Food Services in Hospitals: Expanding the Concept of 'Plate Waste' to 'Tray Waste'. *Sustainability* **2020**, *12*, 6872. [CrossRef]
90. Malefors, C.; Svensson, E.; Eriksson, M. Automated Quantification Tool to Monitor Plate Waste in School Canteens. *Resour. Conserv. Recycl.* **2024**, *200*, 107288. [CrossRef]
91. Gao, H.; Jia, H.; Guo, B. Resources Available for Me Versus Us: Implications for Mitigating Consumer Food Waste. *J. Mark. Res.* **2024**, *61*, 619–637. [CrossRef]
92. Çetin, K.; Süren, T. An Investigation into the Causes of Food Waste by Tourists in All-Inclusive Resorts in Turkey. *J. Qual. Assur. Hosp. Tour.* **2024**, *25*, 677–699. [CrossRef]
93. Sundin, N.; Malefors, C.; Danielsson, M.; Hardiyanti, M.; Persson Osowski, C.; Eriksson, M. Investigating Goal Conflicts in Menu Planning in Swedish School Catering on the Pathway to Sustainable Development. *Resour. Conserv. Recycl.* **2023**, *190*, 106822. [CrossRef]
94. Betz, A.; Buchli, J.; Göbel, C.; Müller, C. Food Waste in the Swiss Food Service Industry—Magnitude and Potential for Reduction. *Waste Manag.* **2015**, *35*, 218–226. [CrossRef]
95. Padeyanda, Y.; Jang, Y.-C.; Ko, Y.; Yi, S. Evaluation of Environmental Impacts of Food Waste Management by Material Flow Analysis (MFA) and Life Cycle Assessment (LCA). *J. Mater. Cycles Waste Manag.* **2016**, *18*, 493–508. [CrossRef]
96. *ISO 14001*; Environmental Management Systems—Requirements with Guidance for Use. ISO: Geneva, Switzerland, 2015.
97. Amicarelli, V.; Bux, C.; Lagioia, G. How to Measure Food Loss and Waste? A Material Flow Analysis Application. *Br. Food J.* **2020**, *123*, 67–85. [CrossRef]
98. Giordano, C.; Piras, S.; Boschini, M.; Falasconi, L. Are Questionnaires a Reliable Method to Measure Food Waste? A Pilot Study on Italian Households. *Br. Food J.* **2018**, *120*, 2885–2897. [CrossRef]
99. Eriksson, M.; Malefors, C.; Bergström, P.; Eriksson, E.; Persson Osowski, C. Quantities and Quantification Methodologies of Food Waste in Swedish Hospitals. *Sustainability* **2020**, *12*, 3116. [CrossRef]
100. Derqui, B.; Fayos, T.; Fernandez, V. Towards a More Sustainable Food Supply Chain: Opening up Invisible Waste in Food Service. *Sustainability* **2016**, *8*, 693. [CrossRef]
101. Amato, M.; Musella, M. Quantification of Food Waste within Food Service in the Historic Centre of Naples: A Case Study. *Qual. Access Success* **2017**, *18*, 22–28.
102. Principato, L.; Pratesi, C.A.; Secondi, L. Towards Zero Waste: An Exploratory Study on Restaurant Managers. *Int. J. Hosp. Manag.* **2018**, *74*, 130–137. [CrossRef]
103. Ioannou, A.; Georgali, P.-Z.; Fokaides, P.A. Quantification of Food Waste in an Insular Island State for All Stages of the Food Supply Chain. *Resour. Conserv. Recycl.* **2022**, *185*, 106486. [CrossRef]
104. Connors, P.L.; Rozell, S.B. Using a Visual Plate Waste Study to Monitor Menu Performance. *J. Am. Diet. Assoc.* **2004**, *104*, 94–96. [CrossRef]
105. Silvennoinen, K.; Nisonen, S.; Pietiläinen, O. Food Waste Case Study and Monitoring Developing in Finnish Food Services. *Waste Manag.* **2019**, *97*, 97–104. [CrossRef] [PubMed]
106. Sehnem, S.; Pereira, L.H.; Junior, S.S.; Bernardy, R.J.; Lara, A.C. Management of Food Waste in Restaurants by Way of Circular Practices. *J. Mater. Cycles Waste Manag.* **2022**, *24*, 1020–1036. [CrossRef]
107. Giboreau, A.; Schwartz, C.; Morizet, D.; Meiselman, H.L. Measuring Food Waste and Consumption by Children Using Photography. *Nutrients* **2019**, *11*, 2410. [CrossRef]
108. Deutekom, E.J.; Philipsen, H.; Ten Hoor, F.; Huyer Abu-Saad, H. Plate Waste Producing Situations on Nursing Wards. *Int. J. Nurs. Stud.* **1991**, *28*, 163–174. [CrossRef]
109. McCray, S.; Maunder, K.; Krikowa, R.; MacKenzie-Shalders, K. Room Service Improves Nutritional Intake and Increases Patient Satisfaction While Decreasing Food Waste and Cost. *J. Acad. Nutr. Diet.* **2018**, *118*, 284–293. [CrossRef] [PubMed]
110. Caldeira, C.; De Laurentiis, V.; Corrado, S.; van Holsteijn, F.; Sala, S. Quantification of Food Waste per Product Group along the Food Supply Chain in the European Union: A Mass Flow Analysis. *Resour. Conserv. Recycl.* **2019**, *149*, 479–488. [CrossRef] [PubMed]
111. Beretta, C.; Stoessel, F.; Baier, U.; Hellweg, S. Quantifying Food Losses and the Potential for Reduction in Switzerland. *Waste Manag.* **2013**, *33*, 764–773. [CrossRef]
112. Smith, S.L.; Cunningham-Sabo, L. Food Choice, Plate Waste and Nutrient Intake of Elementary- and Middle-School Students Participating in the US National School Lunch Program. *Public Health Nutr.* **2014**, *17*, 1255–1263. [CrossRef]

113. Martins, M.L.; Cunha, L.M.; Rodrigues, S.S.P.; Rocha, A. Determination of Plate Waste in Primary School Lunches by Weighing and Visual Estimation Methods: A Validation Study. *Waste Manag.* **2014**, *34*, 1362–1368. [CrossRef] [PubMed]
114. Thamagasorn, M.; Pharino, C. An Analysis of Food Waste from a Flight Catering Business for Sustainable Food Waste Management: A Case Study of Halal Food Production Process. *J. Clean. Prod.* **2019**, *228*, 845–855. [CrossRef]
115. Papargyropoulou, E.; Steinberger, J.K.; Wright, N.; Lozano, R.; Padfield, R.; Ujang, Z. Patterns and Causes of Food Waste in the Hospitality and Food Service Sector: Food Waste Prevention Insights from Malaysia. *Sustainability* **2019**, *11*, 6016. [CrossRef]
116. Pirani, S.I.; Arafat, H.A. Reduction of Food Waste Generation in the Hospitality Industry. *J. Clean. Prod.* **2016**, *132*, 129–145. [CrossRef]
117. Prescott, M.P.; Herritt, C.; Bunning, M.; Cunningham-Sabo, L. Resources, Barriers, and Tradeoffs: A Mixed Methods Analysis of School Pre-Consumer Food Waste. *J. Acad. Nutr. Diet.* **2019**, *119*, 1270–1283.e2. [CrossRef] [PubMed]
118. Filimonau, V.; Ermolaev, V.A.; Vasyukova, A. Food Waste in Foodservice Provided in Educational Settings: An Exploratory Study of Institutions of Early Childhood Education. *Int. J. Gastron. Food Sci.* **2022**, *28*, 100531. [CrossRef]
119. Ghaziani, S.; Ghodsi, D.; Dehbozorgi, G.; Faghih, S.; Rajabpour Ranjbar, Y.; Doluschitz, R. Comparing Lab-Measured and Surveyed Bread Waste Data: A Possible Hybrid Approach to Correct the Underestimation of Household Food Waste Self-Assessment Surveys. *Sustainability* **2021**, *13*, 3472. [CrossRef]
120. Sakaguchi, L.; Pak, N.; Potts, M.D. Tackling the Issue of Food Waste in Restaurants: Options for Measurement Method, Reduction and Behavioral Change. *J. Clean. Prod.* **2018**, *180*, 430–436. [CrossRef]
121. Gustavsson, J.; Cederberg, C.; Sonesson, U.; van Otterdijk, R.; Meybeck, A. Global Food Losses and Food Waste. In Proceedings of the Save Food Congress, Düsseldorf, Germany, 16 May 2011; Food and Agriculture Organization of the United Nations: Rome, Italy, 2011.
122. Garrone, P.; Melacini, M.; Perego, A. Opening the Black Box of Food Waste Reduction. *Food Policy* **2014**, *46*, 129–139. [CrossRef]
123. Teigiserova, D.A.; Hamelin, L.; Thomsen, M. Towards Transparent Valorization of Food Surplus, Waste and Loss: Clarifying Definitions, Food Waste Hierarchy, and Role in the Circular Economy. *Sci. Total Environ.* **2020**, *706*, 136033. [CrossRef] [PubMed]
124. Hopkins, J. *Extreme Cuisine: The Weird and Wonderful Foods That People Eat*; Tuttle Publishing: Clarendon, VT, USA, 2019.
125. Freedman, M.R.; Brochado, C. Reducing Portion Size Reduces Food Intake and Plate Waste. *Obesity* **2010**, *18*, 1864–1866. [CrossRef] [PubMed]
126. Zhao, X.; Manning, L. Food Plate Waste: Factors Influencing Insinuated Intention in a University Food Service Setting. *Br. Food J.* **2019**, *121*, 1536–1549. [CrossRef]
127. Setti, M.; Falasconi, L.; Segrè, A.; Cusano, I.; Vittuari, M. Italian Consumers' Income and Food Waste Behavior. *Br. Food J.* **2016**, *118*, 1731–1746. [CrossRef]
128. Cerrah, S.; Yigitoglu, V. Determining the Effective Factors on Restaurant Customers' Plate Waste. *Int. J. Gastron. Food Sci.* **2022**, *27*, 100469. [CrossRef]
129. Cohen, J.F.W.; Jahn, J.L.; Richardson, S.; Cluggish, S.A.; Parker, E.; Rimm, E.B. Amount of Time to Eat Lunch Is Associated with Children's Selection and Consumption of School Meal Entrée, Fruits, Vegetables, and Milk. *J. Acad. Nutr. Diet.* **2016**, *116*, 123–128. [CrossRef] [PubMed]
130. Young, W.; Russell, S.V.; Robinson, C.A.; Barkemeyer, R. Can Social Media Be a Tool for Reducing Consumers' Food Waste? A Behaviour Change Experiment by a UK Retailer. *Resour. Conserv. Recycl.* **2017**, *117*, 195–203. [CrossRef]
131. Vassiliades, C.; Diemuodeke, O.E.; Yiadom, E.B.; Prasad, R.D.; Dbouk, W. Policy Pathways for Mapping Clean Energy Access for Cooking in the Global South—A Case for Rural Communities. *Sustainability* **2022**, *14*, 13577. [CrossRef]
132. Corrado, S.; Caldeira, C.; Eriksson, M.; Hanssen, O.J.; Hauser, H.-E.; van Holsteijn, F.; Liu, G.; Östergren, K.; Parry, A.; Secondi, L.; et al. Food Waste Accounting Methodologies: Challenges, Opportunities, and Further Advancements. *Glob. Food Secur.* **2019**, *20*, 93–100. [CrossRef] [PubMed]
133. Gerassimidou, S.; Dora, M.; Iacovidou, E. A Tool for the Selection of Food Waste Management Approaches for the Hospitality and Food Service Sector in the UK. *Resources* **2022**, *11*, 80. [CrossRef]
134. Eriksson, M.; Lindgren, S.; Persson Osowski, C. Mapping of Food Waste Quantification Methodologies in the Food Services of Swedish Municipalities. *Resour. Conserv. Recycl.* **2018**, *137*, 191–199. [CrossRef]
135. Gallardo, A.; Edo-Alcón, N.; Carlos, M.; Renau, M. The Determination of Waste Generation and Composition as an Essential Tool to Improve the Waste Management Plan of a University. *Waste Manag.* **2016**, *53*, 3–11. [CrossRef] [PubMed]
136. Kuperberg, K.; Caruso, A.; Dello, S.; Mager, D. How Will a Room Service Delivery System Affect Dietary Intake, Food Costs, Food Waste and Patient Satisfaction in a Paediatric Hospital? A Pilot Study. *J. Foodserv.* **2008**, *19*, 255–261. [CrossRef]
137. Wang, M.; Rasoolimanesh, S.M.; Kunasekaran, P.; Zhao, Y. Understanding Over-Ordering Behaviour in Social Dining: Integrating Mass Media Exposure and Sense of 'Mianzi' into the Norm Activation Model. *Serv. Ind. J.* **2022**, *44*, 1018–1037. [CrossRef]
138. Phasha, L.; Molelekwa, G.F.; Mokgobu, M.I.; Morodi, T.J.; Mokoena, M.M.; Mudau, L.S. Influence of Cultural Practices on Food Waste in South Africa—A Review. *J. Ethn. Foods* **2020**, *7*, 37. [CrossRef]

139. Minton, E.A.; Johnson, K.A.; Vizcaino, M.; Wharton, C. Is It Godly to Waste Food? How Understanding Consumers' Religion Can Help Reduce Consumer Food Waste. *J. Consum. Aff.* **2020**, *54*, 1246–1269. [CrossRef]
140. Doukas, Y.E.; Vardopoulos, I. A Pathway to Sustainable Policy: Unraveling the Interplay of Socio-Cultural Factors, Local Contexts, and Institutional Dynamics in Food Waste Reduction. In Proceedings of the 3rd International Conference RETASTE: Rethink Food Resources, Losses and Waste, Athens, Greece, 27–29 September 2023; Manios, T., Lasaridi, K., Abeliotis, K., Daliakopoulos, I., Eds.; Book of Abstracts; Hellenic Mediterranean University: Heraklion, Greece, 2023; pp. 275–277, ISBN 9786188477476.
141. Synani, K.; Abeliotis, K.; Lasaridi, K. Food Waste in the Catering and Hospitality Sector: Generation Factors and Innovative Practices. In Proceedings of the 8th International Conference on Sustainable Solid Waste Management, Thessaloniki, Greece, 23–26 June 2021.
142. Sigala, E.G.; Gerwin, P.; Chroni, C.; Abeliotis, K.; Strotmann, C.; Lasaridi, K. Artificial Intelligence Reduces Food Waste in the Horeca Sector. Available online: https://papers.ssrn.com/sol3/papers.cfm?abstract_id=4990161 (accessed on 24 October 2024).

MDPI AG
Grosspeteranlage 5
4052 Basel
Switzerland
Tel.: +41 61 683 77 34

Sustainability Editorial Office
E-mail: sustainability@mdpi.com
www.mdpi.com/journal/sustainability

www.ingramcontent.com/pod-product-compliance
Lightning Source LLC
Chambersburg PA
CBHW040817250926
51850CB00007BA/1194
9783725846450